In the Formation of the Classical Islamic World Series

MANUELA MARÍN
The Formation of al-Andalus, Part I: History and Society

MARIBEL FIERRO AND JULIO SAMSÓ
The Formation of al-Andalus, Part II: Language, Religion, Culture and the Sciences

URI RUBIN
The Life of Muhammad

FRANK PETERS
The Arabs and Arabia on the Eve of Islam

And in the Variorum Collected Studies Series

MICHAEL BRETT
Ibn Khaldun and the Medieval Maghrib

G. REX SMITH
Studies in the Medieval History of Yemen and South Arabia

C.E. BOSWORTH
The Arabs, Byzantium and Iran: Studies in Early Islamic History and Culture

C.F. BECKINGHAM
Between Islam and Christendom: Travellers, Facts and Legends in the Middle Ages and the Renaissance

DANIEL MARTIN VARISCO
Medieval Folk Astronomy and Agriculture in Arabia and the Yemen

PAOLO M. COSTA
Studies in Arabian Architecture

ELIYAHU ASHTOR, EDITED BY B.Z. KEDAR
Technology, Industry and Trade: The Levant versus Europe, 1250–1500

NICHOLAS LOWIK, EDITED BY JOE CRIBB
Islamic Coins and Trade in the Medieval World

VARIORUM COLLECTED STUDIES SERIES

Studies in Medieval Islamic Technology

To Pat, with love

Donald R. Hill

Studies in Medieval Islamic Technology

From Philo to al-Jazarī –
from Alexandria to Diyār Bakr

edited by
David A. King

Routledge
Taylor & Francis Group
LONDON AND NEW YORK

First published 1998 by Ashgate Publishing

Published 2016 by Routledge
2 Park Square, Milton Park, Abingdon, Oxon OX14 4RN
711 Third Avenue, New York, NY 10017, USA

Routledge is an imprint of the Taylor & Francis Group, an informa business

British Library CIP Data
Hill, Donald Routledge
Medieval Islamic Technology: From Philo to al-Jazari –
From Alexandria to Diyar Bakr. (Variorum Collected Studies Series: CS555).
1. Technology – Islamic Countries – History. I. Title.
609.1'7671

US Library of Congress CIP Data
Hill, Donald Routledge
Medieval Islamic Technology: From Philo to al-Jazari –
From Alexandria to Diyar Bakr / Donald R. Hill.
p. cm. – (Variorum Collected Studies Series: CS555).
Includes bibliographical references.
1. Technology – Islamic Empire – History. 2. Civilization, Islamic
I. Title. II. Series.
T27.3.I75H55 1998 98–4611
609.17'671–dc21 CIP

Transfered to Digital Printing in 2010

ISBN 9780860786061 (hbk)

CONTENTS

This volume contains xxvi + 368 pages

PUBLISHER'S NOTE

The articles in this volume, as in all others in the Collected Studies Series, have not been given a new, continuous pagination. In order to avoid confusion, and to facilitate their use where these same studies have been referred to elsewhere, the original pagination has been maintained wherever possible.

Each article has been given a Roman numeral in order of appearance, as listed in the Contents. This number is repeated on each page and quoted in the index entries.

EDITORIAL PREFACE

The scholarly world knows of the major publications of Donald Hill: his books on al-Jazarī and the Banū Mūsā, his little book on water-clocks; his overview of Islamic engineering (with A. Y. al-Hassan); and his useful little history of engineering in classical and medieval times. The purpose of this volume is two-fold: first to pay tribute to the man and his contributions to scholarship, and second to gather the most significant of his various articles that are scattered in various journals and collected works.

To achieve that first aim I include below a personal tribute to Donald, modified from my notice about him which appeared in the journal *Arabic Science and Philosophy* in 1995, and supplemented by information from the eloquent obituary that appeared in *The Times* of London on 27 June, 1994. The author of that obituary was Anthony Turner, who graciously consented to my using it and then polished the final version which appears below. I append a list of Donald's publications.

The achievement of the second aim is very much due to John Smedley, editor of Variorum, who sent me occasional faxes to remind me that I promised him to prepare this volume in 1994. It was the awesome prospect of preparing the indexes which kept me waiting until October, 1997, when finally I had a whole week free to devote to the task. The delay at least ensured that paper X was included here.

To what is recorded in the notice that follows I can only add that it was for me a pleasure to know Donald Hill from 1973, when we met in Aleppo, to my last visit to him in his home in Great Bookham in 1989. It was always comforting to know that there was Donald Hill working away on Arabic texts on mechanical devices, and he clearly drew some comfort from the fact that Ted Kennedy was working away at Arabic astronomical texts, with four of his students, Richard Lorch, Julio Samsó, George Saliba, and myself doing our best to pick up where Kennedy left off. And such texts on mechanical devices were all for Donald. So, for example, when I stumbled across a manuscript of an Andalusian treatise on mechanical devices in the Biblioteca Laurenziana in Florence in 1974, the sensible thing to do was to alert Donald of its existence, whereupon he published a preliminary account of it (XVII).[1] His attribution of the treatise to Abū ʿAbdallāh Ibn Muʿādh al-Jayyānī is unfortunately incorrect, as he was soon to realize: the author is Ibn Khalaf al-Murādī, otherwise unknown to us. Likewise when I found a previously-unrecorded manuscript of the treatise on mechanical devices by the Banū Mūsā in the Topkapı Library in 1975, with better diagrams than those in other available manuscripts, I alerted Donald and he, with mixed feelings, had to redraw all of the diagrams for his book on the treatise.[2] It was a particular pleasure for me to be able to help Donald in this limited way. Alas he never consulted with me or any of my colleagues when he was writing on Islamic astronomy and mathematics. As a result, there are a few errors in his writings on those subjects, not committed by him, but simply well-established in the literature that he was citing.[3] When we met for the last time in 1989 my wife and I were travelling by car with my father, from Buckinghamshire to Frankfurt; we stopped off in Great Bookham. It was a most interesting experience to witness the encounter of two men (my father was only seven years years older than Donald) who had both in a sense, by sheer love of their subject, dedication, native wit and hard work, beaten the established academic system in England.

The choice of the articles was rather delicate, but I made a deliberate attempt to provide the reader with access to Donald's early papers on trebuchets and camels and horses (XIX and XX) and his first writings on what he liked to call "fine technology" (I and II). Particularly important studies are his overview of the textual sources for the history of Islamic technology (V) and his findings in the writings of medieval Muslim geographers (VI). There is a preview of the contents of the treatise of the Banū Mūsā (XII), but alas no similar overview of al-Jazarī's treatise for comparison: the reader must simply consult Donald's books. The newly-discovered Florence manuscript enabled Donald to present an overview (XVII) of a work which has completely changed our understanding of Andalusian technology and raised numerous questions that have still to be answered. His papers on mining and hydraulic technology (VII and VIII) show what an Arabist and friend of the Islamic world can do with diverse textual and archaeological sources. The reader should also consult Donald's chapter on Islamic technology published posthumously in the new *Encyclopedia of the History of Arabic Science* (publication B36).

The compilation of the indexes was an extremely delicate task. It is difficult enough to prepare indexes of one's own works, let alone those of a colleague. I took certain liberties, mostly in an attempt to keep down their length. (John Smedley has good reason not to trust me in this regard.) Particularly important research articles like V and VI are indexed as fully as seemed reasonable. Articles of a more general nature have been indexed with less thoroughness; thus for VII on mining I suppressed place-names, preferring to include the names of the materials mined. I have made a deliberate effort to cover technical terms and have put them in a separate index, in the hope that this might be of use to future researchers.[4]

Notes

1 Namely in item no. B4 in the bibliography that follows, reprinted here as no. XVII. I mentioned the treatise in a review of his book on al-Jazarī published as "Medieval Mechanical Devices", *History of Science* 13 (1975), pp. 284–289 (especially n. 8), repr. in my *Islamic Astronomical Instruments*, London: Variorum, 1987, repr. Aldershot: Variorum, 1995, XX. Scholarly reaction was immediate: see, for example, A. I.

Sabra, "A Note on Codex Biblioteca Medicea-Laurenziana Or. 152", *Journal for the History of Arabic Science* 1 (1977), pp. 276–283, and M. V. Villuendas, "A Further Note on a Mechanical Treatise Contained in Codex Medicea Laurenziana", *ibid.* 2 (1978), pp. 395–396. The subsequent literature on this very important manuscript is listed in Josep Casulleras, "El último capítulo del *Kitāb al-asrār fī natā'iŷ al-afkār*", in *idem* and Julio Samsó, eds, *From Baghdad to Barcelona. Studies in the Islamic Exact Sciences in Honour of Prof. Juan Vernet*, 2 vols, Barcelona: Instituto "Millás Vallicrosa" de Historia de la Ciencia Arabe, 1996, pp. 613–653, especially p. 613, n. 2. The reason for the mis-attribution to Ibn Muʿādh is that the following treatise in the manuscript is indeed by Ibn Muʿādh: on this see now M. V. Villuendas, *La trigonometría europea en el siglo XI - Estudio de la obra de Ibn Muʿāḏ: El Kitāb Maŷhūlāt*, Barcelona: Instituto de Historia de la Ciencia de la Real Academia de Buenas Letras, 1979, where, however, the treatise on mechanics is not mentioned.

2 Compare his article on the treatise of the Banū Mūsā listed in the bibliography as B5 and reprinted here as XII, p. 47, note, and Donald's book on the same subject listed as A4, pp. ix and 14–16.

3 One error that should be mentioned is the attribution of the spherical cosine formula to al-Battānī (item B32 of the bibliography, pp. 490–491, and elsewhere) and the prosthapheretical formula to Ibn Yūnus (item A4, p. 29). Just because a medieval author used a procedure which is mathematically equivalent to a modern formula, this does not mean that he knew the formula. In fact al-Battānī did not use spherical trigonometry at all, and Ibn Yūnus proposed equivalent solutions to a problem of spherical astronomy which can be seen to be equivalent by virtue of the prosthapheretical formula. But such claims litter the modern literature and they will continue to be repeated.

4 For technical terms in medieval technological treatises one should consult the copious indexes of Eilhard Wiedemann, *Aufsätze zur arabischen Wissenschaftsgeschichte*, 2 vols, Hildesheim & New York: Georg Olms, 1970; and *idem*, *Gesammelte Schriften zur arabisch-islamischen Wissenschaftsgeschichte*, 3 vols, Frankfurt am Main: Institut für Geschichte der Arabisch-Islamischen Wissenschaften, 1984–85. Donald's main glossary is to be found in his book on the treatise of the Banū Mūsā, pp. 253–265. There are also glossaries in the Arabic editions of the texts of the Banū Mūsā and al-Jazarī (see endnote 4 to the section that follows).

ACKNOWLEDGMENTS

John Smedley's usual tolerance, forbearance and patience were this time somewhat stretched, but his encouragement was incessant. Anthony Turner contributed suggestions, criticism and above all encouragement in the preparation and organization of this volume; this was but a small part of our continuous collaboration over many years. Kurt Maier (Frankfurt) typed the raw indexes from highlighted photocopies of the various articles. After I had organized the raw data into a preliminary version of the final indexes François Charette (Frankfurt) controlled various problematic entries and made numerous corrections and additions. He then prepared the camera-ready copy of the indexes, of the introductory sections and of five of the articles.

For permission to reprint the various articles I thank the following co-authors: Ahmad Yusuf al-Hassan (VII), Kamal Shehadeh (X) and Richard P. Lorch (X), and the following publishers: The American University of Beirut, Beirut, Lebanon (I); the Institute for the History of Arabic Science, Aleppo (II, XVI, XVII); the Société Internationale de l'Astrolabe, Paris (III); Ithaca Press, an imprint of Garnett Publishing, Reading (IV); Cambridge University Press (V); Mansell, an imprint of Cassell of London, for *History of Technology* (VI and XII); the editors of the *Encyclopaedia of Islam* at E. J. Brill, Leiden (VII, VIII, XI, XIII); Les Belles Lettres, Paris (IX); the Institut für Geschichte der Arabisch-Islamischen Wissenschaften, Frankfurt am Main (X); *Annals of Science* and Taylor & Francis, London (XIV); Simon & Schuster, New York, for Charles Scribner's Sons, formerly of New York (XV); Secretaría de Estado de Cultura, Madrid (XVIII); the Center for Medieval and Renaissance Studies, University of California, Los Angeles (XIX); and Oxford University Press (XX).

Donald Routledge Hill, 1922–1994*

> "On my return (from seeing the superb miniatures of al-Jazarī in the Aya Sofya library) I kept telling my lamented colleague Gustave von Grünebaum that the engineering relations between Islam and the West could never be understood until al-Jazarī was edited; and he kept replying, 'Who knows enough both of technology and of Arabic to do it?'."
> *Lynn White, Jr., in his foreword to Donald Hill's book on al-Jazarī.*

Donald was born in London on August 6, 1922. He served in the Middle East as a lieutenant in the Royal Engineers from 1941 to 1946, including two years in the Eighth Army in North Africa and Italy, where he was wounded in action. In 1949 he obtained a degree in engineering from London University and joined the Iraq Petroleum Company, for whom he worked on construction projects in the Lebanon, Syria and Qatar. In 1947 in Beirut he married Patricia Sulmen of Jerusalem. Back in England he held several senior posts in major petro-chemical companies from 1955 until his retirement in 1984.

Donald was a competent linguist. Having already mastered French, German, Spanish and Italian, his sojourn in the Middle East provided the opportunity to add colloquial Arabic to his list. But he also mastered the intricacies of classical Arabic, and ventured into

*Modified from the obituary notice in *Arabic Science and Philosophy* 5 (1995), pp. 297–302. That notice relied to some extent on the eloquent obituary in *The Times* (London), June 27, 1994, more of which has been inserted here. This writer is indebted to the author, Anthony Turner, for his generous permission to use it, as well as for his improvements to the penultimate version of the present notice.

the field of Islamic history. He received an M.Litt. degree in Islamic history from the University of Durham in 1964 and a Ph.D. from the University of London in 1970. His thesis was a comprehensive computer analysis of the termination of hostilities during the early phase of the Arab conquests.

Donald's early study of trebuchets arose out of his awareness of the inadequacy of the Arabs in siege warfare during that early period. This brought him into contact with Lynn White Jr. of the University of California, the leading historian of medieval technology. And it was Lynn White who drew Donald's attention to a medieval Arabic treatise on mechanical devices, known to the scholarly world only through a partial German translation by Eilhard Wiedemann and Fritz Hauser (1915). The subject of medieval Islamic technology had barely progressed since the time of Wiedemann, who died in 1928. By the time Donald published *The Book of Ingenious Mechanical Devices* by Ibn al-Razzāz al-Jazarī in 1974 he was recognized as the leading authority on the subject. For this book he was awarded the Dexter Prize of the American Society for the History of Technology in 1974 (jointly with Daniel Boorstin). But this was just the beginning. Of major importance to drawing attention to Donald's work was the fact that two years after his book was published, the realization of an exhibition "Science and Technology in Islam" in the context of the World of Islam Festival in London provided an opportunity to test the firmly held and stongly iterated opinion that all al-Jazarī's machines were totally impractical. For the exhibition held in the Science Museum in 1976, three of al-Jazarī's machines - a phlebotomy device, a double-acting pump and the "castle" water-clock - were reconstructed. All three were built without great difficulty following al-Jazarī's instructions and all three worked well. Particularly impressive was the mechanically-complex water-clock some four metres high and about 3 metres deep, in which all sorts of things happen at each hour. The reconstruction is described in Donald's book on water-clocks.[1] Some 20 years later the clock was exhibited in Amsterdam.[2] In the same way as watching the norias turn at Hama gives an impression of the life of materials as the wood creaks, groans, shudders and sings, and one beholds the beauty of the water cascading in the sunlight from the overflowing containers, so the rebuilt castle-clock, all its mechanism hidden away, gives a sense of the wonderment such mechanical marvels must have evoked, leaving one to ponder on the nature of the spirits imprisoned within it. The medieval literary equation of mechanics with magic has been well documented. Hearing the castle water-clock and seeing its automatic activities, makes one realize that the literary trope reflects medieval reality.[3]

When justice had been done to al-Jazarī, Donald set about preparing for publication the translation and commentary of two earlier treatises, those of the Banū Mūsā and Pseudo-Archimedes, as well as a survey of Arabic treatises on water-clocks. He became interested in the social context of technological ideas and their transmission from one culture to another. Donald's collaboration over many years with Ahmad Yusuf al-Hassan on a history of Islamic technology took him far away from the Banū Mūsā and al-Jazarī. Also Dr. al-Hassan had the facilities in Aleppo to publish the edited texts of both al-Jazarī and the Banū Mūsā.[4] I can claim some input in this monumental undertaking because I raised the question in my review of Donald's book on al-Jazarī: it is nice to have a new translation, more readable than anything of Wiedemann's, but where is the edited text, or at least a facsimile of the best available manuscript?[5] In 1984 Donald published a *History of Engineering in Classical and Medieval Times*. And he did not shy away, as others of us did, from an invitation by Edinburgh University Press to prepare *Islamic Science and Technology*, alas his last major work.

Over the years Donald retained his affiliation with the School of Oriental and African Studies in London but essentially worked in isolation and without institutional backing. It was he who inspired others to publish the Arabic texts of the treatises of al-Jazarī and

Taqī 'l-Dīn. But it is most unfortunate that he had no students, and he does not seem to have a successor. His papers, which I had the privilege of inspecting in November, 1994, were (at least compared with my own) remarkably tidy and well-organized. This does explain how he was able to publish one book after the other.

Some days before his death on May 30, 1994, Donald signed and dispatched to various colleagues copies of his latest book, *Islamic Science and Technology*. Few scholars have the satisfaction of having devoted many years so fruitfully to a highly technical subject and also having produced a popular book in which the results of this research are made available to the educated public. Donald's little book will surely be for many years to come *the* book in the English language to which people will turn for a readable introduction to the many aspects of our field.

Donald was, in addition to his professional and academic achievements, very much a family man. His wife Pat, and later their three children and more recently their four grandchildren, were never far from his thoughts. The scholarly world shares their loss, for Donald ranks with B. Carra de Vaux, Eilhard Wiedemann and Lynn White among the main contributors to our understanding of medieval technology.

What is there left to do in this field now that Donald is no longer with us? A great deal. First, a colloquium "Aspects of Islamic Technology—in memoriam Donald Hill" was organized by the Society for the History of Medieval Technology and Science at the Warburg Institute, London, on April 1, 1995.[6] Second, there is one topic that Donald was working on in his last years and for which his notes are available, alas not in the best of shape. I refer to wheels and perpetual motion as treated in various Arabic sources. I am looking for a competent young scholar of medieval Arabic with the necessary technical knowledge to complete Donald's study. Third, there is an Iranian tradition of fine technology which Donald left virtually untouched. There are treatises on fine technology in Persian that have never been studied.[7] And there is material in modern Iranian publications on the subject that needs to be incorporated into the corpus of literature that Donald built up.[8] There is even material in modern Arabic publications that appears to have escaped Donald's notice.[9] Fourth, the relevance of the Islamic textual material for the future of the archaeology of the Islamic world has to be taken seriously. There is, for example, still no properly-documented study of the water-operated astrolabic clock in Fez.[10] The mathematics, and to some extent the technology behind such remarkable, truly Islamic, architectural features as the *muqarnas*, a honey-comb-type structure used from one end of the Islamic world to the other to decorate corners of ceilings and dome-foundations, is indeed described in texts.[11] Finally, there are close to a thousand medieval Islamic astronomical instruments, mainly in brass, that have never been seriously studied by a historian of technology. These precious sources are now to some extent under our control, firstly as far as their locations are concerned, and secondly as far as their inscriptions, dating, and provenance are concerned.[12] They have been studied so far only from the perspective of the history of astronomy, but they are also primary sources for the history of very fine technology.[13]

David A. King

Notes

1 *Arabic Water-Clocks* (listed as A5), pp. 103–111.

2 Catalogued in Anthony J. Turner, *Time*, The Hague: Tijd voor Tijd Foundation, 1990, pp. 72–73, no. 88.

3 These three sentences are in their essence taken from Anthony J. Turner, "Donald Hill and Arabic Water-Clocks", unpublished paper delived at the 1995 London Colloquium (see n. 6 on p. xxii).

4 Ahmad Yusuf al-Hassan, ed., *Kitāb al-Ḥiyal ("The Book of Ingenious Devices") by the Banū Mūsā bin Shākir*, Aleppo: Institute for the History of Arabic Science, 1981;

and *idem*, ed., *al-Jāmi' bain al-'ilm wal-'amal al-nāfi' fī ṣinā'at al-ḥiyal ("A Compendium on the Theory and Practice of the Mechanical Arts") by Ibn al-Razzāz al-Jazarī*, Aleppo: Institute for the History of Arabic Science, 1979. See also *idem*, ed., *Taqi-al-Din and Arabic Mechanical Engineering, with The Sublime Methods of Spiritual Machines, An Arabic Manuscript of the Sixteenth Century*, Aleppo: Institute for the History of Arabic Science, 1976.

We should also mention Muḥammad Aḥmad Dahmān, *'Ilm al-sā'āt wa-'l-'amal bihā ta'līf Riḍwān ibn Muḥammad al-Sā'ātī ... ma' majmū'a fi 'l-mīkānīk al-Islāmī* [= Riḍwān ibn Muḥammad al-Sā'ātī on the construction and use of clocks, with notes on Islamic mechanics], [Damascus]: Maktabat al-Dirāsāt al-Islāmiyya, n.d. [*ca.* 1980?].

5 But facsimiles have to be done properly. On the pitfalls see Jan-Just Witkam, "Arabic Manuscripts in Distress: The Frankfurt Facsimile Series", *Manuscripts of the Middle East* 4 (1989), pp. 175–180.

6 The papers presented were the following: Emilie Savage-Smith, "Metallurgy, Magic and Divination"; Charles Burnett, "Islamic Technology and Magic"; Anthony Turner, "Donald Hill and Islamic Water-Clocks"; and Alan Williams, "The Metallurgy of Islamic Arms and Armour".

7 See Storey, *Persian Literature - A Biobibliographical Survey*, 5 vols, London: Luzac, 1927 onwards, II:3, section on mechanics on pp. 445–446.

8 See Parviz Mohebbi, *Techniques et Ressources en Iran du 7e au 19e siècle* (Bibliothèque Iranienne, 46), Tehran: Institut Français de Recherche en Iran, 1996.

9 Notably Muhammad Ahmad Dahmān's edition of Riḍwān al-Sā'ātī's treatise *'Ilm al-sā'āt wa-l-'amal bihā*, Damascus, n.d. [*ca.* 1980].

10 The best description is still Derek J. de Solla Price, "Mechanical Water Clocks of the Fourteenth Century in Fez, Morocco", *Proceedings of the Tenth International Congress of History of Science*, Ithaca, N.Y., 1962, pp. 599–602.

11 Articles "Muḳarbas" by A. Fernandez-Puertas and "Muḳarnas" by Doris Behrens-Abouseif in EI^2, VII, pp. 500–501 and 501–506. For the most recent account of the technical literature see Yvonne Dold-Samplonius, "Practical Arabic Mathematics: Measuring the Muqarnas by al-Kāshī", *Centaurus* 45 (1992), pp. 193–242.

12 A long-awaited, but still unpublished, survey of Muslim instrument-makers, unfortunately listed alphabetically rather than organized in any historical or geographical manner, is Alain Brieux and Francis R. Maddison, *Répertoire des facteurs d'astrolabes et de leurs œuvres*, Paris: C.N.R.S. A catalogue of all medieval Islamic and European instruments, arranged chronologically by region, is described in D. A. King, "Medieval Astronomical Instruments - A Catalogue in Preparation", *Bulletin of the Scientific Instrument Society* (London), no. 31 (1991), pp. 3–7. This catalogue is not about to be published in the near future, but descriptions are already available for most Islamic and European instruments from before *ca.* 1550. On the kind of surprises that reward such detailed studies of instruments see *idem*, "Astronomical Instruments between East and West", in Harry Kühnel, ed., *Kommunikation zwischen Orient und Okzident*, Vienna: Österreichische Akademie der Wissenschaften, 1994, pp. 143–198.

13 Donald wrote (no. B31, reprinted as XVIII — see there p. 11 and n.6 on p. 16) that astronomical instruments belong "more properly to the history of astronomy rather than that of technology". There I would say he erred. He modified his opinion when working on the geared luni-solar mechanism of al-Bīrūnī (no. B16), and in his latest books (nos A6 and A8) he devoted several pages to instruments. Metallurgical considerations, construction techniques, brazing and welding, screws, precision markings and scale-graduations, engraving and inlaying techniques. These are aspects of instruments which could teach us a great deal.
Useful studies of technological aspects of European instruments are to be found in Robert B. Gordon, "Metallography of Brass Made in the Sixteenth Century", *Journal of the Historical Metallurgy Society* (London) (1986), pp. 93–96, and *idem*, "Sixteenth-Century Metalworking Technology used in the Manufacture of two German Astrolabes", *Annals of Science* 44 (1987), pp. 71–84; and Allan Chapman, *Dividing the Circle: the Development of Critical Angular Measurement in Astronomy, 1500–1800*, London, 1990, and New York, 1990, 2nd edn, New York, Wiley & Sons, 1996; *idem*, "The Design and Accuracy of Some Observatory Instruments of the Seventeenth-Century", *Annals of Science* 40 (1983), pp. 457–471; *idem*, "A Study of the Accuracy of Scale Graduation on a Group of European Astrolabes", *Annals of Science* 40 (1983), pp. 473–488.

Donald R. Hill - Publications

Note: Boldface Roman numerals to the right of the entries refer to the reprints in this volume. The following abbreviations are used:

Al-Abhath: Journal published by the American University of Beirut.

DSB: *Dictionary of Scientific Biography*, 16 vols, New York: Charles Scribner's Sons, 1970–80.

EI ^{2}: *Encyclopaedia of Islam*, new edition, 8 vols to date. Leiden: E. J. Brill, 1960 to present.

IHAS: Institute for the History of Arabic Science, Aleppo.

JHAS: *Journal for the History of Arabic Science*, Aleppo: IHAS.

UR: Journal published by the Iraqi Cultural Centre, London.

A. Books

1 *The Termination of Hostilities in the Early Arab Conquests*, London: Luzac, 1971.

2 *The Book of Knowledge of Ingenious Mechanical Devices by Ibn al-Razzāz al-Jazarī*, Dordrecht (NL): Reidel, 1974, reprinted Islamabad: Pakistan Hijra Council, 1989.

3 *On the Construction of Water-clocks*, Occasional Paper No. 4, London: Turner & Devereux, 1976.

4 *The Book of Ingenious Devices by the Banū Mūsā ibn Shākir*, Dordrecht: Reidel, 1979, reprinted Islamabad: Pakistan Hijra Council, 1989. A Persian translation has been published in Mashhad, Iran, 1993.

5 *Arabic Water-Clocks*, (Sources and Studies in the History of Arabic-Islamic Science, History of Technology Series - 4), Aleppo: IHAS, 1981.

6 *A History of Engineering in Classical and Medieval Times*, London: Croom Helm, and La Salle, Ill.: Open Court, 1984, reprinted in paper-back, London and New York: Routledge, 1996. Divided into three parts as follows. I: Civil engineering: irrigation and water supply, dams, bridges, roads, building construction, surveying. II: Mechanical engineering: Water-raising machines, power from water and wind. III: Fine technology: Instruments, automata, clocks.

7 *Islamic Technology: An Illustrated History* (with Ahmad Y. Al-Hassan), Cambridge: Cambridge University Press and Paris: U.N.E.S.C.O., 1986. Contains chapters on: mechanical engineering; civil engineering; military technology; ships and navigation; chemical technology; textiles, paper and leather; agriculture and food technology; mining and metallurgy; engineers and artisans.

8 *Islamic Science and Engineering*, Edinburgh: Edinburgh University Press, 1993. Contains: Mathematics (arithmetic, algebra, geometry, trigonometry); Astronomy (folk astronomy, sources of Islamic astronomy, spherical astronomy, planetary theory, instruments, observatories, astrolagy); Physics (mechanics and optics); Chemistry (alchemy, industrial chemistry); Machines (water-raising machines, water-mill, windmills, siege engines); Fine Technology (Islamic sources, clocks, trick vessels, automata, fountains, miscellaneous); Bridges and Dams (bridges, dams); Irrigation and Water Supply (irrigation systems, measurement of irrigation waters, qanats, water-supply); Surveying (levelling, triangulation, quantity surveying); Mining (Islamic mines, maining techology, non-ferrous metallurgy, iron and steel metallurgy); Transmission of Islamic Knowledge to Europe (science, engineering).

B. Articles, chapters in books

1 "Trebuchets", *Viator - Journal of the Center for Medieval and Renaissance Studies* (University of California, Los Angeles) 4 (1973), pp. 99–114. **XIX**

2 "The Camel and the Horse and the Early Arab Conquests", in V. J. Parry and M. E. Yapp, eds, *War, Technology and Society in the Middle East*, Oxford: Oxford University Press, 1975, pp. 32–43. **XX**

3 "Mechanical Technology" in John R. Haynes, ed., *The Genius of Arab Civilization: Source of Renaissance*, 1st edn, New York: New York University Press, 1975 / Phaidon Press, 1976, pp. 175–187; 2nd edn, Cambridge, Mass.: MIT Press / Eurabia Ltd., 1983, pp. 203–215; Arabic edn, Cambridge, Mass.: MIT Press, 1978, pp. 187–199.

4 "A Treatise on Machines by Ibn Muʿādh Abū ʿAbdallāh al-Jayyānī", *JHAS* 1 (1976), pp. 33–44. **XVII**

5 "The Banū Mūsā and their Book of Ingenious Devices", *History of Technology* 2 (1977), pp. 39–76. **XII**

6 "Al-Jazarī", *DSB*, vol. XV, suppl. 1 (1978), pp. 253–255. **XV**

7 "Notice of an Important al-Jazarī Manuscript", *JHAS* 2, (1978), pp. 291–298. **XVI**

8 "Medieval Arabic Mechanical Technology", *Proceedings of the First International Symposium for the History of Arabic Science*, Aleppo, April 5–12, 1976, Aleppo: *IHAS*, 1976 (published 1979), pp. 221–237. **II**

9 "Water-clocks", in Alan Smith, ed., *The Country Life International Dictionary of Clocks*, New York: Putnam, 1979, pp. 129–135.

10 "Al-D̲jazarī", *EI*², Supplement, fasc. 5–6 (1982), pp. 266–267.

11 "Ḥiyal" [= Automata], *EI*², Supplement, fasc. 5–6 (1982), pp. 371–374.

12 "Traditional Islamic Engineering", *UR* 1 (1982), pp. 37–39.

13 "Traditional Islamic Technology: Water Power", *UR* 2/3 (1982), pp. 89–92.

14 "Traditional Mesopotamian Irrigation", abridged translation of "Techniques hydrauliques en Mésopotamie ancienne" by R. van Laere, *Orientalia Lovaniensis Periodica*, University of Leuven Press, 1980, pp. 11–53, in *UR* 2 (1984).

15 "Information on Engineering in the Works of Muslim Geographers", *History of Technology* 9 (1984), pp. 127–142. **VI**

16 "Al-Bīrūnī's Mechanical Calendar", *Annals of Science* 42 (1985), pp. 139–163, repr. in J. V. Field, D. R. Hill and M. T. Wright, eds, *Byzantine and Arabic Mathematical Gearing*, London: Science Museum, 1985. **XIV**

17 "Ḳusṭā b. Lūḳā", *EI* [2], vol. V (1986), pp. 529–530. **XIII**

18 "Hydraulic Machines" (Section 3 of the article "Māʾ" [= Water]), *EI* [2], vol. V (1986), pp. 860–862. **VIII**

19 "Mining Technology" (with A. Y. Al-Hassan), (Section 2 of the article "Maʿdin" [= Mines]), *EI* [2], vol. V (1986), pp. 967–973. **VII**

20 "Engineering in the Medieval Arab World", *Arab Affairs* 2 (Winter 1986/7), pp. 38–46.

21 "Islamic Fine Technology and its Influence on the Development of European Horology", *Al-Abhath* (Beirut) 35 (1987), pp. 9–28. **I**

22 "Engineering", in *Cambridge History of the Middle East*, Cambridge: Cambridge University Press, 1988, pp. 269–270.

23 Introduction to *Héron d'Alexandrie: Les Mécaniques ou L'Élévateur des corps lourds*; facsimile of the 1984 edn by Bernard Carra de Vaux, Paris: Les Belles Lettres, 1988, pp. 9–20. **IX**

24 "I precedenti greci dell'ingegneria europea del Medioevo", in *Ars et Ratio dalla torre di Babele al ponte di Rialto*, Palermo:

Sellerio, 1990, pp. 225–244.

25 "Miḳyās" [= the Nilometer], *EI*^{2}, vol. VII (1990), pp. 39–40. **XI**

26 "Mathematics and Applied Science", chapter 14 of M. J. L. Young, J. D. Latham and R. B. Serjeant, eds, *Religion, Learning and Science in the Abbasid Period*, (Cambridge History of Arabic Literature), Cambridge: Cambridge University Press, 1990, pp. 248–273.

27 "The Literature of Arabic Alchemy", *ibid.*, chapter 18, pp. 328–341.

28 "Mand̲j̲anīḳ" [= Trebuchets], in *EI*^{2}, vol. VI (1991), pp. 405–406.

29 "Mechanical Engineering in the Medieval Near East", *Scientific American* 264:5 (May 1991), pp. 100–105.

30 "Arabic Mechanical Engineering: Survey of the Historical Sources", *Arabic Sciences and Philosophy* 1 (1991), pp. 167–186. **V**

31 "Tecnología andalusí", in Julio Samsó et al., eds, *El legado científico andalusí*, Madrid: Ministerio de Cultura, 1992, pp. 157–172 and 298–309. **XVIII**

32 "Science and Technology in Ninth-Century Baghdad", in P. L. Butzer and D. Lohrmann, eds, *Science in Western and Eastern Civilization in Carolingian Times*, Basel: Birkhäuser, 1993, pp. 485–502.

33 "Arabic Fine Technology and its Influence on European Mechanical Engineering", in D. A. Agius and R. Hitchcock, eds, *The Arab Influence in Medieval Europe*, Reading: Ithaca, 1994, pp. 25–43. **IV**

34 "From Philo to al-Jazarī", in W. D. Hackmann and A. J. Turner, eds, *Learning, Language and Invention: Essays Presented to Francis Maddison*, Aldershot: Variorum and Paris: Société Internationale de l'Astrolabe, 1994, pp. 188–206. **III**

35 "Construction of a Fluting Machine by Apollonius the Carpenter" (with Kamal Shehadeh and Richard P. Lorch), *Zeitschrift für Geschichte der Arabisch-Islamischen Wissenschaften* (Frankfurt am Main) 9 (1994/95), pp. 326–356. **X**

36 "Engineering", in Roshdi Rashed and Régis Morelon, eds, *Encyclopedia of the History of Arabic Science*, 3 vols, London and New York: Routledge, 1996, vol. 3, pp. 751–795, with French translation "Technologie", in *eidem*, eds, *Histoire des sciences arabes*, 3 vols, Paris: Éditions du Seuil, 1997, vol. 3, pp. 11–54.

C. Reviews

1 Thorkild Schiøler, *Roman and Islamic Water-Lifting Wheels*, Odense (DK): Odense University Press, 1973; in *Technology and Culture* 15 (1974), pp. 628–630.

2 Seyyed Hossein Nasr, *Islamic Science: An Illustrated Study*, London: World of Islam Publishing Company Ltd., 1976, in *The Times Literary Supplement*, 30 April, 1976.

3 Ahmad Yusuf al-Hassan, *Taqī al-Dīn and Arabic Mechanical Engineering*, Aleppo: University of Aleppo Press, 1976, in *ISIS* 69 (1978), pp. 117–118.

4 Charles Singer et al., eds, *A History of Technology*, vol. VIII, Consolidated Indexes, compiled by Richard Raper, in *The Times Literary Supplement*, March 1985, p. 366.

5 Atilla Bir, *The Book "Kitāb al-Ḥiyal" of Banū Mūsā bin Shākir: Interpreted in Sense of Modern System and Control Engineering*, Istanbul: Research Center for Islamic History, Art, and Culture, 1990, in *ISIS* 83 (1992), p. 480.

I

ISLAMIC FINE TECHNOLOGY AND ITS INFLUENCE ON THE DEVELOPMENT OF EUROPEAN HOROLOGY

It is generally accepted that the invention of the mechanical clock is one of the most important events in the history of technology. It was the first machine to use a steady gravitational pull as its motive force, and the means of controlling and transmitting that force embodied principles that were to be of great significance in the development of machine design. It is all the more unfortunate, therefore, that the ancestry of the mechanical clock has been largely neglected by western historians of technology and horology. One of the few scholars to attempt to make up for this neglect was the late lamented Derek de Solla Price and his comments on the subject are worthy of attention:

> Now, the history of the mechanical clock is as peculiar as it is fundamental. Almost any book on the history of time measurement opens with a pious first chapter dealing with sundials and water clocks, followed by a chapter in which the first mechanical clock described looks recognizably modern. The beginning is indeed so abrupt that it often seems to me that the phrase "history of time measurement" must have been expressly coined to conceal from the public the awful fact that the clock (as distinct from other time-telling devices) had no early history. It appears to spring forward at birth fully formed and in healthy maturity, needing only a few improvements such as the substitution of a pendulum for the foliot balance and the refinement of the tick-tocking escapement into a precision mechanism.[1]

The mechanical clock does, of course, have an early history and no-one has done more than Professor Price in tracing its origins; a little later we shall meet one of the devices which he identified as being among its ancestors. Although he acknowledged the Islamic contribution to the history of time-measurement, however, his researches were mainly directed elsewhere. Indeed, only in quite recent times has some of the important material in Arabic become easily

1 Derek de Solla Price, *Science since Babylon* (Yale University Press, 1975), p. 30.

accessible. My purpose in this paper is to attempt to trace the origins, development and diffusion of the various ideas and components which eventually coalesced into the mechanical clock. It should be borne in mind, however, that almost all these separate parts have identities of their own, either as parts of other machines or as machines in their own right.

The essentials of the mechanical clock in its early form were the weight-drive, the 'going train' of gears and the verge-and-foliot escapement. Some form of marking the passage of the hours was also necessary: the tower clocks of medieval Europe, in addition to dials, usually incorporated elaborate biological and celestial automata; the mechanism for striking the hours was activated by the 'striking train' of gears, which included segmental wheels, i.e. cogwheels having teeth on only part of their perimeters. If we examine some of the significant developments in fine technology, particularly in Islam, before the fourteenth century we shall be able to see how almost all this array of mechanisms came to be available when it was needed by the clockmakers.

It is not possible in the space available to enter into a detailed discussion about the origins of fine technology — the kind of engineering that is concerned with delicate mechanisms and sophisticated controls. It is known that mechanical devices were constructed in India as early as the third century B.C. and that elaborate timepieces incorporating biological automata were in use there in the eleventh century A.D.[2] In China, inflow clepsydras were in widespread use for timekeeping from about 200 B.C. onwards, and steelyard clepsydras are first mentioned in a Chinese book of A.D. 450, with the implication that they had been in use for some time before this date. The great water-clock described by Su Sung in a book completed in A.D. 1090 was probably derived from this relatively simple type of clepsydra. This was a monumental clock, driven by a scoop-wheel; the wheel was provided with a very ingenious escapement system, in essence two steelyards upon each of which the scoops acted in turn. When a scoop was full, its weight overcame the balancing system, and it fell freely for a given distance until checked, without recoil, by a locking device. The next scoop then came under the delivery jet and the cycle repeated itself. The clock was about 40 feet in height and included a rotating armillary sphere, together with numerous jackwork figures with both audible and visible effects.[3]

There is no evidence that Chinese or Indian ideas were transmitted westwards, and indeed the chronology would seem to preclude this possibility. For about 250 B.C., in the Fertile Crescent and Eastern Mediterranean, a tradition of fine technology was founded that flourished and grew from that time onwards. This

2 V. Ragavan, "Yantras or Mechanical Contrivances in Ancient India," *The Indian Institute of Culture* (Bangalore, 1952), Transaction No. 10.

3 Joseph Needham, *Science and Civilisation in China* (Cambridge, Cambridge University Press, 1965), vol. 4, pt. 2, pp. 446-66.

tradition, founded by Greeks, Egyptians and Syrians, was continued by their successors and descendants in the world of Islam, and later became a fertile source of inspiration to the peoples of other cultures.

Our earliest information about the origins of fine technology occurs in the writings of Vitruvius, who attributes the invention of the organ and the complex water-clock to an Egyptian engineer called Ctesibius. In Chapter 8 (vol 2, book 9) of *De Architectura,* Vitruvius describes two clocks constructed by Ctesibius, both which were operated by the steady rise of a float in the receiver of an inflow clepsydra. In one of these clocks there was a vertical rod soldered to the top of the float; the rod was toothed and meshed with a toothed drum. We are not told about the transmission mechanisms, although evidently there were other gears, but the end result was the activation of automata and the discharge of pellets. In the same chapter Vitruvius describes another clock, not attributed to Ctesibius, in which a chain attached to the float is passed round a horizontal axle. Fixed to the axle is a disc upon which a star map is engraved together with a Zodiac circle in which a small image of the sun can be set to its correct position for the day in question. An hour scale consisting of a network of bronze wires is placed close to the disc. As the disc rotates the 'sun' moves in a circular arc through the unequal hours for that day.[4] The fragment of a plate from a clock of this type, dated to about the second century A.D., was discovered near Salzburg, evidence that this device continued in use in the Roman Empire.[5]

Another remarkable product of the so-called Hellenistic Age is the geared calendar known as the Antikythera Mechanism because of its discovery as part of the cargo of a sunken vessel in 1900 near the island of Antikythera (between Crete and Kythera). It was probably made in Rhodes about 87 B.C. It consists of a large number of gear wheels, all of which were turned by hand by the rotation of a single gear wheel. Among its purposes were the finding of the positions of the sun and moon, and the risings and settings of the circuit of notable fixed stars through the cycle of years and months. A very interesting part of the mechanism is a differential turntable, the purpose of which may have been to produce as output a function of the Metonic cycle of the moon.[6]

Treatises by two important writers have survived from the Hellenistic period. The *Pneumatics* of Philo (fl.c. 230 B.C.) exists only in a number of Arabic versions, all of which contain Islamic additions to Philo's original text. It is likely, for example, that several of the water-raising machines are of Islamic origin. Nevertheless, a number of machines can be positively attributed to Philo, and

4 Vitruvius, *De Architectura,* ed. F. Granger, 2 vols. (London: Loeb Classics, 1970) vol. 2, pp. 255-67.

5 H. Diels, *Antike Technik* (Leipzig and Berlin, 1914) pp. 213 ff.

6 Derek de Solla Price, *Gears from the Greeks* (New York: Science History Publications, 1975), *passim.*

indicate that he had a good grasp of the principles of pneumatics.[7] From Hero of Alexandria (fl.c. 62 A.D.) we have the *Pneumatics* and the *Mechanics,* the latter extant in a good Arabic version by Qusṭā b. Lūqā, made in Baghdad towards the close of the 3rd/9th century.[8] The *Pneumatics* is clearly derived from Philo's work, which it extends and amplifies. The *Mechanics* demonstrates a mastery of theoretical mechanics, and includes discussions on mechanical advantage and the use of complex gear-trains for moving heavy loads, although it is unlikely that these were ever incorporated in practical machines.

Another treatise, *On the Construction of Water-clocks,* carries the name of Archimedes (c. 250 B.C.), and this too exists only in Arabic versions.[9] The water-clock was driven by the descent of a heavy float in a cylindrical reservoir, the speed of descent being kept constant by a feed-back control system: water from the reservoir discharged through a conical valve, the plug of which was soldered to the top of a small float in a float-chamber, which had an outlet pipe at the bottom. When water issued from this outlet the valve opened momentarily, water ran into the float-chamber from the reservoir and the valve closed momentarily — and so on. By this means an almost constant head was maintained in the float-chamber, and this ensured that the rate of discharge was also constant. Power from the float was transferred to a ball-release device through a pair of gears. The treatise describes a number of other automata, some of which may have been added by Philo, while others were certainly Arabic additions. Arabic writers are unanimous in attributing the invention of the feed-back control system and the ball-release mechanism to Archimedes and there is no reason to doubt this attribution, since Arabic engineers were usually very scrupulous in acknowledging their sources.[10]

We know that the tradition for building monumental water-clocks, with their associated automata, continued into Byzantine times and beyond. A clock of this type in Gaza was described by Procopius in the sixth century: there was a Gorgon, whose eyes rolled fearsomely at the end of each hour, and two rows of twelve doors each. The upper row was for the night, one being illuminated each hour by a light which moved at constant speed in front of the doors. The lower row was for the daytime; each door had two leaves. At the end of the first hour an eagle above the

7 Philo of Byzantium, "Le Livre des Appareils Pneumatiques et des Machines Hydrauliques par Philon de Byzance," Arabic text ed. with French translation by Carra de Vaux, *Paris Académie des Inscriptions et Belles Lettres,* 38 (1903), pt. 1.

8 *Pneumatics,* 1st ed. B. Woodcroft (London, 1851); facsimile of Woodcroft, ed. with introduction by Marie Boas Hall (London and New York: MacDonald, 1971); *Mechanics,* ed. with French trans. by Carra de Vaux, "Les Mechaniques ou l'Elevateur de Heron d'Alexandrie sur la version de Qosṭā ibn Lūqā," *Journal Asiatique,* ninth series, 1 (1893), pp. 386-472; 2, pp. 152-269, 450-514.

9 Donald R. Hill, *On the Construction of Water-clocks* (Occasional Paper no. 4, Turner and Devereux, London, 1976).

10 For example, al-Jazarī acknowledged his derivation of the water machinery in his first clock from 'Archimedes' and also makes respectful mention of the Banū Mūsā's work on machines.

first door unfolded its wings and moved forward. The door opened, while a figure of Helios, moving along a ledge in front of the doors, pointed to it. The figure of Hercules emerged from the door carrying the spoils of his first labour, the lion's skin, and the eagle covered his head with a victor's wreath. The hero then moved back into his niche and the door leaves closed upon him. A similar action occurred every hour, Hercules carrying in succession the spoils of his twelve labours. The hours were also marked by the activation of other automata and by the sounding of a gong. Unfortunately, Procopius tells us nothing at all about the machinery used to drive the clock and activate the automata.[11] Writing in 600/1203, Riḍwān b. al-Sāʿātī tells us that the basic water-machinery of Archimedes was transmitted to Persia, where the feed-back control system was retained, whereas the other arrangements were changed and doors for the day and the night were added. This design was transmitted to Damascus in the days of the Byzantines and continued to be constructed in Umayyad times.[12] The description of the Gaza clock is a confirmation of Riḍwān's statement.

It should always be borne in mind, when discussing fine technology, that some of its ideas came from everyday machines. Of these the saqiya, the noria and the geared water mill are most relevant to our theme. The saqiya, a chain-of-pots driven through a pair of gears by an animal moving in a circle, was almost certainly invented in Egypt in the third century B.C.[13] The noria, the compartmented water-driven wheel, probably originated in Syria or al-Jazīra, about 200 B.C.[14] The question of the origin of the water mill driven by a vertical paddle-wheel through a pair of gears remains unresolved. Since Vitruvius implies, however, that it was invented by analogy with the noria, it may well be that its invention occurred in the same area some years later.

There is strong evidence in the foregoing summary, and from other material, to suggest that the main centre for the inventions of fine technology and utilitarian machines lay in Alexandria from about 300 B.C. to 300 A.D., although new ideas also saw the light in Greece, Asia Minor, Syria and Iran. In any case, although the language of culture in Seleucid Syria and Ptolemaic Egypt was Greek, the engineers themselves came from all the countries of the Middle East and the Eastern Mediterranean. Although the stream of invention lost much of its force in the four ensuing centuries, the tradition of fine technology and mechanical engineering remained intact until the advent of Islam.

The story of the establishment of Islam and the diffusion of the Arabic language

11 H. Diels, "Über die von Prokop beschriebene Kunstuhr von Gaza," *Abhandl. d. preuss. Akad. Wiss. Berlin* (Teubner: Phil. Hist. Klasse, 1917), No.7.

12 Donald R. Hill, *Arabic Water-clocks,* Institute for the History of Arabic Science, University of Aleppo, 1981, pp. 12-3.

13 Donald R. Hill, *A History of Engineering in Classical and Medieval Times* (London: Croom Helm, 1984), pp. 138-9.

14 *Ibid.,* 140-1.

from the Atlantic to Central Asia, in the two centuries following the Arab conquests, has been well documented. After the establishment of a homogeneous Islamic culture, the science and technology of earlier times were absorbed and then developed into recognisably Muslim forms. In some fields — irrigation for example — the process occurred very smoothly, simply because the indigenous engineers passed on their skills to their sons who were by then Arabic-speaking Muslims. Lack of space precludes detailed discussion of Muslim achievements in fields such as irrigation, milling and water raising, achievements that were the result of a combination of innovativeness and a large increase in the scale of operations. In undertakings of this kind, in which the machines and structures were durable, craftsmen could learn their professional skills on site, without recourse to written manuals. With fine technology things were rather different. We have seen that the art of constructing monumental water-clocks continued from Byzantine to Islamic times without interruption, and there may well have been existing water-clocks available for inspection. As in the case of heavy engineering, it is also probable that the skills of fine technology were passed on from master to pupil. It is certain, however, given the complexity of some of the smaller devices, that documents such as the treatises of 'Archimedes', Philo and Hero played an important role in transferring this technology into the world of Islam. Indeed, we know this to be so because, as already mentioned, Muslim engineers acknowledged such works as part of their inheritance.

Before proceeding to discuss the purely Muslim contributions to fine technology, it will be as well to recall the purpose of this paper, and to see which of the discrete elements of the mechanical clock were already present in the Muslim inheritance. In the saqiya and the water mill we have simple gears capable of transmitting high torque; in the Antikythera Mechanism we have complex gears, including a differential but *not* segmental or epicyclic gears. Neither in this device, nor in any other pre-Islamic examples, do we have the combination of complex gears and high torque. With reservations, because it is not freely suspended, the heavy float in the 'Archimedes' clock may be regarded as part of weight-drive, and the feed-back control system as a type of escapement. Biological automata were quite advanced and fairly simple celestial analogues had been introduced. Audible, as well as visual, signals were used for marking the passage of the hours.

The first substantial evidence for the development of fine technology in Islam is provided by the Banū Mūsā's *Book of Ingenious Devices,* written about the middle of the 3rd/9th century in Baghdad.[15] Although these three brothers undoubtedly took the works of Philo and Hero as their starting point, their work exhibits a greater mastery of physical media than does that of either of their predecessors. In

15 *The Book of Ingenious Devices* has been published in both English and Arabic: the Arabic edition is edited by Aḥmad Y. al-Ḥasan, Institute for the History of Arabic Science, University of Aleppo, 1981; the annotated English translation is by Donald R. Hill, Reidel, Dordrecht, 1979.

particular they display an astonishing skill in the manipulation of small variations in aerostatic and hydrostatic pressures. They also made considerable use of automatically-operated conical valves in control systems, and were the first of a succession of Arabic engineers to demonstrate a pre-occupation with 'in-line' automatic controls. In many ways they were ahead of their time, and none of their successors attempted to improve upon their results, although their work was well known and appreciated in the Muslim world. Perhaps the words of Ibn Khaldūn explain why no advance was made again before the introduction of pneumatic instrumentation in modern times: 'there exists a book on mechanics that mentions every astonishing remarkable and nice mechanical contrivance. It is often difficult to understand, because the geometrical proofs occurring in it are difficult. People have copies of it. They ascribe it to the Banū Shākir'.[16] There is nothing of direct relevance to our present theme in the Banū Mūsā's book, no addition to our collection of clockwork elements. Their importance in the present context is nevertheless great, because they exemplify a recurring characteristic of Arabic engineers — a pre-occupation with finding means for controlling the release of energy. And uncontrolled energy is both useless and dangerous.

We now have to make an apparent digression, and consider a chapter called *Wheels which turn by themselves,* which is to be found in six different Arabic manuscripts.[17] There are variations from one manuscript to another, but there are enough similarities to suggest that they all came from a common source. The devices have often been dismissed as unintelligible or absurd or both, but our written sources are not so abundant that we can afford to dispense with any document, however fanciful it may seem. Two devices, in particular, are worthy of attention.[18] In one of these a large wheel has a number of S-shaped pipes, partially filled with mercury, fixed to one of its sides, in such a manner that when the wheel is at rest in any position it is statically in balance. The wheel is mounted on an axle, which also carries, at either side of the main wheel, a chain-of-pots wheel. According to the text, once the large wheel has been set in motion, it will keep on rotating because of the action of the mercury sliding up and down in the pipes. The second device consists of a pair of water-wheels driven through gears by the descent of two lead weights attached to ropes wound around the main axle of the machine. It is reasonably certain that I have given here the basic principles described in the two passages,although there are obscurities in the text and the accompanying illustrations add little to our understanding. Both concepts are, of course, nonsense — the first because there is no input of energy and the second because the descent of the weights is uncontrolled. It is assumed that this treatise

16 *Muqaddima,* English translation by F. Rosenthal, Routledge and Kegan Paul, 2nd ed. (London, 1967), vol. 3, p. 132.

17 See Thorkild Schioler, *Roman and Islamic Water-Lifting Wheels* (Odense University Press, 1973), pp. 72-8.

18 Folios 6R to 8R in MS 954 of the Bodleian Library, Oxford.

was written for the amusement of courtly circles and not for practical application, but nevertheless it is difficult to understand why it was written at all. We can conjecture that the writer knew quite well what he was about and that he omitted certain essential components in order to safeguard 'trade secrets'. Weight is lent to this supposition by the existence of a practical machine which combines the main principles of both machines — mercury escapement and weight-drive.

The *Libros del Saber* were prepared in Spain in A.D. 1277 under the direction of Alfonso X of Castile, with the express purpose of making Arabic knowledge available in Castilian.[19] It consists of translations and paraphrases of Arabic works some with known authors some anonymous. In the latter category come the five clocks described in the last section of the work, only one of which concerns us for the moment. This consisted of a large drum made of walnut or jujube wood tightly assembled and sealed with wax or resin. The interior of the drum was divided into twelve compartments, with small holes between the compartments through which mercury flowed. Enough mercury was enclosed to fill just half the compartments. The drum was mounted on the same axle as a large wheel powered by a weightdrive wound around the wheel. Also on the axle was a pinion with six teeth that meshed with 36 oaken teeth on the rim of an astrolabe dial. The mercury drum and pinion made a complete revolution every 4 hours and the astrolabe dial made a complete revolution in 24 hours.

This type of timepiece had been known in Islam since the eleventh century — at least 200 years before the first appearance of weightdriven clocks in the West. The *Libros del Saber* was translated into Italian in 1341, but it was not until 1598 that a similar clock was described in a treatise by Attilio Parisio. It seems likely, although it remains to be proved, that Parisio's clock was derived from the one in the *Libros*. A few years later, Solomon de Caus, a German specialist in water mechanisms (1576- c.1635) described a compartmented cylinder clock that was to be the prototype of many examples constructed in exactly the same form which were to appear again and again in the seventeenth and eighteenth centuries.[20] Clearly, the compartmented drum, whether filled with mercury or with water, was an effective escapement for weightdriven clocks.

In our search for the precursors of the mechanical clock, no single piece of evidence is more important than an Arabic treatise written in Muslim Spain in the 5th/11th century by a certain al-Murādī. Unfortunately, the only known manuscript copy is so badly defaced that it is impossible to deduce from it precisely how any of the machines was constructed, but enough remains for us to make an assessment of its value. Most of the devices were water-clocks, but the five automata machines are of greater relevance to our subject. The water-clocks, although they are not

19 Ed. Rico y Sinobas (5 vols. Madrid, 1963).

20 Silvio A. Bedini, "The Compartmented Cylindrical Clepsydra," *Technology and Culture,* 3 (1963), pp. 114-41.

without points of interest, do not show any essential advances upon the clock described in the 'Archimedes' treatise. The automata machines are similar to clocks without precise timing, in that the machinery causes events to take place at intervals, such as the opening and closing of doors, the discharging of balls and the movement of figures — in other words actions that we have already encountered in Hellenistic water-clocks. Al-Murādī's devices are exceptional, however, on several counts, not the least of which is their sheer size: they contain gear-wheels up to about 70 cm in diameter, counterweights of over 3 kg, and ropes rather than cords or wires are used in the transmission systems. The transmission also includes complex gear trains with segmental gears and epicycles. In one of the devices, for example, the main central gear-wheel had 64 teeth on half its perimeter and meshed with two outer cogwheels, each of which had 32 teeth around its complete perimeter. None of the descriptions of the gearing is complete, but taking the surviving text with the illustrations, there seems little reason to doubt that the machines contained epicycle gears. So here, for the first time we have the transmission of high torque by complex gears, which were far more sophisticated than anything known in Europe before the fourteenth century.[21] It is also worthy of note that these five machines were powered by water-wheels, the same method that was used in the clock of Su Sung, who was roughly a contemporary of al-Murādī. It is therefore just as likely that the Chinese clock was inspired by Islamic ideas as that transmission was in the reverse direction. For our purposes, however, the gears are the most important feature of al-Murādī's machines.

Some well-known Arabic scientists built water-clocks, but these cannot detain us for long, because they did not incorporate any of the mechanisms that we are interested in. Al-Khāzinī tells us that the famous Egyptian scientist Ibn al-Haytham (d. 430/1039) built a water-clock, but gives us no details of its construction.[22] In about 442/1050 the astronomer al-Zarqāll constructed a large water-clock on the banks of the Tagus at Toledo in Spain.[23] This not only told the hours of the day and the night, but also recorded the phases of the moon. It consisted of two large basins, in a building, fed by underground conduits. As soon as the new moon appeared the basins began to fill, at the rate of one fourteenth of their capacity every day. When the moon began to wane the basins emptied at the same rate, until on the twenty-ninth day both were empty. There was some kind of compensating mechanism, so that if any water were extracted from or poured into one of the basins, the water immediately returned to its correct level. This clock was still in operation when the Christians occupied Toledo in 1085 and one of the clepsydras continued in service until after 1133. The methods of control were

21 Hill, *Arabic Water-clocks*, pp. 36-46.

22 Al-Khāzinī, *Kitāb Mīzān al-Ḥikma*, ed. Hāshim al-Nadwa (Hyderabad, Dn, 1940), 8th Treatise, pt. 1.

23 J. M. Millas Vallicrosa, *Estudias sobre Azarquiel* (Madrid, 1950), pp. 6-9.

probably hydraulic: compensating systems occur in some of the Banū Mūsā's devices, and al-Jazarī, in his fountains, uses various methods for changing the flow from one channel to another. Al-Khāzinī himself made steelyard clepsydras which are remarkable for the close attention paid to accuracy of results rather than for mechanical ingenuity.[24]

It has already been noted that the tradition for building water-clocks continued without a break in Syria from the days of the Byzantines into Umayyad times. It seems likely that one of the places where a clock had traditionally been located was by the Jayrūn gate in Damascus: writing late in the 4th/10th century, al-Muqaddasī refers to the 'Gate of the Hours' in a portico by the Jayrūn Gate,[25] and it was here that Muḥammad al-Sāʿātī built a monumental water-clock about 555/1160. The clock was described by Muḥammad's son, Riḍwān, in a treatise written in 600/1203.[26] It was based upon the 'Archimedes' water machinery and contained some interesting mechanical elements, together with some inaccuracies and structural weaknesses. A similar clock, but much improved in its engineering, is described in al-Jazarī's great machine book, composed in Diyār Bakr in 602/1206.[27]. This book is one of the most important engineering documents to have come down to us from any cultural area before modern times, not only for the importance of the fifty machines described in it, but also because the construction of each machine is described with scrupulous care with the avowed (and successful) purpose of enabling later craftsmen to reconstruct any of the machines. There is a wealth of information about techniques and components, many of which were not to appear in Europe until the Renaissance, or even later. Many of these are of great significance in the history of machine technology, including calibration techniques, methods of casting metals, cranks, pumps, control systems, etc. Restricting ourselves, however, to the subject of this paper we should first of all mention a most ingenious water-clock, driven by a submersible bowl called a *ṭarjahār*. This clock, which had a large number of automata, operated on a closed-loop system whereby it continued to work as long as balls were loaded into a magazine at the top of the clock – there was no need to refill the reservoir at the end of the day. Furthermore, as in the small mechanical clocks introduced in the fifteenth century, a large number of interrelated mechanisms were fitted into a confined space, in this case a metal box about 24 cm square in plan by 30 cm high. (Category I, Ch. 4).

24 Al-Khāzinī, 8th Treatise.

25 Al-Muqaddasī, *Aḥsān al-Taqāsīm,* ed. M.J. de Goeje (Leiden: Brill, 1906), p. 158.

26 Hill, *Arabic Water-clocks,* pp. 69-88.

27 *A Compendium on the Theory and Practice of the Mechanical Arts,* ed. Aḥmad Y. al-Ḥasan (Institute for the History of Arabic Science, University of Aleppo, 1979); annotated English translation, *The Book of Knowledge of Ingenious Mechanical Devices,* Donald R. Hill (Dordrecht, Reidel, 1974).

The first clock, which is a most impressive piece of engineering, was reconstructed full-size for the Science Museum, London as part of the 1976 World of Islam Festival: it worked perfectly. At the top of the clock is a Zodiac circle about one metre in diameter made from sheet copper. The outer rim is divided into twelve equal segments for the signs of the Zodiac, and inside this are two movable annuli, set eccentrically to the centre of rotation of the circle. The outer annulus carries a roundel of yellow glass to represent the sun and the inner one has a roundel of white glass for the moon. Every day the 'sun' and the 'moon' are set to their correct positions in the Zodiac for that day. At daybreak the sun rises above an artificial horizon and the circle turns at constant speed, one Zodiacal signs rising while another sets, until sunset. In the water machinery of the clock there is a sophisticated flow regulator which varies the working period of the clock from day to day according to the length of the hours of daylight for each day (Category I, Ch. 1).

In his introduction to the section on musical automata (Category IV, Ch. 7), al-Jazarī mentions an old machine that he had examined. This was driven by a water-wheel and he comments: 'even if the wheel caused a number of rods to fall in succession it would not be slow enough to display the changes adequately'. This is a clear indication that al-Jazarī knew about escapement systems for wheels. It is also evidence that al-Murādī's use of water-wheels in automata machines was not unique. Segmental gears, another of al-Murādī's ideas, recur in two of al-Jazarī's water-raising machines (Category V, Chs. 1 and 2).

We should now turn briefly to touch upon the subject of the construction of astronomical instruments in Islam. The great scientist al-Bīrūnī (d. after 442/1050) has left us a short treatise on the construction of a geared calendar. This is not so complex as the Antikythera instrument, although Muslim skills in astronomy and metalwork would certainly have been equal to the construction of such a calendar. Al-Bīrūnī's instrument was probably 'made to order' and included only the displays required by his client. It had seven gear-wheels of various sizes and its output comprised the phase of the moon for a given day, together with the positions of the sun and the moon in the Zodiac on that day.[28]

Al-Bīrūnī says that the wheels on his calendar are like those fitted to the backs of astrolabes. Geared astrolabes have survived in a very few cases, but the survivors are later than the time of al-Bīrūnī. His remark, however, assures us that they were in use before his time. The normal planispheric astrolabe, which is both a computational and an observational instrument, originated in Hellenistic Egypt but its construction and applications were perfected in the Islamic world. Another type of astronomical instrument, the equatorium, was first developed in Muslim Spain early in the 5th/11th century. It is essentially a computational instrument for obtaining astronomical results without recourse to laborious calculations. The

28 Donald R. Hill, "Al-Bīrūnī's Mechanical Calendar," *Annals of Science*, 42 (1985), pp. 139-63.

spheres, epicycles, deferent circles etc. of Ptolemaic astronomy are represented by metal discs and lengths of cord.[29]

There is sufficient evidence to show that water-clocks equipped with some kind of alarm device were in use in western Europe during the Middle Ages, where it was important in the monasteries for the brethern to be called for the services at the correct times. For example, there is a Latin text of the Cistercian rule, dating to the early part of the twelfth century, that prescribes rules for the sounding of the alarms for services.[30] About 80 years later we find similar instructions in Dante: 'One says *mezza Terza* before ringing for that part of the day, and *mezza Nono* after that part is rung; similarly *mezza Vespro'* (Convito IV, Ch. 23). The earliest description we have of a European water-clock appears in Latin in Ms Ripoll 225 from the Benedictine monastery of Santa Maria de Ripoll at the foot of the Pyrenees. Estimates for its date vary from the middle of the tenth century to some time in the eleventh.[31] The description of the main water machinery is missing but the section describing the striking train is complete. Although this is not very clearly described, it appears that cams on a wheel driven by the water-machinery released weights at intervals, and the release of the weights caused the rotation of iron arms that struck bells. An illustration in a manuscript dated about 1285 shows a water-clock in a monastery in northern France. It is not possible to determine the precise operation from the illustration, but it seems that there was a reservoir below which an inflow clepsydra was suspended. A chain or cord from this receiver passes round the axle of a wheel. The wheel is divided into 15 segments, and between each pair of segments is a hole and a projection on the perimeter. A row of bells above the wheel indicates that this was a chiming clock.[32] The existence of a considerable market for water-clocks in Europe in the twelfth century is implied by the establishment of a guild of clockmakers in Cologne in 1183.[33]

The word *horologium* can mean either water-clock or mechanical clock, an ambiguity that has caused difficulties in establishing a date for the invention of the latter. It could not have been earlier than 1271, since in a treatise written in that year, Robertus Anglicus tells us that the makers of horologia were trying to make a weightdriven clock. 'But they cannot quite complete their task which if they could, it would be a really accurate horologium.' This is a clear indication that the invention of the escapement was approaching at that time, and also implies that the

29 E.S. Kennedy, "The Equatorium of Abu al-Ṣalt," *Physics,* Year 12, fasc. 1 (1970).

30 C.B. Drover, "A Medieval Monastic Water-clock," *Antiquarian Horology,* 1 (1954), pp. 54-8 and 63.

31 Francis Maddison, Bryan Scott and Alan Kent, "An Early Medieval Water-clock," *Antiquarian Horology,* 3 (1962), pp. 348-53.

32 Drover, *op. cit.*

33 Lynn White Jr., *Medieval Technology and Socail Change* (Oxford, Oxford University paperback, 1964), p. 120.

invention was made by a clockmaker who was already familiar with the construction of water-clocks.[34]

It is certain that the invention of the mechanical clock, with verge escapement occured at the end of the thirteenth or the beginning of the fourteenth century. Between 1286 and 1340 a number of tower clocks were built in England, France and northern Italy and in the same period there are references in the works of Jean de Meung and Dante which indicate that the poets had seen a new and exciting machine. The first positive evidence we have is the Dover clock (now in the London Science Museum) which carries the date 1348. This is also the year in which Giovanni de' Dondi began to build his astronomical clock, which is a machine of astonishing sophistication, incorporating complex gearing and displaying not only the time, but the movements of the planets and the dates of the movable feasts.[35] About the middle of the fourteenth century the mechanical clock spread with astonishing rapidity throughout western Europe. To quote Lynn White:

> Something of the civic pride which earlier had expended itself in cathedral-building now was diverted to the construction of astronomical clocks of astounding intricacy and elaboration. No European community felt able to hold up its head unless in its midst the planets wheeled in cycles and epicycles, while angels trumpeted, cocks crew, and apostles, kings, and prophets marched and countermarched at the booming of the hours.[36]

It can hardly be doubted that the verge escapement was invented in Western Europe and that this made possible the rapid developement of the mechanical clock. We have shown, however, that all the other elements of the mechanical clock already existed in Islamic machines such as water-clocks, ingenious devices and astronomical instruments. We have therefore to consider whether this array of elements entered Europe from Islam, and if so how the transmission occurred. it seems very unlikely that the Hellenistic tradition of fine technology persisted in northern Europe in the turbulent and uncultured centuries that followed the dismemberment of the Roman Empire and even less likely that improvements were made to it during those Dark Ages. In all aspects of civilised life, northern Europe did not begin to advance until the beginning of the eleventh century, after the dangers and disruptions of the Viking incursions were over. It seems reasonable to assume, therefore, that a nascent European technology was fuelled by imports from the more advanced culture of Islam from that time onwards, and that the rapid development of all types of technology in Europe from the twelfth century onwards

34 Lynn Thorndike, "Invention of the Mechanical Clock about 1271 AD," *Speculum,* 16 (1941), pp. 242-3.
35 H.A. Lloyd, *Giovanni de' Dondi's Horological Masterpiece* (Limpsfield, Oxted, Surrey, 1956).
36 White, *op. cit.,* p. 124.

was due in no small measure to the impetus of borrowings from Islam.

In the matter of astronomical instruments we do not need to look far to identify the route by which Muslim knowledge entered Europe. It has now been established that the first European treatises on the astrolabe were of Arabic inspiration and were produced in the monastery of Ripoll in Catalonia. These early treatises were full of unassimilated, half-understood Arabic words, and it was not until the middle of the twelfth century that properly constructed treatises on the instrument began to appear in Latin.[37] Similarly, the treatises of Ibn al-Samḥ (c. 1025) and al-Zarqāll (c. 1050) were translated into Castilian in the *Libros del Saber.*

It cannot be said that the diffusion of other Arabic ideas on fine technology can be so readily identified. Certainly, the Christians examined al-Zarqāll's water-clock when they occupied Toledo in 1085 but these were operated by hydraulic means and have little relevance to machine technology. A water-clock and a candle clock appear in the *Libros del Saber,* but these are fairly crude devices when compared to those described by al-Jazarī, for example. We must assume that the translator of these two chapters had only some inferior Arabic material at his disposal. We know, however, that more advanced devices were constructed by the Muslims in Spain, as attested by the mercury clock in the *Libros* and by the devices described by al-Murādī. It is significant that al-Murādī, in the introduction to his treatise, says that he was writing it in order to revive a subject that was in danger of being forgotten, implying that this kind of technology had long been known in Muslim Spain. We therefore have the situation in which sophisticated water-clocks were made in the Iberian peninsula, coupled with the fact that the first known European description of a water-clock was written in the monastery of Ripoll, the very place from which the knowledge of the astrolabe had passed into northern Europe. Moreover, knowledge of the astrolabe was probably disseminated into Europe by Gebert d'Aurillac, later Pope Silvester II, after he had visited Ripoll about the year 967. Since the church was keenly interested in finding a means for timekeeping, it is likely that information about the new Arabic water-clocks was carried to Europe by a churchman – either Gebert himself or a later visitor to Ripoll.

Spain was probably not the only source of European information about Arabic fine technology, although it may have been the earliest. In the Muslim East, the writings of Riḍwān and al-Jazarī, among others, assure us that there had been no falling-off in the tradition of fine technology. Al-Jazarī makes it clear that he regards himself as in a succession of craftsmen and engineers, and we should regard his work as the summation of a long line of anonymous engineers stretching back over the centuries. We have no reason to doubt that this tradition was flourishing at the time of the Cursades, and we should bear in mind that the Crusaders were not

37 Emmanuel Poulle, "Les instruments astronomiques de l'Occident latin au XIe et XIIe siècles," *Cahiers de civilisation medievale,* 15 (1972), pp. 27-40.

all illiterate knights and men-at-arms but included in their numbers churchmen, merchants and scholars. They became acclimatised to the more civilized life of the East, not only on the Syrian coast, but also, for over forty years, in the region of Edessa on the Upper Euphrates – an area where al-Jazarī was to pick up the threads of a continuing tradition only a few decades later. There is little reason to doubt that Crusaders took back with them to Europe, along with other Islamic ideas, some elements of the developed technology of Islam.

It is never easy to trace the diffusion of technological ideas and innovations. In the past, engineers have seldom put their results down on paper, and when they have done so we have no records of their works having translated into other languages. Transmission occurs by word of mouth, by the reports of travellers or by the inspection of older constructions by practising craftsmen. Nevertheless, the weight of evidence is in favour of European fine technology having been strongly influenced by Islamic exemplars. In particular, it has been shown that all the elements of the mechanical clock, except the verge escapement, had been familiar for centuries to Islamic engineers. Most, if not all, of these were diffused from Islam to Europe from the tenth century to the twelfth.

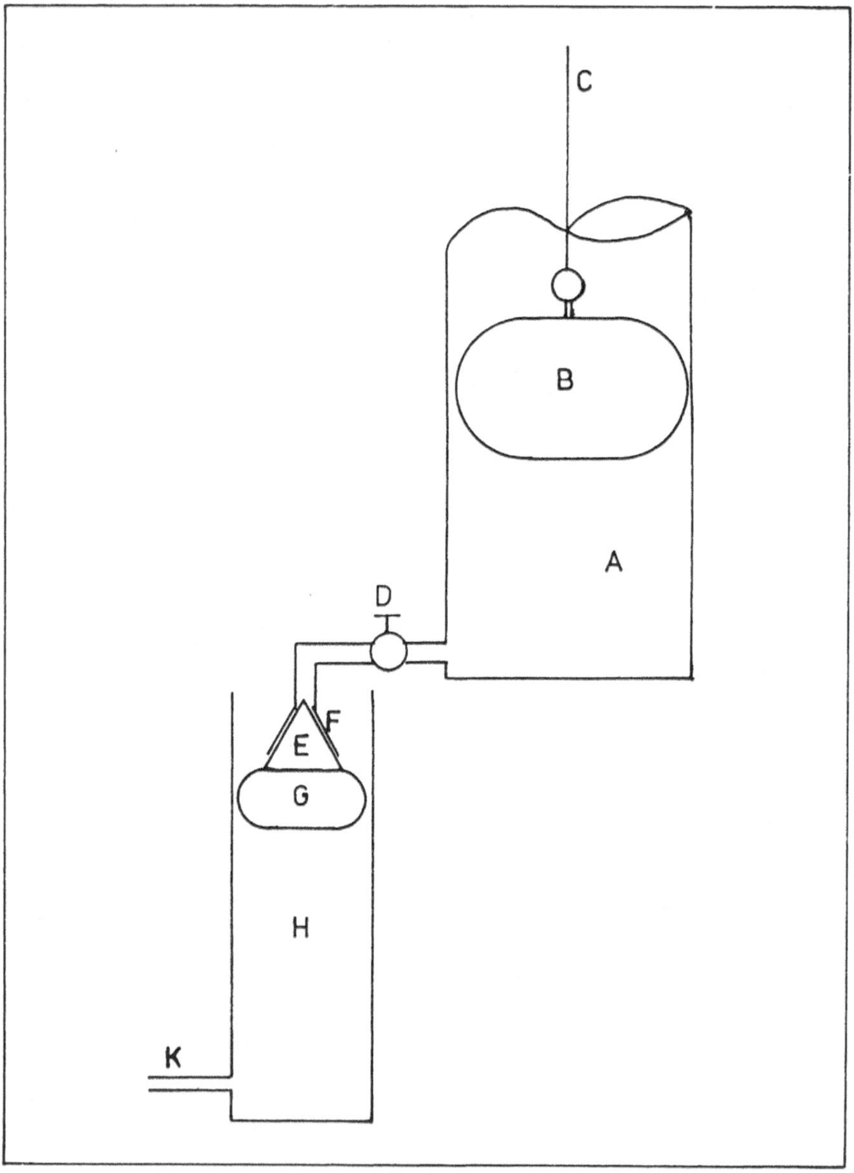

Figure 1

'Archimedes' Water-machinery as used by al-Jazarī and Riḍwān

A-Reservoir; B-Main float; C-Cord to actuate mechanisms; D-tap; E-valve-plug; F-valve-seat; G-float; H-float-chamber; K-outlet pipe.

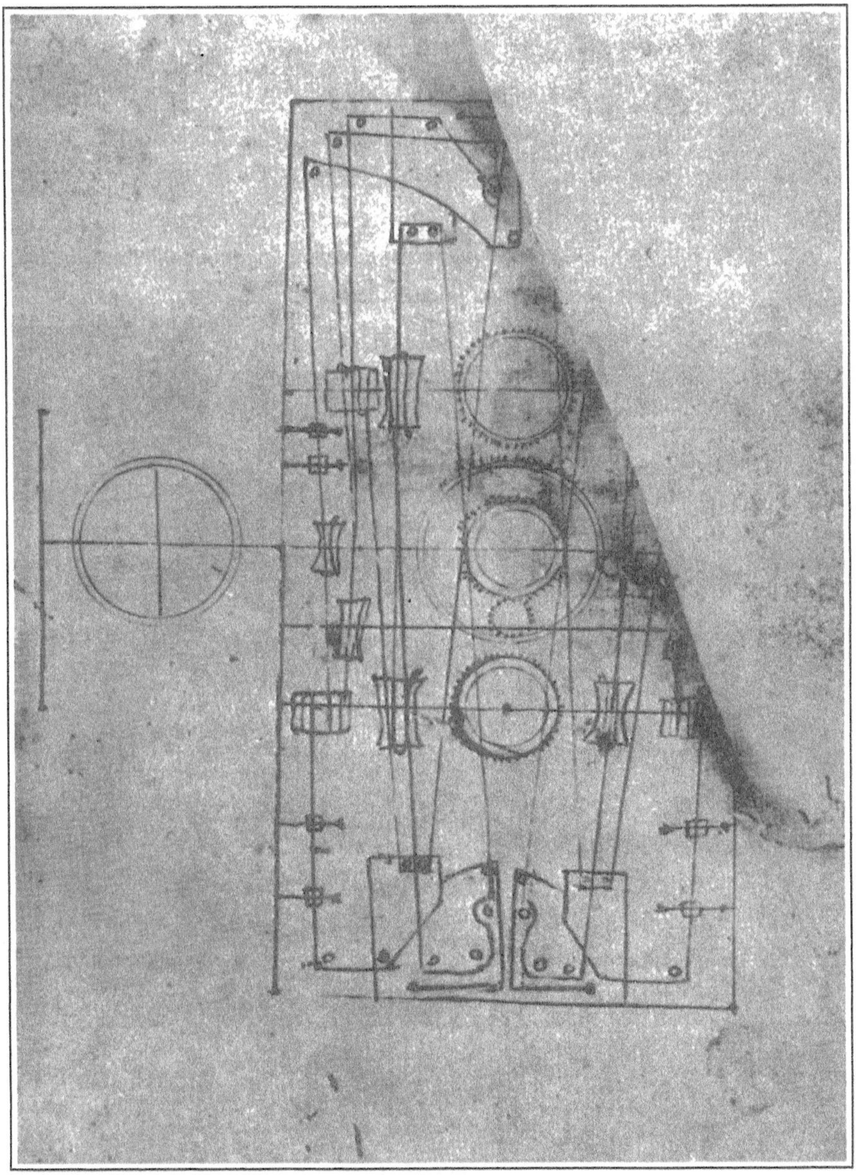

Figure 2

Al-Murādī automata machine

The water-wheel is shown as the concentric circles at the right.

Figure 3

Al-Jazarī's boat clock (from Arabic edition, ed. A.Y. Ḥasan, Aleppo, 1979, p. 101)

The scribe is made to rotate and indicate the passage of the minutes by a pulley system (not shown) driven by the *ṭarjahār*. For the main mechanism see Figure 4.

Figure 4

Detail of boat clock

The balls are loaded into the magazine by removing the detachable dome; the ball-release mechanism is in the 'castle', level with the bird's head. Operation: when the bowl *a* submerges after one hour, the wire *h* pulls the ball-release mechanism and the ball drops from the bird's beak into the serpent's mouth; the serpent rotates, while chain *d* raises the *ṭarjahār*. When the serpent's head reaches the bottom, the ball drops on to the cymbal, and the combined action of the chain and the links *b* tilts the bowl, which empties and settles on the surface of the water in reservoir *n*. The serpent's head rises and returns to its former position, and the cycle starts again. This is therefore a closed-loop system - it continues to operate as long as there are balls in the magazine.

Plate 1

Working face of al-Jazarī's first clock

This clock is about 1·3 metres wide by over 3 metres high.
(Courtesy, Museum of Fine Arts, Boston)

II

Medieval Arabic Mechanical Technology

This paper is concerned with the flourishing of the mechanical arts in the Arab world between the third/ninth and the sixth/thirteenth centuries, as embodied in a number of surviving treatises, with especial reference to those of the Banū Mūsa and al-Jazarī. It must be emphasised, however, that the study of Arab technology (and indeed the history of technology in general) is a relatively neglected field, so that a full assessment of the Arab achievement cannot be made until a comprehensive programme of research has been completed. We must also take into account the existence of a long tradition of utilitarian engineering comprising the construction of water-raising machines, mills, war-engines, buildings, domestic appliances, and articles for personal and household decoration. The names of the craftsmen who constructed and maintained these appliances are unknown to us, and their results were not committed to writing, but their work, quite apart from its necessary rôle in society, was a fruitful source of ideas for the engineering writers. The methods, components and vocabulary of millwrights, carpenters, jewellers and painters were well understood by the best engineering writers, and the influence of humbler craftsmen is readily discernible in their books.

The treatises we are about to consider deal with *ḥiyal*, usually translated as 'ingenious devices'. These depend very largely upon the use of water pressure and air pressure for motive power to produce visual effects, such as the discharge of liquids from nozzles or fountainheads, or the activation of automata. These results depended upon their ability to use an empirical knowledge of statics in design, and their ability to cut, form and assemble the components from the available materials. Their materials included all the common metals and alloys with the exception of zinc, together with small quantities of gold and silver; timber; glass; leather; cords of hemp, cotton and silk; materials for decoration. The commonest materials were wood and sheet copper, the latter being used for the manufacture of vessels, pipes, floats and jackwork figures. Liquid containers were thoroughly tinned internally to inhibit corrosion. Wire was usually made from copper, occasionally from iron, which was also used for the manufacture of chains. Mechanical transmission was achieved by gears, cams, and pulley systems; the larger gearwheels and pulleys were made from wood, smaller ones from copper. Liquids were conveyed in pipes and channels, and the flow was controlled by siphons,

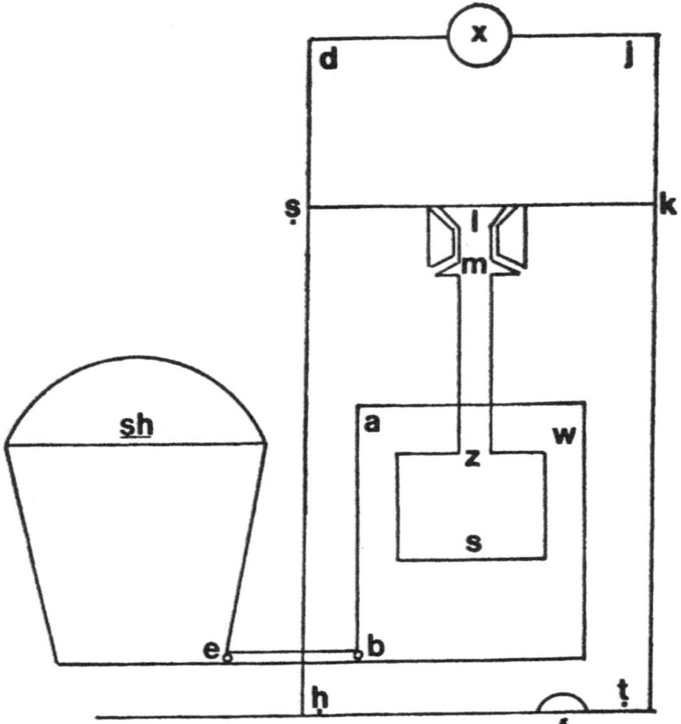

77

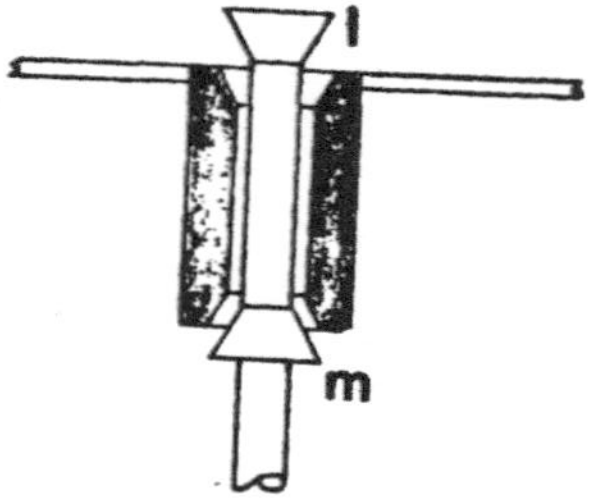

77a

Fig. 1

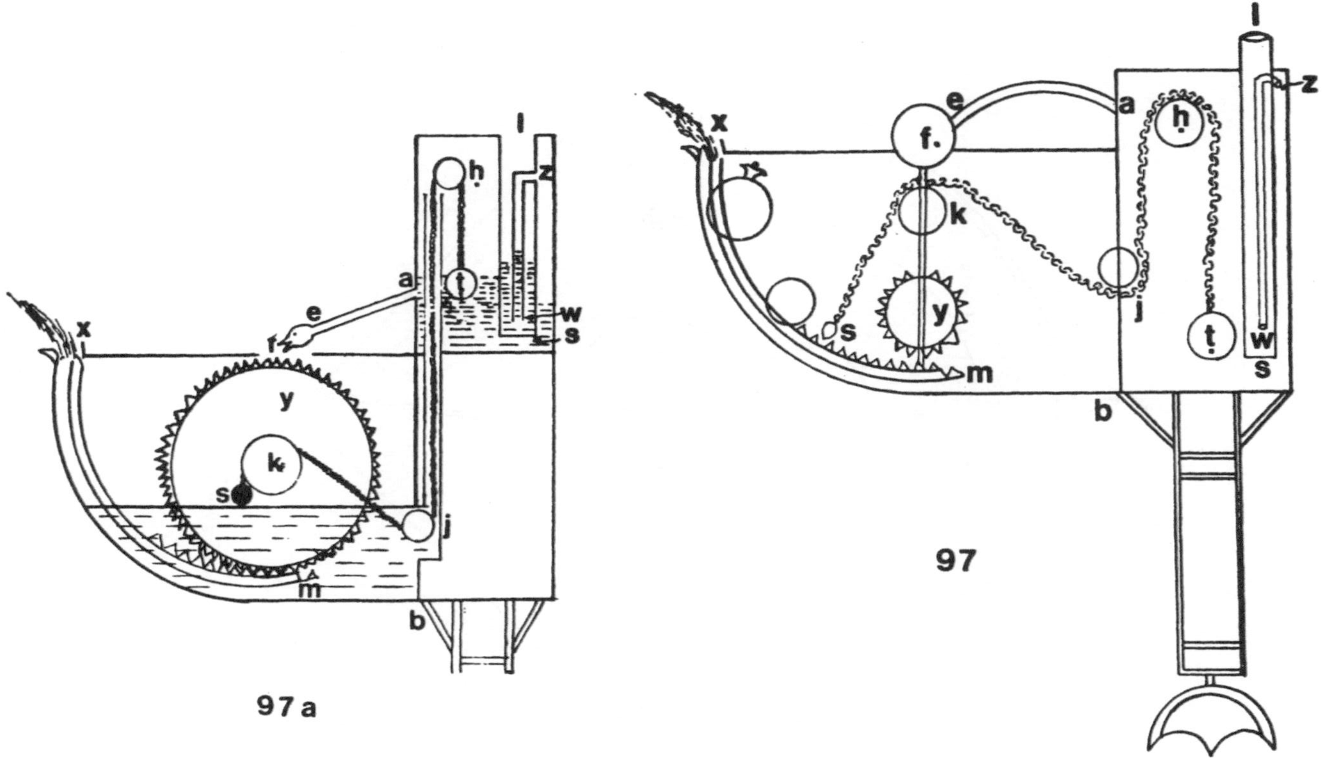

Fig. 2

II

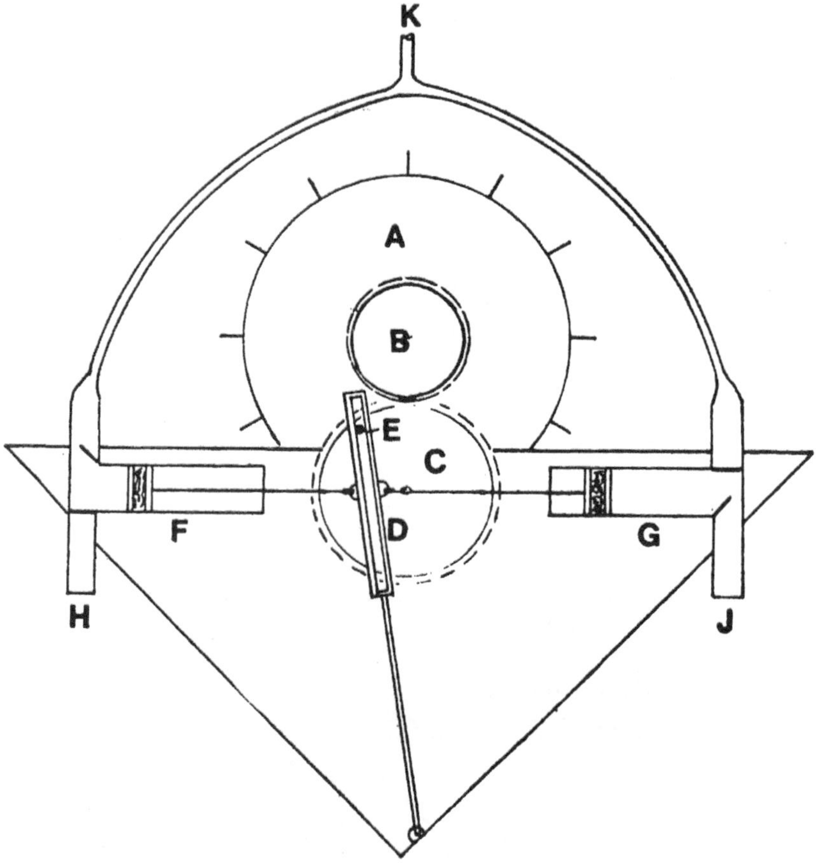

Fig. 3

Water Driven Pump with Twin Cylinders

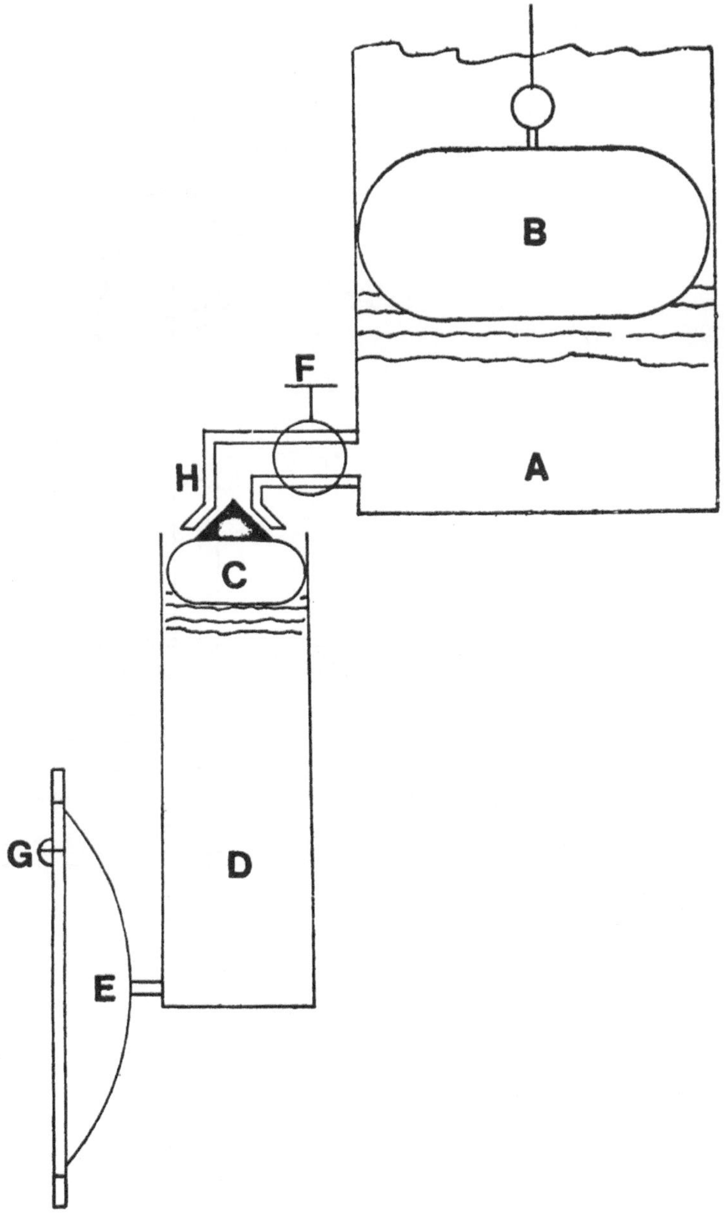

Fig. 4

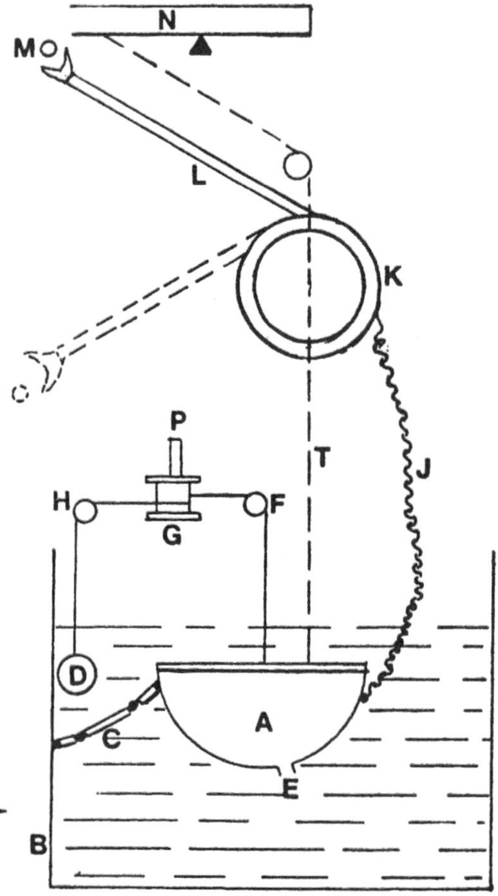

Fig. 5 Principle of the Ṭarjahār Clock →

orifices, taps and valves. Taps and valves were made from cast bronze, the plugs and seats ground together with emery to obtain a watertight fit. Taps were often made with multiple borings in plug and seat, so that different liquids could be drawn from compartments within the main vessel. Fixing and jointing was a problem, since screws, nuts and bolts were not available, and much ingenuity was required to obtain the correct fit: the three types of fit used in modern workshop practice, push-fit, sliding-fit and running-fit were fully understood and differentiated. The commonest jointing medium was solder, iron nails were used frequently, and also lugs, cotters, and male-female joints. Al-Jazarī secured the caps of his candle-clocks to the main body by bayonet fittings. The flow of water was used to turn miniature scoop-wheels and vaned wheels, to which components the terminology of the mill wrights was applied. Wheels of all kinds were mounted on wooden or iron axles, the ends of which were housed in bearings, although bearings and lubrication are unfortunately one of the few cases where descriptions are inadequate, probably because it was assumed that engineers were sufficiently familiar with the techniques.

The same inadequacy is to be found in the descriptions of the use of tools, probably for the same reason. Most of our knowledge must therefore be inferred from the operations performed on the various materials. Wood was handled with great confidence: it was made smooth and flat, grooves were cut in it, and it was cut to circular profiles. We can therefore assume a range of saws, planes, chisels and gouges. The ways in which metal was formed and jointed presumes a variety of metalworking tools, including hammers of various sizes, soldering irons, files and scrapers. The file was used for removing surplus metal from castings and for cutting the teeth of small gear wheels. The lathe was mentioned more than once by al-Jazarī; in addition to the normal turning operations it must have been used for the more intricate drillings. For measuring and setting-out the engineers used plumb-lines, squares, rules and dividers. Apart from inscribing circles, dividers were also used for cutting circular plates from sheet metal.

The mathematical equipment of the Arab engineers is not easy to assess. The manufacture and assembly of the devices demanded some competence in plane geometry, arithmetic and mensuration, although units of weights and measures must have caused problems in the absence of consistent standards. As would be expected al-Jazarī is completely familiar with numerals, whereas the Banū Mūsa always write out their numbers in words. The mathematical bases of many physical phenomena were not established in the period under discussion, but throughout history men have made sophisticated use of the phenomena before the scientific explanations were known. The concept of engineering as 'applied science' is very recent. Al-Jazarī's calibration of his flow-regulator (see below) is a fine example of the empirical approach, and one can find similar examples in all the *ḥiyal* treatises. For instance, the Banū Mūsa did not really understand the concept of differential pressue but this did not prevent them from using aerostatics and hydrostatics with astonishing skill and confidence. They simply used a set of assumptions that constituted a reliable model for the conditions they were dealing with, in much the same way as the astronomers used the Ptolemaic model for arriving at satisfactory solutions of their problems.

Although some of the ideas incorporated in the *ḥiyal* treatises came from the Far East, India and Iran, Arab engineering can be firmly identified as the continuation of an Eastern Mediterranean tradition. The Egyptians and the Romans made significant advances in mechanics. But the greatest contribution was undoubtedly made by the Greeks, and it was from the Hellenistic world that Arab mechanical engineering received its main impetus. During the era of the great ᶜAbbasid caliphs in Baghdad a large number of Greek works were systematically translated into Arabic, sometimes through the medium of Syriac. Among these were the *pneumatics* of Philon of Byzantium (2nd century B.C.)[1] and the *Mechanics* of Heron of Alexandria (60 A.D.)[2], and a treatise on a water-clock, attributed to Archimedes[3]. The last named is important because it was acknowledged as a source by al-Jazarī but its derivation cannot be established with certainty: it probably incorporated ideas from Archimedes and Philon, together with later Arab additions. Apart from these written sources, the Arabs would have been able to examine the water-clocks, or the remains, of them, that had been constructed in Syria. There was a notable example in Gaza[4], and we know that there was a long-established tradition in Damascus for the construction of water-clocks, a tradition that persisted from classical and Byzantine times through into the Umayyad period, as Riḍwān informs us.[5]

1. Carra de Vaux, 'Le Livre des Appareils Pneumatiques et des Machines Hydrauliques par Philon de Byzance', *Paris Académie des Inscriptions et Belles Lettres*, 8 (1903), pt. 1.

2. Woodcroft, B., *The Pneumatics of Hero of Alexandria*, London 1851. The new edition (1971) with introduction by Marie Boas Hall is to be preferred.

3. Wiedemann, E. and Hauser, F., 'Uhr des Archimedes und zwei andere Vorrichtungen' *Nova Acta. Abh. der Kaiserl. Leop. Carol Deutschen Akademie der Naturforscecher*, 103 (Halle 1918), No. 2, 164-202. A new English edition 4: D.R. Hill *On the Construction of Water-Clocks* Turner and Devereux. London 1976.

4. Diels, H., 'Über die von Prokop beschriebene Kunstuhr von Gaza', *Abh. der Königlich Preussischen Akademie der Wissenschaften*, Philos-Hist. Klasse, No. 7 (Berlin 1917).

5. The only known manuscript of Riḍwan's treatise is Ms. Arab. 1348. Forschungsbibliothck, Gotha. The reference is on folio 4R. See also: Wiedemann, E. and Hauser, F., 'Über die Uhren in Bereich der Islamischen Kultur', *Nova Acta*, 100 (Halle 1915), 176-266.

As far as we know, the flourishing of the mechanical arts in the Arab world began with the Banū Mūsa in Baghdad about 236/850 and reached its summit with al-Jazarī in Diyārbakr in 602/1206, and I propose to illustrate and estimate the Arab achievement from these exemplars. This selectivity is not random since from the frequent references to the Banū Mūsa and al-Jazarī in the works of other Arab writers it can be inferred that they were regarded as key figures in the field of mechanical technology, and moreover al-Jazarī himself mentions the Banū Mūsa as being his most eminent predecessors. By confining most of my attention to these major figures I can at least, in the space available, do a certain amount of justice to their achievements. This should not be taken as implying, however, that we can afford to neglect other fields of study – research for new written material, archaeological investigations, and the examination of known works by Arab writers on mechanics or allied subjects.

First, however, two other works in particular deserve mention, one of these being a kind of dictionary of the sciences entitled *Mafatīh al-ᶜUlum* (Keys of the Sciences), written by Abū ᶜAbd Allah al-Khuwārizmī between 365/975 and 381/991.[6] The chapter on *ḥiyal* lists the various components and techniques used by the constructors of these devices together with brief descriptions and some etymological information. This therefore provides us with a valuable guide to the state of development reached by Arab engineers at the close of the fourth/tenth century. The other work was composed by Riḍwan b. al-Saᶜati in 600/1203, and describes how he undertook the repair of the clock built by his father Muḥammad over the Jayrun gate in Damascus.[7] Riḍwan was not primarily an engineer and his lack of technical knowledge is clearly demonstrated by his presentation. The writing is discursive and imprecise, the drawings are crude freehand sketches. Nevertheless his treatise tells us a great deal about the construction of the clock, which must have been very similar to the large water-clocks whose remains can still be seen in Fez, Morocco.[8] Many other Arabic works, not dealing specifically with engineering matters contain material of interest to the historian of technology, although much research and reading is required to identify and extract such material. I have in mind the geographers, the historians, and, above all, the great Arabic scholars whose minds embraced all the aspects of the natural world – men such as al-Bīrūnī, al-Khāzini, Ibn al-Haytham, and many others. When the

---

6. al-Khuwārizmī, Abū ᶜAbd Allāh Muhammad b. Ahmad, *Liber Mafatih al Olum*, Arabic text. Edited with critical apparatus in Latin by G. van Vloten, Leiden 1895.

7. See 5 above.

8. Derek de Solla Price, 'Mechanical Water Clocks of the 14th Century in Fez, Morocco', *Proceedings of the Xth International Congress of the History of Science*, Ithaca N. Y. and Philadelphia 1962.

full story of Arab mechanical engineering comes to be told, full account must be taken of all the available evidence. For the present, however, I shall confine myself to the two works mentioned above: they are worthy representatives of a great tradition.

The Banū (sons of) Mūsa are among the most important and significant figures in the development of Arabic science and mechanics. Their father, Mūsa b. Shākir, was a notable astronomer and a close friend of the Caliph al-Maʾmūn. After the death of their father the three brothers, Muḥammad, Aḥmad and al-Ḥasan, were brought up and educated as the wards of al-Maʾmūn and afterwards were accepted in the courtly circles of his successors. They became great patrons of scientists and men of letters, and it was largely due to their inspiration that the renowned translation bureau was established in Baghdad. Eminent scholars such as Thābit b. Qurra and Ḥunayn b. Iṣhāq laboured to produce faithful translations of the works of classical authors – philosophers and scientists – many of which now exist only in Arabic, the originals having been lost. The world therefore owes an incalculable debt to these translators, and to the Banū Mūsa, for the preservation of many priceless works. But translation is, of course, only part of the story. Many of the translators, and indeed the Banū Mūsa themselves, were eminent scientists in their own right, and wrote works on mathematics, astronomy and natural science that formed the basis for the great flowering of Arabic science in the succeeding centuries.

A number of works are accredited to the Banū Mūsa but only one deals directly with mechanics. Called *Kitāb al-Ḥiyal* (the Book of Ingenious Devices), it was probably composed about 236/850. At present we know of only complete, or almost complete, manuscript copies of this work and there is no full edition in any language, although valuable studies in German have been made.[9] One hundred devices are described, over eighty of which are trick vessels of various kinds. The remainder include fountains that change shape at intervals, self-trimming and self-feeding lamps, a bellows for removing polluted air from wells, and a mechanical grab for excavating in the beds of seas and rivers. The descriptions are fairly brief, between one and two pages, and each model *(shakl)* is provided with a drawing in sectionalised elevation. Letters of the alphabet are written on the drawings and these letters are referred to in the text. In both the descriptions and the illustrations

9. One Ms is in the Vatican (No. 317) and the other is in two parts: Gotha (Arab 1349) and Berlin (von Ahlwardt No. 5562). See also: Hauser, F., 'Über das Kitāb al-Ḥiyal–das Werk über die sinnreichen Anordungen der Banū Mūsà', *Abh. zur Gesch. der Naturwissenschaften* und der Medizin. And Wiedemann and Hauser, 'Über Trinkgefässe und Tafelaufsätze nach al-Gazarī und den Banū Mūsa', *Der Islam*, 8 (1918), 268-291.

the lettering is often haphazard and inconsistent. but this may be the fault of the copyists. Furthermore, the letters are used in the manner of a geometrician to identify spatial positions on the drawings, whereas an engineer would use them to identify individual components. The trick vessels, which occupy so large a part of the book, produce a number of unusual effects. For instance, a pitcher that will accept nothing more the pouring is interrupted; jars into which two liquids are poured in succession, and which discharge those liquids alternately in given quantities; troughs which are constantly replenished with water unless an excessive amount is withdrawn, whereupon they are not replenished; and so on. These effects were produced by a sophisticated, if empirical, use of the principles of hydrostatics, aerostatics, and mechanics. The components used included tanks, pipes, floats, siphons, lever arms balanced on axles, taps with multiple borings, cone-valves, rack-and-pinion gears, screw-and-pinion gears. The large vessels described by the Banū Mūsa often contain a number of these components linked together in various ways to produce the desired effects. Two devices, less complex than many of the others, will serve to give an idea of the methods employed. The first consists of a large basin or trough, placed beside a cylindrical vessel and connected by a pipe to a small tank inside the vessel. In this tank is a float, carrying a vertical valve rod, which has (Fig. 1) on its ends the plugs of two conical valves. The rod passes through a double-valve seat, so that the upper valve opens upwards, the lower valve downwards. The upper valve seat is located in a hole in a plate that divides the vessel into two compartments, the upper compartment being the water reservoir. When a small amount of water is taken from the basin the lower valve opens and a quantity of water flows through the valve into the small tank and through the pipe into the basin. The float rises and closes the valve. As long as only small amounts of water are removed from the basin its contents will always be replenished. If, however, a large amount of water is withdrawn at one time, the upper valve closes and no further flow is possible. The most interesting feature of this device is the feed-back control effect of the lower valve and the use of a double-valve on one valve-rod. Indeed, the use of the cone-valve, which permits a much greater degree of control than plate- or flap-valves, is the feature that mainly distinguishes the work of the Banū Mūsa from that of their Greek predecessors, with whom otherwise they show many points of resemblance. Cone valves do occur once or twice in Heron's treatise on pneumatics, but these were simply push-pull 'bath plug' types of component and did not embody the concept of automatic control that is a common feature in the work of the Banū Mūsa. Cone-valves are an essential component in many modern machines, but they did not enter the general vocabulary of European engineering until the tenth/sixteenth century.

The second device is a lamp with automatic feed and self extruding wick. The wick is attached to the end of a curved, toothed (Fig. 2) rack that rests in guides in the bottom of the lamp, which is soldered to the side of the oil reservoir. A large pinion engages the rack, and this pinion is connected, over two pulleys, to a float in the oil reservoir. Toward the lower part of the lamp is a hole that carries an air pipe into the reservoir, the reservoir being otherwise airtight – it has a special filling arrangement that allows the oil to be poured in while making it impossible for air to enter the vessel. When the wick is ignited the oil in the lamp is consumed until the air vent is uncovered, whereupon oil from the reservoir flows through a pipe into the top of the lamp, until the air vent is again blocked. When the oil level in the reservoir sinks the float descends, rotating the pinion which moves the rack carrying the wick, and the wick is therefore extruded. Dimensions must have been carefully worked out, so that the wick was entruded at a rate that produced a flame of the required size.

The work of the Banū Mūsa carries a strong flavour of theoretical mechanics, as opposed to practical enginering. They seldom give dimensions, usually omit any mention of the materials used, and give only the vaguest and most general directions about construction. It is highly likely that they were not themselves skilled mechanics, and that the devices were constructed by craftsmen according to the brothers' designs. Doubtless this was not a one-way communication, since at all times craftsmen have been ready to advise designers, often in the bluntest terms, about the practicability of their designs. The delicacy and complexity of many of the devices implies the existence of a body of highly skilled metalworkers in Baghdad at this time. Our greatest admiration, however, must be reserved for the devices themselves, which demonstrate a deep empirical understanding of the effects of differential fluid pressures. It is unfortunately impossible to demonstrate this quality in a paper such as this, since a full appreciation of the achievement of the Banū Mūsa can only be attained by a close examination of their complete work.

It is even more difficult to do justice to our second subject within the confines of an essay. In 602/1206 Ibn al-Razzāz al-Jazarī completed his book on machines entitled *Kitāb fi ma'rifat al-ḥiyal al-handasiyya*[10] – 'The Book of Knowledge of Ingenious Mechanical Devices'.All we know of his life is what he himself tells us in the introduction to his work, namely that at the time of writing he had spent twenty-five years in the service of the Artuqid princes of Diyārbakr, and that it was his master Naṣr al-Dīn who commanded him

10. Ibn al-Razzāz al-Jazarī, *The Book of Knowledge of Ingenious Mechanical Devices*. Translated and annotated by Donald R. Hill, Reidel, Dordrecht, 1974.

to compose the book. The Artuqids were a Turcoman dynasty who had maintained a precarious autonomy in al-Jazīra throughout the sixth / twelfth century, although by the time al-Jazarī entered their service they had become vassals of Salāh al-Dīn.[11] An eloquent testimony to the importance attached to al-Jazarī's work is the number of manuscript copies that was made. At least twelve are known ranging in date from the seventh/thirteenth to the thirteenth/nineteenth century.[12]

The book is divided into six categories *(naw')*: I. Water-clocks and Candle-clocks, II. Trick vessels, III. Vessels for hand-washing and phlebotomy, IV. Fountains and perpetual flutes, V. Water-raising machines, VI. Miscellaneous. The value of the book lies not only in the importance of the methods, techniques and components, but also in the fact that the construction of each device is described in scrupulous detail by a man who was obviously a master craftsman. So accurate are these descriptions that a modern craftsman can reconstruct the machines from them with comparative ease (see below). In each case the presentation is clear and orderly. First comes a description of of the appearance of the machine and a general explanation of its operation and use. Then the manufacture and assembly of individual sections are carefully described, with close attention to the sequence of operations. Finally the operation of the machine is described in complete detail. Illustrations – both general arrangements and detailed sketches – are included for each model, and provided with letters of the alphabet to which reference is made in the text. These letters are placed on individual components and correlate closely with the letters in the descriptions.

Only one of al-Jazarī's complete machines can be said to have direct significance in the development of modern engineering. This is a pump driven by a paddle-wheel, which operates, through a system of gears, an oscillatihg slot-rod to which the connecting (Fig 3) rods of two pistons are attached. The pistons work in horizontally opposed cylinders, each provided with valve-operated suction and delivery pipes. The delivery pipes are joined above the centre of the machine to form a single outlet into the irrigation system. This pump is remarkable for three reasons: the application of the double-acting principle, the conversion of rotary to reciprocating motion, and the earliest known use of true suction pipes. It therefore has an important place in the development of the steam engine and of modern reciprocating pumps.

Many of the components and techniques used by al-Jazarī are essential features of modern engineering. Conical valves, which occur frequently,

11. Cahen, Claude 'Le Diyar Bakr au Temps des Premier Urtulides', *Journal Asiatique* (1935) 219-276.

12. Hill, *op. cit.*, 3-6.

have already been mentioned. we can do no more than list some of the others: tipping-buckets that discharge their liquid contents at set intervals of time, used nowadays in rain-gauges and other measuring instruments; orifices calibrated for a given discharge; static balancing of wheels; use of paper models to establish a design; lamination of timber to minimise warping; cams; segmented gears. In one of the water-raising devices a crank is part of the power transmission. This is the earliest known use of a crank in a machine, although hand-operated cranks had been known for centuries in small mills and lifting machinery. In describing the construction of a pair of doors for the palace at Amid, al-Jazarī clearly describes the casting of brass in closed mould-boxes with green sand, a method not adopted in Europe until the tenth/sixteenth century.

Three of al-Jazarī's machines have recently been built for the 1976 World of Islam Festival in London. These comprise the pump, one of the phlebotomy devices and the monumental water-clock described in Chapter I of Category I. All of them were built according to al-Jazarī's instructions and all of them work perfectly. The water-clock is a most impressive sight, standing about twelve feet high with a working face about five feet wide. As al-Jazarī specified, it records the passage of the solar or unequal hours: that is to say, the hours of daylight and darkness are divided by twelve to give 'hours' that vary in length from day to day, and from daytime to night-time, on a given day. Every hour two falcons drop balls from their beaks on to a cymbal, door leaves open to reveal a standing figure, another door turns over to show a different colour, and one of twelve glass roundels becomes completely illuminated. The clock is crowned by a zodiac circle, which rotates at a steady speed throughout the day. All the devices mentioned so far are operated by a float that sinks at a steady rate in a cylindrical reservoir. A string attached to the top of the float passes through a pulley system which actuates the tripping mechanisms, and rotates the zodiac circles. At the sixth, ninth and twelfth hours a band of musicians – two drummers, a cymbalist and two trumpeters–play their instruments. This effect is obtained by allowing the outflowing water to collect in a special container from which it is suddenly released at the due time. The released water runs first over a water-wheel, upon whose axle are cams that operate the arms of the percussion players. The water then runs into a closed air-vessel, driving out the air into a mechanical whistle to simulate the sound of the trumpeters.

The water machinery which ensures the steady fall of water in the reservoir is a masterpiece of careful engineering. The reservoir itself was made of perfectly uniform cross section by (Fig 4) the use of a wooden template and water levels. A tap made of cast bronze leads out from the bottom of the reservoir, its outlet end forming a vertical valve-seat. The conical plug of the valve is soldered to a float which is in a float-chamber placed at the side of and slightly below the bottom of the reservoir. The float-chamber has an outlet pipe at the bottom. When, therefore, the tap from the reservoir is open water runs through the valve into the float chamber, the valve closes momentarily, water flows from the float-chamber, whereupon the valve closes momentarily. An almost constant head of water is therefore maintained in the float-chamber – indeed, the fluctuation in level is not discernible to the naked eye. (Level control by a closed-loop system of this nature was re-invented (?) in England in the eighteenth century A.D. for water supply to a boiler.) Upon leaving the float-chamber the water enters a flow regulator, essentially a device which permits the orifice to be rotated in a full circle and hence vary the static head over the orifice from day to day. Al-Jazarī's method for graduating the flow-regulator is of great interest. He first calculated the required hourly rates of flow for each month in the year. He then set the orifice – a piece of drilled onyx to the topmost position, for the day of the summer solstice, and enlarged the orifice with copper wire and emery until the required flow was achieved. All the other positions were obtained by painstaking trial-and-error. The calibration of the orifice is a very delicate operation, since its diameter is less than one millimetre and a small variation in diameter makes a considerable difference to the rate of flow.

Finally we should mention another type of water-clock that occurs only in al-Jazarī's work and may have been his own invention. This consists basically of a bowl with an orifice in its underside, (Fig. 5) called a *ṭarjahār*. The bowl is placed in a reservoir and connected to its side by a hinged chain. From the side of the bowl a chain goes up to a large ring from which a hollow-pipe with a serpent's head is extruded – the ring and serpent are in fact a balance-arm. From a rod across the open top of the *ṭarjahār* a cord runs up to a ball-release device situated above the balance-arm. The balls are loaded in a magazine from which an outlet channel runs into the head of a falcon. The reservoir is filled with water, the *ṭarjahār* sinks slowly, submerging completely in either thirty minutes or sixty minutes. When it submerges the ball-release mechanism ejects one ball into the falcon's head. This ball drops into the mouth of the serpent, the serpent's head sinks, and at the end of its travel drops the ball on to a cymbal. Meanwhile the *ṭarjahār* has been raised and tilted by the rotation of the ring: the water is evacuated from it and it settles back on to the surface of the water, and the cycle is repeated.

It is impossible to over-emphasise the importance of al-Jazarī's work in the history of engineering. Until modern times there is no other document, from any cultural area, that provides a comparable wealth of instructions for the design, manufacture and assembly of machines. No doubt this is partly due to the fact that there was usually a social and cutural divide between those who made and those who wrote. When a scholar described a machine that had been constructed by an illiterate craftsman he was usually interested mainly in the finished product; he neither understood nor cared about the messy business of construction. A secondary factor was that a craftsman, even when literate, might have been reluctant to impart his knowledge to potential competitors, preferring to pass it on to his sons by word-of-mouth. Al-Jazarī tells us that he hesitated about committing his results to paper for another reason, the fear of adverse criticism, and that it was written only in obedience to his master's command. We therefore owe a great debt to this obscure prince for our possession of a unique document. We rely so heavily upon al-Jazarī for information about Arab mechanical technology that it would be only too easy to lose a sense of perspective. After all, over three centuries separate al-Jazarī from the Banū Mūsa and we know from the *Mafatīh* and other evidence that the intervening period was in no way sterile and undoubtedly witnessed the flourishing of mechanical arts, whenever and wherever conditions were favourable.

In order to arrive at a just assessment of the achievements of the Arabs in mechanical technology, it is first necessary to deal with certain criticisms that have been levelled at them. The first of these is concerned with the triviality of most of the devices which they constructed. It cannot be denied that there is some truth in this viewpoint, although the engineers were in most cases fulfilling the requirements of their princely paymasters, but it ignores some important factors, namely the continuing and parallel tradition of utilitarian machines, the useful machines included in the *ḥiyal* treatises, the aesthetic worth of many of the devices, and the importance of many of the components and techniques. The second two points are discussed more fully below, but one cannot omit the observation that many historians of technology, particuarly in the West, are too ready to consider early machines solely in the light of their relevance to material progress.

The princely patrons of the Arab engineers were men of international tastes, and presumably demanded constructions that would astonish the senses, and embellish their palaces' formal gardens. Hence we have the water-clocks, with their representations of human figures, birds, animals, and astronomical phenomena, and the fountains with alternating shapes. The first have their successors in the figures on the cathedral clocks of Europe and perhaps in the stories of 'live' puppets such as Pinocchio and Petrouchka, the second are the precursors of the water gardens of rich Europeans. Man has probably always had a desire to attempt to reproduce the biological and astronomical phenomena that he observes around him and it is therefore hardly surprising to find the Arab engineers, with their patrons' encouragement, following this ancient tradition. Nevertheless, they handled the theme with great delicacy, never using this art iconographically, or to show earthly rulers in an exalted status. The human figures are those of servants and courtiers, the animals are shown performing their natural functions. It is beyond the scope of this paper to enter deeply into this subject, but it must be pointed out that the representation of biological and astronomical phenomena was one of the factors that led men to regard the universe, for good or ill, in rationalistic, mechanistic terms.

We have now reached the crucial question of the position of the Arab engineers in the development of mechanical technology, involving comparison with their predecessors, particularly the Greeks, and their successors in Western Europe. Any conclusions must be tentative, not only because so little research has been undertaken on Arab technology, but also because much of the evidence of the Greek achievement in mechanics has been lost. Nevertheless, we must do our best, in the light of present knowledge, to identify the techniques that differentiate Arab from Greek engineering, bearing in mind that not the least of the Arabs' achievements was the synthesis of earlier engineering with their own additions and improvements, and the transmission of this integrated corpus of knowledge to other cultural areas. We can identify the following as specifically Arab contributions: al-Jazarī's pump; al-Jazarī's *tarjahar* clocks; use of valves and taps – single-acting and double-acting – for feed-back control; casting of metals in closed mould-boxes with green sand; use of the crank in a machine; the tipping-bucket in its perfected form; segmental gears. There were also a number of ingenious mechanisms such as one-way hinges, ball-release mechanisms, and tripping devices, which were designed and constructed for individual machines. An important feature is the frequent occurrence of delayed-action mechanisms, which delayed the opening or closing of valves, for instance, until a set period had elapsed. To pass from specific instances to more general attitudes, the main distinguishing feature of the Arabs was a constant striving after control in order to construct machines which would work automatically for long periods without human intervention. This was a conscious endeavour, as indicated by a remark of al-Jazarī's about machines 'with marvellous methods of control'. The examples given earlier – the Banū Mūsa's basin and lamp, al-Jazarī's clocks – all demonstrate clearly this preoccupation with control. Indeed the techniques devised for given purposes were often more sophisticated than were strictly necessary. It is simpler, for example, to maintain a static head by fixing an overflow pipe to the supply vessel, rather than using a valve-operated feed-back control. It cannot be doubted that there is aesthetic satisfaction in seeking for elegant solutions to engineering problems, a kind of inspired curiosity that impels men beyond the limits imposed by strict necessity. The Arab engineers felt and exercised this curiosity, and in so doing left behind them a whole range of concepts that were ready to hand when, centuries later, engineers were sufficiently advanced to put them to practical use. To sum up, the Arabs are distinguished from the Greeks by their addition to the engineering vocabulary of vital components and techniques, and more so by their fuller application of the principles of automatic control.

European technology in the Middle Ages was not retarded by the general intellectual stagnation as were the arts and sciences, and by the sixth/twelfth century the western machine technology had developed a momentum that was partly self-impelled, partly fed by ideas from other areas. With some over-simplification, we can identify three major events in which Arabic ideas were important for the development of modern machinery, namely the invention of the mechanical clock, the Scientific Revolution of the eleventh/seventeenth century and the Industrial Revolution in the twelfth/eighteenth. At the start of the eighth/fourteenth century the mechanical weight-driven clock with escapement made its entry on the European scene and before long every major city in Europe boasted its cathedral clock complete with striking bells and automata. A remarkable example of this tradition is the astrarium completed by Giovanni de' Dondi in 1365 A. D.[13] This is weight driven and, through a complex system of gearing, records the movements of sun, moon and planets, including the retrograde motions. One is struck by the essential similarity between the Arabic water-clocks and the later, mechanical clocks of Europe: both were weight driven, although in the water-clocks the controlled fall of the water-level provided the escapement; both included cosmographical representations; both made use of biological automata. Furthermore, the two types of clock made use of the same range of mechanisms such as gears, drums, cams and so on. We know that water-clocks were used in monasteries and that their use was known in the court of Alfonso X of Castile,[14] undoubtedly with stimulus from Arabic sources. We can be sure, therefore, that the Arabic water-clocks were a direct influence on the development of the mechanical clock, and hence indirectly upon the growth of mechanical technology in general, since so much present-day machinery derives from the clock.

The Scientific Revolution was founded partly upon changes in philosophical attitudes, but the reasons behind it can be ascribed, perhaps mainly to the existence of a numerous group of mathematical practitioners – writers

---

13. Silvio A. Bedini and Francis R. Maddison, 'Mechanical Universe, the astrarium of Giovanni de' Dondi', *Transactions of the American Philosophical Society*, n. s. lvi, 5 (1966).

14. Von Bassermann-Jordan, Ernst, *The Book of Old Clocks and Watches*, 4th Edition. Translated into English by H. Alan Lloyd, London 1964, 337.

of books on the useful scientific crafts, makers of mathematical and scientific instruments.[15] Thus it can be said that the first effective co-operation between technologists and scientists began with the former supplying the latter with the tools of their trade, rather than the modern situation whereby the technologist applies scientific results to achieve practical ends. Foremost among the instrument artisans were the clockmakers, so here we have a transmission from the Arab water-clocks through the mechanical clock to the Scientific Revolution. Another potent influence on instrument making during this period was the astrolabe,[16] the most important scientific instrument of the Middle Ages, and still a useful aid in the teaching of spherical astronomy. The tradition for the design and construction of astrolabes came to the West from the Arabs, but Arab astrolabists were never surpassed in skill by Europeans. Individual components were also of significance: for instance segmental gears, which are used by al-Jazarī in two of his water-raising machines, are first found in Europe in de' Dondi's clock, and were used by Blaise Pascal (1623-62 A.D.) in his calculating machine, the precursor of modern calculators.

It would not be difficult to fill several pages with examples of medieval Arabic components that were embodied in machinery from the start of the Industrial Revolution to the present day, particularly in fine mechanisms, instrumentation and control systems, and linkages. The conical valve, as mentioned earlier, is an essential component in much modern machinery, and it is note-worthy that the Banū Mūsa's twin valve works on the same principle as the slide-valve in a double-acting steam engine. The continuous-recording gas calorimeter of C. V. Boys, perfected about forty-five years ago, but still in use today, has a clock powered by a water-wheel incorporated in a closed-circuit water system. The measuring of the water is done by a tipping-bucket so arranged that the jet passes to the water-wheel when the bucket is not receiving. Alternating tipping-buckets are still to be seen working at the present day in production machinery.[17] Arabic components and ideas are also embodied in measuring devices, vending machines and pneumatic instruments. Even more importantly, the modern concepts of feed-back control and indeed automation itself can be traced back to the pre-occupation of the Arab engineers with the idea of self-operating devices.

We have established that the Arabs took over a corpus of engineering knowledge, mainly from the Greeks but from other peoples, also, that they refined and developed this knowledge by the addition of new ideas and techniques, and that Arabic mechanical technology gave important stimuli to modern

15. Derek de Solla Price, *Gears from the Greeks*, New York, 1975, 52.

16. Article 'Asṭurlāb' in *Encyclopaedia of Islam*, New Edition, Vol. I, 722-728.

17. Needham, Joseph, *Science and Civilisation in China*, Vol. 4, Part 2, 537.

machine design. When we come to consider how this knowledge was transmitted we enter the realms of conjecture. While there is ample evidence that Arabic science, medicine, mathematics and philosophy were received in Europe in written form, the transmission of engineering ideas cannot be so readily identified. This is partly due to the lack of fundamental research on technological subjects: we certainly do not yet have adequate knowledge of the Arabic writings and there may also be medieval translations of some Arabic documents into European languages. After all, it is only comparatively recently that we were made aware of the existence of the possible influence of the translation of an Arabic work upon the plan and imagery of Dante's Divine Comedy[18] and it is not improbable that the Renaissance engineers had translations from the Arabic at their disposal. But even if the written evidence does not come to light, this by no means disproves the existence of a cultural penetration of Europe by Arabic engineering ideas. Technological ideas have frequently been carried from one cultural area to another by travellers' reports, by observations of commercial agents, by direct contacts between craftsmen, and by the inspection of earlier machines. Until modern times, cross-fertilizations of this nature were probably more frequent and more fruitful than written communications. Over the period from the third/ninth century to the seventh/thirteenth century and beyond, close commercial and cultural relationships between the Arab world and Europe were established in various places. Among others we may mention the exchange of embassies between the courts of the Abbasid Caliphs and the Carolingians; Sicily under Arabs, Normans and Hohenstaufens; the trading arrangements between Italian mercantile cities and the Eastern Mediterranean; and above all, the centuries-long interpenetration of Islam and Christendom in the Iberian peninsula. Moreover the High Middle Ages was a golden age for travel – of both Muslim and Christian travellers. A number of these great travellers were many sided men who took pains to obtain information and to meet famous scholars. We can reasonably conjecture that it was by such means, with the possible addition of written transmission, that Arabic ideas came to be included in the writings of Renaissance engineers such as Giovanni de' Dondi. Leonardo da Vinci, Antonio da San Gallo the Younger, and Agostino Ramelli. Thanks to the adventurous, non-utilitarian spirit of these engineers a whole repertory of fine mechanisms and control devices many of them of Arab ancestry were available when the Scientific and Industrial revolutions required them.[19]

---

18. Cerulli, E., *Il 'Libro della Scala'e la questione della fonti arabo-spagnole della Divine Commedia*, Vatican 1949.

19. White, Lynn, 'The Flower of Early Renaissance Technology', *Developments in the Early Renaissance*, ed. B. A. Levy (Albany 1972), 51-52.

In conclusion I would like to return to a theme that has been mentioned several times in the course of this paper, namely the need for a proper programme of research to be established as soon as possible. This should include archaeological investigations, the preparation of Arabic editions of the known treatises, the discovery and editing of new material, and an examination of the historical and sociological content of medieval Arabic mechanical technology. I look forward to the time when a fully integrated history of the subject can be submitted to the attention of scholars and thinking people throughout the world. In such a programme Arab scholars can, and indeed must, play a leading rôle, not only for the invaluable contribution that they can offer, but because only they can make the Arab people aware of their technological heritage. I am not one of those who believe that there can be any retreat from technology – one has to take a very short view of history to believe that to be possible. On the contrary, I believe that technology must be used intelligently, together with a number of other disciplines, to repair past mistakes and improve mankind's well-being. Too often modern western technology has been regarded as an alien import in the rest of the world, a one-sided view that has had largely harmful and negative effects. It is my belief that when the Arabs recognise and appreciate their own contribution to technology, they will again emerge in the forefront of technological advance.

III

From Philo to Al-Jazarī

There are a number of technical treatises from Hellenistic and Islamic times that provide us with the best available information about early mechanical engineering. Of the machines that they describe some – water-clocks, water-raising machines, specialized machines – served utilitarian purposes, while others – trick vessels, automata, fountains – were designed to give amusement and aesthetic pleasure to the members of courtly circles. In general it is convenient to use the term 'fine technology' for the machines described in the treatises. (The expression also embraces instrumentation, but this subject is not included in the scope of the present article.) The range of mechanisms, components and techniques incorporated in the treatises is wide. Much of this array was to enter the vocabulary of European engineering and to be of importance in the genesis of the Industrial Revolution. Moreover, the best of the treatises provide us with information about techniques of manufacture and construction that are to be found in no other pre-Renaissance sources.[1]

The pre-history of fine technology begins with such devices as the simple outflow clepsydra, which was known in Egypt and Babylonia before 1500 BC An important development was the inflow clepsydra, whose time and date of origin are unknown. It may have been a Greek invention. At first this consisted of two vessels, an outflow clepsydra that discharged into a receiver having an hour-scale marked on its interior. Clearly, this was no more accurate than the simple outflow type, but the next innovation made it a much better timekeeper. The reservoir vessel was provided with an overflow pipe and a constant supply of water, flowing in at a greater rate than the outflow from the orifice. A constant head was therefore maintained in the reservoir.

[1] Donald Hill, 'Arabic Mechanical Engineering: Survey of the Historical Sources', *Arabic sciences and philosophy*, i, 1991, pp. 167–186.

From *Learning, Language and Invention: Essays presented to Francis Maddison*, ed. W.D. Hackmann and A.J. Turner. Published by Variorum, Ashgate Publishing Ltd, Gower House, Croft Road, Aldershot, Hampshire, GU11 3HR, Great Britain.

The water supply could be from a large vessel, a stream or the city mains.[2] Both Plato (d. 347) and Aristotle (d. 322) are said to have had 'alarm clocks'.[3]

The earliest types of automata were probably the articulated figures such as those from the Egyptian tombs from the Twelfth Dynasty onwards. The next stage of development is also found in ancient Egypt: talking statues worked by means of a speaking trumpet concealed in hollows leading down from the mouths. The beginnings of sophisticated, mechanized constructions, however, are to be found in Hellenistic Egypt.[4]

The earliest Hellenistic engineer about whom we have any information is Ctesibius, who is known only by passages from the writings of Vitruvius.[5] According to Vitruvius he constructed two inflow clepsydras, in one of which a float in the receiver rose at constant speed. A rod soldered to the top of the float carried a figure whose finger moved over an hour-scale. In the other clock the rod was toothed and engaged a pinion which transmitted the movement to various automata. Ctesibius is also credited with the invention of the water-organ and the force-pump.[6] Two celestial globes are said to have been constructed by Archimedes.[7] There is also a work attributed to him on the construction of water-clocks, which will be discussed in connection with monumental Islamic water-clocks. The bulk of our information on fine technology, however, comes from four sources, two Greek and two Arabic. These, in their modern editions, are:

Philo of Byzantium, *c.* 200 BC, *Le Livre des appareils pneumatiques et des machines hydrauliques par Philon de Byzance*, Arabic text with French trans. by Carra de Vaux, *Paris Académie des Inscriptions et Belles Lettres*, 38, 1903, part 1.

Hero of Alexandria, fl. mid-first century AD, *Heron Alexandrinis Opera*, vol. 1, *Pneumatica and Automata*, edited Greek text with German trans., G. Schmidt, Leipzig, 1899.

Banū Mūsà, *The Book of Ingenious Devices*, trans. and annotated by Donald R. Hill, Reidel, Dordrecht, 1979.

Banū Mūsà, *Kitāb al-ḥiyal*, Arabic text of the above, ed. A. Y. Hassan, Institute for the History of Arabic Science, Aleppo University, 1981.

---

2 Donald Hill, *A History of Engineering in Classical and Medieval Times*, London, 1984, p. 225.

3 A. J. Turner, *The Time Museum, Catalogue of the Collection*, vol. 1, *Time Measuring Instruments*, Part 3, 'Water-clocks, Sand-glasses, Fire-clocks', Rockford, Ill., 1984, p. 6.

4 One is preserved in the Louvre, one, a large limestone bust of the god Re-Harmarkhis of Lower Egypt, is in the Cairo Museum. Jointed and talking figures were not confined to Egypt, but probably occurred in early civilizations and are widespread. See Derek de Solla Price, *Science since Babylon*, New Haven and London, 1975, pp. 50–51.

5 Vitruvius, *De Architectura*, trans. F. Granger, London, 1934, vol. 2, pp. 255–267.

6 *Ibid.*, pp. 311–314.

7 Emilie Savage-Smith, *Islamicate Celestial Globes*, Washington, 1985, p. 5.

Al-Jazarī, ibn al-Razzāz, *The Book of Knowledge of Ingenious Mechanical Devices*, annotated and trans. by Donald R. Hill, Reidel, Dordrecht, 1974

Al-jami' bain al-'ilm wal-'amal al-nāfi' fi ṣinā' at al-ḥiyal, Arabic text of the above, ed. A. Y. Hassan, Institute for the History of Arabic Science, Aleppo University, 1979.

These are the only major treatises on fine technology to have come down to us.[8] It might be argued that there are other works of comparable importance awaiting discovery, but if such works had been written they would surely have been mentioned by other writers, particularly scientists and the compilers of biographical dictionaries. But no other comparable works are referred to. Philo is mentioned by Vitruvius[9] and by Hero.[10] The *Pneumatics* exists only in an Arabic translation probably dating from the ninth century. None of Philo's treatises is known to have circulated in Europe in the Middle Ages, but in 1615 Solomon de Caus listed him among his predecessors.[11]

Hero was also known to the Arabs.[12] His *Mechanics* has survived in an Arabic translation made by Qūsta b. Lūqā (d. 912). But although some of his works were known in the Middle Ages only one Greek manuscript from the late fifteenth century is known to have survived. From the sixteenth century onwards, however, nearly one hundred copies of the *Pneumatics* are extant. In most of these the *Automata* is also included. Hero's influence on Renaissance writers on mechanics, hydraulics and pneumatics was considerable.[13]

All the major Arabic biographical dictionaries dealing with scholars, from Ibn al-Nadīm (d. 995) onwards, devote entries to the Banū Mūsà and to no other writers on machines.[14] The great historian Ibn Khaldūn (d. 1406) refers to their machine book in glowing terms,[15] and al-Jazarī mentions them as the most illustrious of his predecessors in the field of fine technology.

Al-Jazarī lived too late to be included in the *Fihrist* of Ibn al-Nadīm and later biographers do not mention him, either because they usually copied from the *Fihrist* or because his reputation did not spread beyond the limits of the Jazīra until some time after his death. He is, however, mentioned by the Egyptian scholar al-Qalqashandī

[8] This does not preclude the possibility of shorter works having been written. There are, for example, short works on musical automata, extant in Arabic but probably from Greek originals. There is also a chapter on perpetual motion machines to be found in several Arabic manuscripts. See note 1 above.

[9] Vitruvius, *De Architectura* (n. 5), vol. 2, p. 75.

[10] *Heron Alexandrinis Opera*, vol. 1, *Pneumatica and Automata*, edited Greek text with German trans., G. Schmidt, Leipzig, 1899, pp. 405–407.

[11] The following works are valuable for a study of the Hellenistic engineers: A.G. Drachmann, *The Mechanical Technology of Greek and Roman Antiquity*, Madison, Wisconsin and Hafner, London, 1963; by the same author, 'Ktesibos, Philon and Heron: a Study in Ancient Pneumatics', *Acta historica scientiarum et medicinalium*, ed. Biblioteca Universitatis Hauensis, Copenhagen, 4, 1948, pp. 1–107; Albert de Rochas, *La Science des philosophes et l'art des thaumaturges dans l'Antiquité*, 2nd edn, Paris, n.d.,. 1930.

[12] E. Wiedemann, *Aufsätze zur arabischen Wissenschaftsgeschichte*, Hildesheim, 1970, vol. 1, pp. 69–70.

[13] Turner *Op. cit.* (n. 3), pp. 31–33.

[14] The Banū Mūsà, trans. Hill, p. 3.

[15] Ibn Khaldūn, *Muqaddimah*, trans. F. Rosenthal, 2nd edn, London, 1967, vol. 3, p. 132.

(d. 1418) as pre-eminent in the science of machines,[16] a pre-eminence that is emphasized by his having been given the soubriquet *Badi' al-Zamān* ('The wonder of the age'). At least 17 manuscripts of his work, from the thirteenth century to the eighteenth, also testify to the esteem in which he was held in the world of Islam.

There were other shorter treatises, some of which embody ideas that are not found elsewhere. These will be referred to as occasions arise, to ensure that this survey includes all the important techniques, but in general attention will be concentrated on the works of the four major writers.

Certain techniques were common to all writers on fine technology. Above all, nearly all their devices rely upon water or other liquids as their prime movers. The water might be used directly so that its pressure, or flow, or both, produced the desired results. Very frequently, however, the water was used to produce pneumatic results – it is significant that the main treatises of Philo and Hero are called *Pneumatica*. Pneumatic effects are caused by differences in aerostatic pressure, usually when the pressure of the atmosphere (1 atm = 101.325 kN/m^2 or 14.7 lbs/in^2) is greater than that in an adjacent container. For example, if a pitcher with a narrow spout is filled with liquid and then made airtight, nothing can be poured from the spout once a tiny drop has emerged, reducing the internal pressure to slightly less than 1 atm. The pressure of the surrounding air prevents any further discharge. If, however, there is a concealed hole in the handle leading to the body of the vessel, then once the hole is uncovered the internal and external pressures are equal and outpouring can occur. Siphons work on the same principle. When water reaches the top of the bend, the pressure there drops below atmospheric. Provided the outer leg is longer than the inner, discharge will occur until the water drops below the opening of the inner leg. Another device that makes use of the pneumatic effect is the occluded orifice. The orifice is let into an otherwise airtight vessel. The hydraulic system can be so arranged that the orifice is opened and closed at very short intervals, so producing a kind of feedback control.

All the early writers believed that pneumatic effects were caused by the impossibility of a vacuum, but this did not prevent them from using these effects in a variety of sophisticated ways. Provided the desired results could be produced by the engineer, using his empirical knowledge, the actual final velocity of the fluids was unimportant. It was not usually necessary to be able to quantify pneumatic phenomena. With notable exceptions, which will be discussed in the appropriate places, liquid flows were slow.

The liquid flows were themselves often the end-product of a particular device. For example three differently coloured liquids might be poured into a vessel and issue from its tap in the reverse order from which they were poured in.

The general formula for flow of liquid in pipes is known as 'Bernoulli's Equation' and is given by:

[16] Al-Qalqashandi, *Subḥ al-a'sha*, ed. M. A. R. Ibrahim, 14 vols, Cairo, 1913–20, vol. 1, p. 477.

$$\frac{p}{w} + \frac{V^2}{2}g + h = A \text{ constant}$$

where p = pressure, w = specific (unit) weight, V = velocity, g = the gravitational constant and h = height. The early engineers did not, of course, know this equation, nor did they understand the concepts of pressure and gravitational constant. They knew that water under gravity flows downwards and this knowledge was sufficient in the many cases where the rate of flow was unimportant. When, as with water-clocks, a precise rate of flow had to be maintained, empirical methods were resorted to.

The following basic hydraulic components were used by all those concerned with fine technology, from Philo onwards. When each writer is discussed the additional components and techniques that he introduced will be described separately.

Pipes	usually made by bending copper sheet round a wooden former and soldering along the seams. Pipes were occasionally made from brass or lead.
Siphons	bent-tube and concentric types.
Pipettes	
Vessels	jars, pitchers, cups, basins.
Funnels	
Orifices	plain and occluded.
Partitioned vessels	
Taps	with single and multiple borings.
Clack valves	
Floats	
Human and animal figures	

Philo of Byzantium

Nothing is known of Philo's life. By his own account he wrote a large number of works dealing with different branches of mechanics, of which only two have survived: two books of a compilation on war machines called *Belopoiika*[17] and the *Pneumatics*. In the former he mentions, among other works he had written, treatises on levers and on the void. Philo is mentioned by Vitruvius[18] as having written on machinery. Hero criticizes Philo's work in his *Automata*[19] saying that his constructions were too cumbersome. This is the only evidence we have of Philo having written about automata.

[17] Carra de Vaux, p. 37.

[18] See n. 9 above.

[19] See n. 10 above.

Philo was known to the Arabs. He is mentioned, for example, by the Egyptian historian al-Kindī (d. 961), who says that he built water-wheels, mills and ingenious devices.[20] He was, however, known earlier than this. His *Pneumatics* was translated into Arabic, probably in the ninth century. This text was edited and translated into French by Carra de Vaux, using MS. 3713 of Santa Sophia as the foundation of the work, with other Arabic texts and a Latin fragment as supporting documents.

It should be mentioned that the drawings which accompany the text in the version of Carra de Vaux are somewhat unsatisfactory. They are derived from the text and the original drawings, but appear as Renaissance-style illustrations in perspective, with the internal parts shown. This sometimes makes the exact layout of the equipment difficult to follow, although in general these drawings are a useful adjunct to the text. The Greek letters are marked on the various components in accordance with Philo's instructions. Nevertheless, it would have been preferable if the original drawings had also been shown. After all, the adapted drawings are themselves translations.

There are 65 devices described in the *Pneumatics.* The first 16 of these are demonstrations of the principles of pneumatics as understood at the time. Examples include siphons – bent-tube and concentric – and pipettes. Although the theory expounded by Philo is faulty, his descriptions of the operation of the components are correct. Of the 65 devices, 41 are operated by pneumatic means, two by steam and the remainder by hydraulic/mechanical power. All except the last five devices can be classified as trick vessels of various kinds, some of them including human or animal figures, or both.

A quadruple lamp that replenishes itself is fairly typical of the pneumatic devices (No. 20). Four lamps are grouped symmetrically around an air tight vessel filled with oil. A pipe runs vertically up the centre of this vessel and from its protruding lower section a pipe is led out into one of the lamps – say, A. Pipes lead out from the centre of the reservoir, one over the centre of each lamp. When the lamps are lit and the fire consumes the oil, and the end of the pipe in A is uncovered, oil then runs into all the lamps from the reservoir until the pipe in lamp A is again covered – and so on. This is an example of feedback control using an occluded orifice.

The hydraulic devices include a pitcher from which only a measured quantity can be drawn each time the tap is opened (No. 28). There is also an automatic lavabo (No. 31) which Philo says is 'like the clocks'. The machine incorporates a tipping-spoon, which is clearly a precursor of the tipping-buckets used to such effect by al-Jazarī.

Heat was used to produce steam in only two of Philo's models, Nos 57 and 58. The first directs steam on to burning incense to produce a draught. The second has steam forced through pipes in the bodies of model birds in order to simulate a whistling sound.

There are several taps with two or more borings. Using such a component attached to a vessel divided into two compartments, turning the tap to one side produces water, turning it to the other side produces wine. Another tap, more complex than this, is

[20] Wiedemann, *Aufsätze* (n. 12), vol. 1, pp. 71–73.

able to throw up a vertical jet of water that turns a small paddle-wheel located in the extended shank of the tap. The vocabulary is the most interesting part of this chapter. Philo calls the wheel a *nā'ūra,* i.e. a noria, and later says that the jet rotates the mill (*raḥā*). This is a clear indication that water-mills were in use, at the latest, by 200 BC.

The final five of Philo's devices are water-wheels. The first is a tympanum in reverse; when water is directed into its compartments it whistles (No. 61). No. 62 is a miniature noria with a similar purpose. No. 63 is also a small noria, this time intended to produce a stream of water for ablutions. No. 64 is a manually operated suction pump and the last machine is a water driven *sāqiya*. There has been considerable discussion as to the possibility of these five chapters being later interpolations, but there seem to be no valid reasons for excluding them from Philo's own work.

Later Philo's work can be evaluated in the light of the writings of his successors. For the moment it is worthy of note that he used no transmission systems – neither gears nor pulleys.

Hero's works on mathematics, mechanics, war machines and surveying are not within the scope of the present chapter, which is concerned only with the *Pneumatics* and the *Automata.* The *Pneumatics* may originally have been intended as notes for pupils, since after the initial didactic chapters there is no order and the chapters appear haphazardly. Nevertheless, the chapters themselves are not haphazard. All are written in the same concise style and according to a fixed plan, beginning with a description of the apparatus, then a description of how it works and finally an explanation (if necessary).[21] They are very clear and it might be possible from the descriptions to reconstruct the devices. To attempt this, however, one would be severely hampered by the lack of dimensions.

The *Pneumatics* is divided into two books, containing in all 80 devices. Many of these are copied more or less exactly from similar constructions in Philo's work. It will therefore be profitable to discuss only those devices and components in Hero's work that are not found in Philo's.

After describing the bent-tube and concentric siphons, Hero describes (I 4 & 5) a siphon that passes through a float to which it is soldered with watertight joints. As the water-level falls or rises the discharge from the siphon remains unaltered, since the distance from the water-level to the end of the longer leg of the siphon is constant [Figure 4]. Three double siphons are described in II 14. These have a bent-tube siphon over the top of the tube and an inverted concentric siphon at the bottom. These siphons were used to produce artificial airlocks. Hero simply says that they could be used in a variety of applications, but gives no examples. As will be seen, double concentric siphons were frequently used to great effect by the Banū Mūsà.

In I11 the manufacture of a clack-valve is described. Philo used the same component but does not describe it, presumably assuming that his readers were familiar with its construction.

---

[21] A. G. Drachmann, *Hero of Alexandria,* in *Dictionary of Scientific Biography,* New York, 1972, vol. 6, pp. 310–313.

In I 16 pulley systems are described for the first time in the *Pneumatics*; far more complex systems are described in the *Automata* (see below).

Hero described the use of gears in his *Barulkos*, a machine describing how a load of 1000 talents can be lifted by a force of five talents, that is, the power of a single man. The machine consists of a number of parallel meshing gears, but it is only a theoretical solution; parallel-toothed wheels were not known to have been used in antiquity for the transmission of heavy torque.[22] In the *Pneumatics* (II 32) Hero has a pair of gears meshing at right angles.

Model I 41 contains a large conical valve. This is a 'bath-plug' type of valve that is lifted in and out of its seat by a chain. This is the only example in Hero's work of this important component.

Of the complete machines described by Hero, the force-pump (I 28) and the water-organ (I 42) were probably copied from a lost work of Ctesibius.[23] The pump is a twin-cylinder, manually operated machine. The cylinders are vertical and stand directly in the water; there are plate-valves on the entry ports.

The water-organ is supplied with air from a pump from which the air is forced into a vessel containing water. From the vessel a pipe leads vertically into an air-chest. There are eight pipes fitted vertically into the air-chest; these can be opened and closed from a keyboard. A second organ is wind operated. A vertical wind-wheel (*anemouria*) with paddle-like vanes is mounted on an axle which carries four cams at 90° to one another. As the *anemouria* is turned by the wind the cams bear down in turn on one end of a rocking arm. This is pin-jointed at the other end to the connecting rod of a vertical piston that moves in a cylinder. A pipe leads from the underside of the cylinder, goes horizontally through a plinth then turns vertically and enters the air-chest which rests on top of the plinth. Eight vertical pipes are fitted to the air-chest. There is no keyboard (I 43).

In models I 38 and 39 temple doors are made to open by the action of heat. In I 38 for example, the temple doors rest on a hollow plinth, into which their hinges are extended as circular axles, resting in bearings at the bottom of the plinth. A pulley system is rigged around these axles: on one end of it is a counterweight, on the other end a container is suspended. A bent tube connects this container to a sphere half filled with water. A pipe leads from the sphere vertically into an altar. All joints are leak-proof. When a fire is lit on the altar the water in the sphere is heated and some of it flows through the bent tube into the vessel, which becomes heavier than the counterweight. The pulley system therefore opens the doors. If the fire is allowed to go out the water returns to the globe and the doors close.

There are several lamps in the *Pneumatics*. A self-trimming lamp is described in I 34. As the level of the oil in the lamps falls a pair of pinions are turned and one of these engages a toothed, curved rack on the end of which is the wick. The wick is

[22] *Ibid.*, p. 11.

[23] See nn. 5, 6 and 10 above. Vitrivius' mentions of the writings of Ctesibius in general and of the organ and the force-pump in particular lend credence to the likelihood of Hero having known these machines.

therefore gradually extruded. Three other lamps are described (II 22, 23, 24), all of which can be replenished without the intervention of an operator.

Of those devices of Hero's that use heat; the small steam-driven machine called the Äolipile has always intrigued historians of technology. On top of a boiler with a flat lid two arms in the shape of inverted L's were erected. A metal sphere was suspended between these. One of them turned in a bearing fixed to the side of the sphere, but the other was a pipe that entered the sphere and the boiler. Protruding from the sphere were two L-shaped pipes, the two outer legs facing in opposite directions. The fire was lit under the boiler. When the water boiled the steam entered the sphere and emerged from the two pipes, causing a turning moment that made the sphere revolve. Although this is a toy steam turbine, it did not, of course, inspire the invention of a practical machine. Nor was this to be expected: the other techniques and materials needed to construct a working steam engine lay far in the future.

The *Automata* has attracted far less attention than it deserves. Whereas Hero's *Pneumatics* is in a direct line of descent from Philo's *Pneumatics* to the Banū Mūsà's *Book of Ingenious Devices*, the *Automata* is in a category of its own. Admittedly, Hero speaks of the ancients having made automatic theatres and indeed he criticizes Philo's designs as being too cumbersome and as having omitted important features,[24] but we have no survivals of any writings on automatic theatres before Hero. In any case, it would seem that his designs were considerably in advance of those of his predecessors. Apart from the intrinsic interest of the constructions themselves, Hero's introductory remarks are noteworthy, since they give an insight into the motives of those who made automata. 'The automatic theatre' says Hero 'enjoyed great popularity among the ancients [*proteros*], first because a variety of skills were developed from it, and secondly because the action of the play was a cause of wonder'. Also, he continues, the manufacture of automata demands a knowledge of every branch of mechanics. It does not appear that any further enquiry into the purposes of fine technology needs to be added to these remarks of Hero.

There were two types of automatic theatre: moving and static. In the first type the whole theatre moved on wheels in front of the audience, various actions were performed, and the theatre then returned to its original position. In the static type the theatre itself did not move, but the scenes were somewhat more complicated than in the moving type.

The moving theatre in Hero's version consisted of a plinth about 46 cm long by 31 cm wide by 23 cm high. The wheels were almost hidden at the base of this plinth. At each corner of the plinth was a pilaster about 62 cm high and 15.5 cm wide, supporting an architrave some 7.7 cm high. (It is unusual for Hero to give measurements. He did so in this case, he says, so that onlookers could see that the theatre was too small to conceal a human operator!) The architrave was decked over and on the deck was a little circular temple, with six columns and a dome surmounted by a spike. On the spike stood Nike with outspread wings, holding a wreath. A Bacchus-figure

[24] See n. 10 above.

stood in the centre of the temple holding a thyrsus in the left hand and a goblet in the right. At his feet sat a small panther and close by was an altar. At each column stood a Bacchante.

When the theatre stopped the altar fire ignited, milk or water flowed from Bacchus' Thyrsus, wine flowed from the goblet, the Bacchantes danced, drum rolls and cymbal strikes were heard. When the noise died down both the Nike and Bacchus turned outwards. All these actions were repeated, then the theatre returned to its starting point.

The motor for all the movements consisted of a heavy weight in a circular chamber filled with millet or mustard seed. The motor was started by pulling a string that moved a shutter from an orifice at the bottom of the chamber, whereupon the seed began to trickle into a container below the weight chamber and the weight began gradually to descend. Cords from the top of the weight led to pulley systems that caused the movements of the theatre and the actions of the automata.

These systems were very ingenious. To take but one example, the forward and reverse movements of the theatre. Between the two front wheels of the theatre's plinth was a wooden cylinder, i.e. a very thick axle. A cord from the weight passed over a pulley then down to this cylinder. A length of this cord was wound around the cylinder in one direction, then a slack section was tacked to the cylinder with some wax, then the rest of the cord was coiled round the cylinder in the reverse direction. The theatre moved forward as the cord uncoiled, remained static as the slack was taken up, then went into reverse when the cord uncoiled in the opposite direction. More complex actions are described: for instance Hero tells us how the theatre can be made to move in a circle or round the sides of a rectangle.

The static theatre was similar in many ways. Dry sand instead of seeds was used, since lightness was unimportant. The movements of the actors were more prolonged and varied than in the static version.

Hero's specifications are a curious mixture of the meticulous and the sketchy. He says that the surface upon which the theatre rolls must be perfectly smooth and level. He gives very careful instructions for preparing the materials. The cords must be pre-stretched so that they do not stretch further while in use. The wheels and axles are to be carefully made and balanced and they are to be oiled before they are used. *This is the only mention of lubrication in any of the treatises on fine technology.*

On the other hand, apart from the overall measurements of the theatre, no dimensions at all are given. There are also no clear directions as to how the various parts are to be assembled. We are not told, for example, how the weight-chamber is located in relation to all the moving parts. It would therefore be very difficult to make a reconstruction of the theatre. It would seem that for a complex construction the early engineers found it difficult to give clear instructions for manufacture. Clarity was not assisted by the rudimentary state of technical drawing. Even so, the *Automata* is a remarkable achievement. It is doubtful whether there is a more sophisticated use of pulleys before modern times. And of course the moving theatre was the first 'horseless carriage', albeit a miniature one.

With the Banū Mūsà we return to the mainstream of fine technology, but with important innovations. These three brothers were named, in order of seniority, Muḥammad, Aḥmad and al-Ḥasan. Their father, Mūsà b. Shākir, was a noted astronomer who was a close companion of the future Caliph al-Ma'mūn when the latter was living in Khurasan. After Mūsà's death the brothers were entrusted to the care of al-Ma'mūn (d. 833) and became members of the courtly circles of this Caliph and his successors in Baghdad. They took part in the turbulent palace politics of the time, undertook major public works, and extended their patronage to a number of eminent scientists and translators. They also made journeys to Byzantium and were instrumental in bringing back to Baghdad copies of scientific works, which were then translated from Greek or Syriac into Arabic. They were also eminent scientists in their own right and composed a number of works on astronomy and mathematics, one of which, *On the Measurement of Plane and Spherical Figures*, was translated into Latin by Gerard of Cremona in the twelfth century. There are entries on the Banū Mūsà in all the major Arabic biographical dictionaries and their political activities are mentioned by the great historian al-Ṭabarī (d. 923), who was alive when they were prominent in public affairs.[25] Apart from the mathematical work, only two of the Banū Mūsà's works have survived. One of these is a short treatise on a musical automaton, describing a very advanced machine that embodies a number of delicate mechanisms and control systems. Unfortunately, space does not permit any discussion of this machine here. It can be studied in two scholarly publications, one an edition of the Arabic text, the other an English translation.[26]

Their major surviving work is *The Book of Ingenious Devices*, containing descriptions of 100 devices. It was probably mainly the work of Aḥmad and it was widely appreciated in medieval Islam. Ibn Khaldūn says of it: 'There exists a book that mentions every astonishing remarkable and nice mechanical contrivance. It is often difficult to understand, because the geometrical proofs occurring in it are difficult. People have copies of it. They ascribe it to the Banū Shākir'.[27]

The Banū Mūsà, unlike Philo and Hero, give no exposition of the general principles of pneumatics, nor do they describe a number of experiments that demonstrate those principles. The first model describes a concentric siphon, but thereafter there are descriptions of various devices, each accompanied by an explanation and an illustration. The illustrations are sectionalized elevants of the complete devices and are in general a useful aid to understanding the construction and operation of the devices. As with their Greek predecessors, the various components on the illustrations are marked with identifying letters and these are referred to in the text.

Models 2–74 are trick vessels producing a bewildering variety of effects. To take but three:

---

[25] The Banū Mūsà, trans. Hill, pp. 3–6.

[26] Arabic text, ed. L. Cheikho, *al-Machriq*, ix, Beirut, 1906, p. 9; English trans. by H. G. Farmer, The *Organ of the Ancients*, London, 1931, pp. 80–85.

[27] See n. 12 above.

Model 17 A boiler with an outlet tap. Hot water issues from the tap only so long as cold water is poured in at the top of the boiler.

Model 43 A jar with a tap, into which three different liquids can be poured without mixing. When the tap is opened the liquids discharge in the sequence in which they were poured in.

Model 62 A jar with an outlet pipe from which, after the jar has been filled with wine, wine discharges as long as water is poured in; then water discharges as long as wine is poured in.

Models 74 to 87 are self-replenishing vessels.
Models 88 to 94 are alternating fountains.
Model 99 is a bellows for removing foul air from wells.
Model 100 is a small clamshell grab.

It would clearly be impossible to describe the construction of even a selection of the trick vessels in detail. Instead, it may give some insight into the methods of the Banū Mūsà if the components that they used frequently are described. In the first place, of course, they used components that appear in the *Pneumatics* of Philo and Hero. These include the bent-tube and concentric siphons [Figures 1 and 2] (not the floating siphon, which is unique to Hero), occluded orifices, human and animal figures, pipes and vessels of various kinds. The new components are shown in Figures 3, 5, 6, 7, 8.

Figure 3 shows a component very frequently used by Banū Mūsà, a double concentric siphon. Apart from a rudimentary drawing in Hero's *Pneumatics*, this device appears in no other treatise on fine technology. It acts like a switch. When liquid is introduced it behaves like a normal siphon. When filling is interrupted the level in the upper tank sinks to position A and some air enters the upper siphon. A state of equilibrium is reached such that $x = y$. Inpouring cannot be resumed through this siphon unless $H > a + c$. When inpouring is restarted the liquid will take an alternative route, through another siphon or a pipe, if such a route has been provided.

Figure 5 shows a simple but effective way of separating two liquids when they are poured into a vessel. The first liquid (say) is poured in forcefully and runs from the bent end of the funnel into tank c. The second liquid is poured in gently and trickles into tank d. The liquids can be evacuated from the tanks through valves, siphons or by some other means.

Figures 6 to 8 show different applications of a component that was used frequently and with great confidence by the Banū Mūsà – the conical valve. This type of valve occurs in the pseudo-Archimedes treatise as part of the feedback control system in a water-clock and it was used for the same purpose by Riḍwān and al-Jazarī (see below). Nowhere else, however, do we find it used so effectively and in such a wide variety of applications. The valves were made of cast bronze: first the plug and the seat were cast together in the same mould, then ground together to a watertight fit with emery powder. They were used by the Banū Mūsà mainly as switching devices. In Figure 6 water runs into tank k, in which there is a float a. This float is connected by rods, as

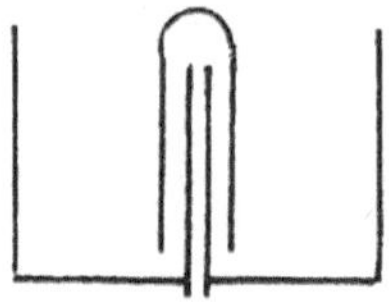

Figure 1. *Concentric siphon*

Figure 2. *Bent-tube siphon*

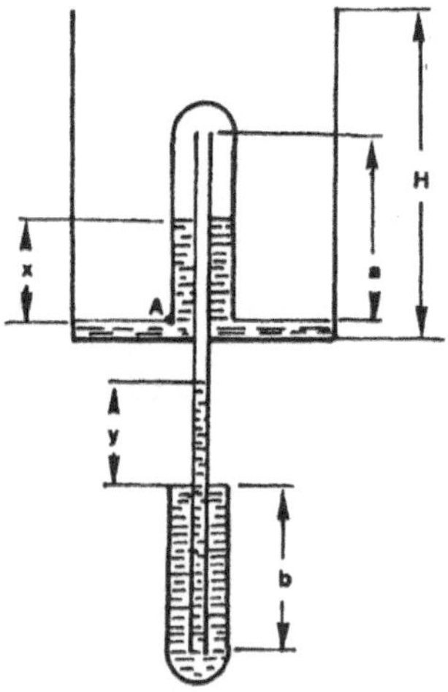

Figure 3. *Double concentric siphon*

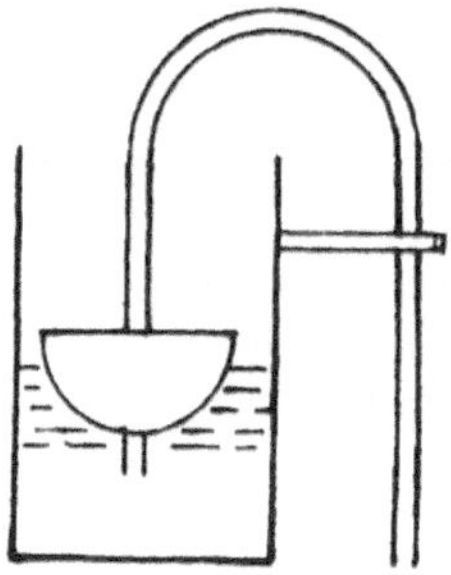

Figure 4. *Floating siphon*

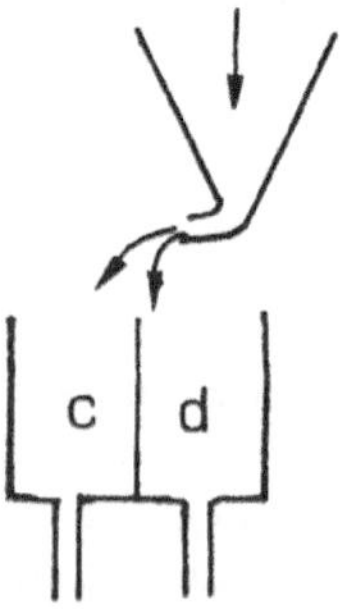

Figure 5. *Method of separating two liquids*

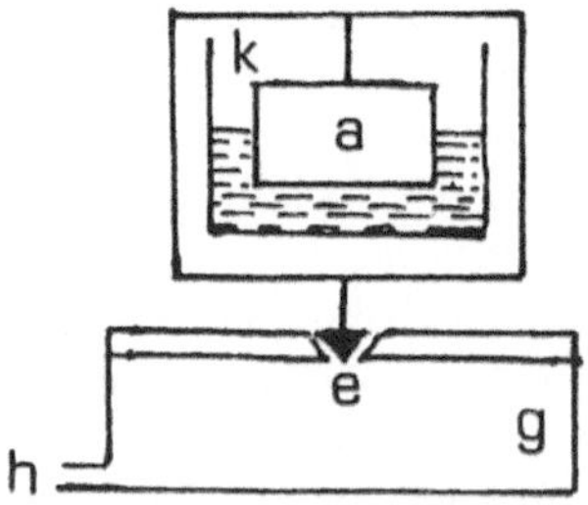

Figure 6. *Hydraulic switches for conical valves*

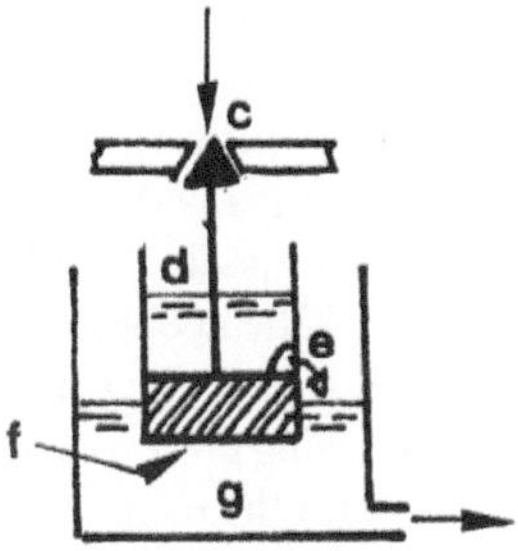

Figure 7. Hydraulic switches for conical valves

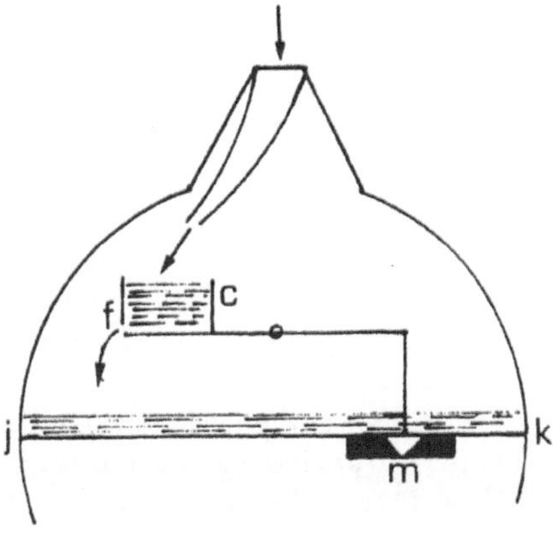

Figure 8. Hydraulic switches for conical valves

Figure 9. Double-acting conical valve

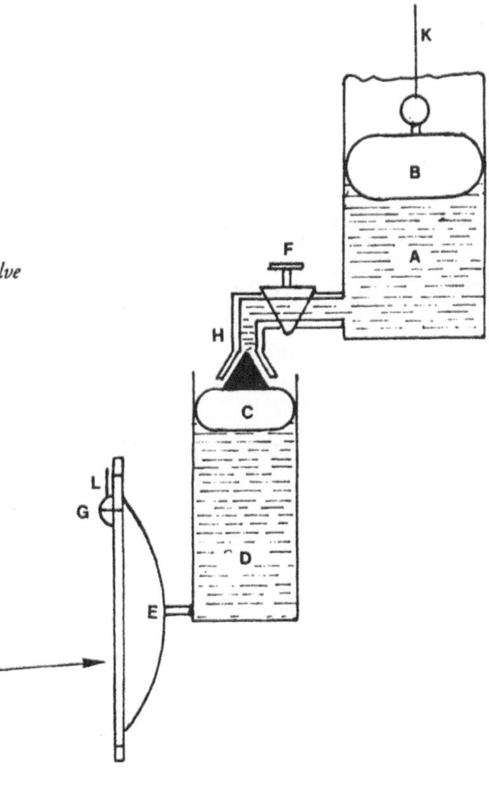

Figure 10. Water-machinery for al-Jazarī's first clock

shown, to a conical valve in a tank g, under tank k. When the water in tank k reaches a certain level, valve e opens and the water runs into tank g and issues from outlet h. In Figure 7 the water runs through the open valve c into tank d, which is on the top of float f, and through hole e into tank g. When filling is stopped tank d evacuates into tank g. Float f rises and closes valve c and no more water can be poured in. Figure 8 shows a valve with the reverse effect of the valve in Figure 7. When poured in the water runs into tank c which fills, the water leaking from hole f on to the partition jk of the vessel. When pouring is stopped, tank c evacuates, allowing the lever arm to tilt to the right and open valve m. The water therefore runs into the main body of the vessel.

The additional components used by the Banū Mūsà, especially the conical valves, allow them to produce a greater variety of effects in their trick vessels than either Philo or Hero. There is also a range of devices, Models 75–87, that have no counterpart in the Greek works. These are all self-replenishing vessels. In the simplest of these the vessel is located outside a large cylindrical container and is connected to a tank on the floor of this container by a pipe, also close to floor level. The top compartment of the container is a reservoir from which a pipe leads down towards the tank. There is a tap in this pipe and this is connected by a rod to the float in the tank. When water is removed from the outer vessel, the level in the inner tank also falls, the tap opens, and the inner tank and the outer vessel are replenished to their original level. The tap then closes (Model 75).

Model 77 is a more sophisticated version. In this case the rod from the float carries two conical valve plugs on its upper end, separated by a short length of rod, which passes through a hole in the floor of the reservoir. This hole is fitted with two valve seats, one just below to fit valve plug m, the other inside the reservoir to fit valve plug l [Figure 9]. When a moderate amount of water is taken from the outer vessel the lower valve opens and the tank and the outer vessel are replenished, as in Model 75. If, however, a large amount of water is withdrawn, valve l drops into its seat and no further replenishment can occur. This is an early example of a fail-safe mechanism and also of a valve rod with twin valves.

Models 88 to 94 are alternating fountains; for example, the fountain-head emits a single jet for a while, then a spray then a jet again and so on. The most interesting feature is that one of these fountains incorporates a miniature horizontal water-wheel driven by axial flow. When a component occurs in a miniature form it is almost always the case that a full-size practical version was already in use. The earliest known axial-flow wheels in Europe are the so-called 'tub-wheels' introduced in the sixteenth century.[28]

Models 95 to 98 are lamps, the first two being self-filling types. This type of lamp had been known for centuries – Philo says that there are many types of self-filling lamps (p. 143). Lamp No. 97 of the Banū Mūsà combines the self-replenishing motif with the self-trimming variety, as in Hero's *Pneumatics* I 34. Indeed, so close is the

---

[28] Norman Smith, *Man and Water*, London, 1976, p. 165.

Banū Mūsà's design to Hero's that it is certain that it is a direct copy. Model 98 is a 'hurricane' lamp with a shield that always turns into the wind to protect the flame. This lamp together with the last two devices – the bellows and the clamshell grab – were probably developed by the Banū Mūsà to meet certain needs in their civil engineering works.

The Banū Mūsà do not include in their work anything similar to the water-wheels of Philo or the pump and organs of Hero. These machines must have been well-known in their time and it is likely that they did not feel that it was appropriate to include them in the *Book of Ingenious Devices.* The end results of most of their devices were in many ways more sophisticated than those of Philo or Hero, but they were equally trivial. Their mastery of control systems and delicate mechanisms, their use of conical valves and switching mechanisms and of small variations in aerostatic and hydrostatic pressures were, however, unique in their time and indeed in the ensuing centuries. Not until relatively recent times, with pneumatic instrumentation, were their achievements in these fields surpassed.

In the interval between the Banū Mūsà's *Book of Ingenious Devices* written about 850 and al-Jazarī's great machine book, completed in 1204 or 1206, there were only two treatises on fine technology of any significance. The first was written in Muslim Spain in the eleventh century by a certain Ibn Jalaf or Ibn Khalaf al-Murādī. It consists of brief descriptions of water-clocks and automata-machines. Although the unique manuscript is badly defaced, the important aspects of the machines can be discerned. First is their ruggedness, with counterweights up to 3 kg and gear-wheels up to 72 cm in diameter. Secondly is the use of complex gear-trains, including segmental gears, i.e. gears with teeth on only part of their circumference in order to allow intermittent working. Although manually operated gear-trains had been known in Hellenistic times, al-Murādī's complex gears are the first to have carried heavy torque. As such they are important in the history of clockwork.[29]

The second treatise was written in 1203 by Riḍwān b. al-Sa'ātī (son of the clockmaker), describing repairs he made to the large water-clock built by his father Muḥammad in Damascus in the reign of Nūr al-Dīn b. Zenkī (1154–1174). This was a large clock with a working face about 4.23 m wide by 2.78 m high. It was very similar to al-Jazarī's first clock but not so well designed; it suffered at least two breakdowns. Also, Riḍwān was a doctor, not an engineer and hence his descriptions often lack clarity. His lack of expertise, however, is occasionally an advantage, since he sometimes describes the manufacture of components in detail, whereas an engineer would take them for granted.[30]

We know nothing of the life of al-Jazarī except what he tells us in the introduction to his book, namely that when he completed his book he had been for 25 years in the service of the Artuqid princes of Diyar Bakr. The book is divided into six categories:

---

[29] Donald Hill, *Arabic Water-clocks*, Aleppo, 1981, pp. 36–46.

[30] *Ibid.*, pp. 69–88.

I. Water-clocks and candle-clocks; 10 chapters
II. Trick vessels; 10 chapters
III. Water dispensers and phlebotomy measuring devices; 10 chapters
IV. Fountains and musical automata; 10 chapters
V. Water-raising machines; 5 chapters
VI. Miscellaneous; 5 chapters

Category I occupies about 40 per cent of the entire work. The first of the water-clocks was also the most elaborate and indeed this clock alone occupies about 12 per cent of the book. A description of the operation and construction of the most important parts of this machine will serve to illustrate, in large part, the unique features of al-Jazarī's methods and skills.

The working face of the clock was a screen of bronze or wood about 1.35 m wide by 2.25 m high, with another 0.75 m for the semidiameter of the zodiac circle. Below the top of the screen was a row of 12 double-leaved doors and below this was a row of 12 single-leaved doors. In front of this row was a small crescent moon. Beneath this was an inverted semicircle of 12 glass roundels and at either side of these was the figure of a falcon below which was a vase with a cymbal in it. Finally, on a platform at the bottom of the screen were the figures of five musicians – two trumpeters, two drummers and a cymbalist. The screen was in the centre of the front wall of a roofless wooden 'house' about 3 m square which contained the water-machinery.

Throughout the day and the night the zodiac circle moved at constant speed at the top of the clock, one sign rising above the 'horizon' – the top of the screen – as another set. The crescent moon moved, also at constant speed, in front of the lower row of doors. At every hour one of the upper doors opened to reveal a standing figure, the lower door rotated to show a different colour and the falcons leant forward, spread their wings and dropped balls from their beaks on to the cymbals. One roundel had become illuminated. At the end of the day all the upper doors were open, all the lower doors had changed colour and all roundels were illuminated. The musicians played at the sixth, ninth and twelfth hours. They were operated by the discharge of water from the orifice.

All the other mechanisms were operated by the water-machinery, attributed by both Riḍwān and al-Jazarī, probably correctly, to Archimedes. This machinery incorporated a feedback control system [Figure 10]. The drive was from the steady descent of the heavy float B in the reservoir A (only the lower part is shown). The cord k went through pulley systems that activated the time-recording automata. A tube made of cast bronze, with a tap F, led out from the bottom of the reservoir, its end H bent down to form the seat of a conical valve. The plug of the valve was soldered to the top of the small float C in float-chamber D. The outlet pipe from the float-chamber entered the flow regulator (see below). When the tap was opened the water entered the float-chamber through the conical valve. The water-level in D rose and closed the valve momentarily, the water flowed through the outlet pipe, the valve opened momentarily, and so on. A virtually constant head was maintained in the float-chamber and the rate of discharge from the reservoir was therefore constant.

The purpose of the flow regulator was to produce the unequal hours, which were used in early clocks. The unequal hours are obtained by dividing the hours of daylight (or darkness) by 12 to give 'hours' that vary in length from day to day. This was done by varying the head of the orifice G below the water-level in the float chamber by rotating it, a piece of drilled onyx, inside an annulus graduated into the 12 zodiacal signs, each subdivided into 30 degrees. The pointer L, on the same radius as the orifice, was set to the correct degree for that day. Al-Jazarī graduated the annulus by painstaking trial-and-error methods, having ascertained that earlier flow regulators, including that of Archimedes, were inaccurate, since they simply divided the annulus into 12 equal divisions of 30° each.

Only one of al-Jazarī's complete machines is of significance in the history of machine technology. This is a water-driven piston pump with horizontally opposed cylinders and true suction pipes. In Greek and Roman pumps the cylinders stood vertically in the water (Category V, Chapter 5).

Throughout the book, however, there are components and techniques that appear for the first time in any treatise. The tipping-bucket occurs in the second clock and in a number of other devices. This was a vessel like 'half a boat' from which stub axles protruded at the top, close to the broad end. These rested in bearings. Water dripped into the vessel from an orifice and after a set time the bucket tipped and discharged all its contents before returning to the horizontal. The sudden tipping of the bucket activated various mechanisms. In one of al-Jazarī's water-raising devices (Category V, Chapter 4) there is a crank as an integral part of the machine. This is the first known use of a non-manually operated crank. Another mechanism used by al-Jazarī was the submersible bowl (*ṭarjahār*); this had an orifice in its underside which caused it to submerge in a pre-determined time. This was used in two of the water-clocks (Category I, Chapters 3 and 4), both of which operated on the closed-loop principle. The *ṭarjahār* was used before this to time the allocation of irrigation water.[31]

A number of techniques that were to be of importance in the history of machine design first appear in the pages of al-Jazarī's book. This does not mean, of course, that they were all invented by him; earlier craftsmen did not usually commit their results to writing. All the same, it is noteworthy that al-Jazarī almost invariably used the best technique available, whether it was his own invention or not. Among these practices were the accurate calibration of orifices; the static balancing of large pulley wheels by fixing small lead weights to their perimeters, much as is done nowadays; the lamination of timber to prevent warping; the use of paper models to establish designs. When manufacturing large vessels from beaten copper, circular wooden templates were used to ensure that the vessel was of uniform circular cross-section along its entire length. In

[31] Al-Muqaddasī, *Aḥsān al-taqāsim*, ed. M. J. de Goeje, Leiden, 1906, p. 357. Al-Muqaddasī (late tenth century) refers to the use of the submersible bowl, which he calls the *ṭarjahara*, for allocation of irrigation water in northern Iran.

the construction of a large ornamental door for the royal palace at Amid (Category VI, Chapter 1) al-Jazarī describes how he cast some of the brass sections first used in Europe at the end of the fifteenth century.[32]

The machines, components and techniques used in al-Jazarī's book would alone be sufficient to guarantee it a foremost place in the history of engineering writing. When to these features his method of presentation is added, however, it becomes no exaggeration to say that this is the most important engineering document, from any cultural area, before the Renaissance. He was writing with the avowed intention of enabling later craftsmen to reconstruct his machines, and in this he was successful.[33] His descriptions, whether of large machines or small devices, always follow the same pattern. First a description of the outside appearance of the machine, then its operation, then a step-by-step description of its manufacture and assembly. Finally, a repetition of its operation. Illustrations include general assemblies of the complete machines, with detailed drawings of various components and sub-assemblies. Although these semi-perspective drawings may at first look strange to our eyes, they are in fact an essential aid to understanding the instructions in the text. As with the other fine technology treatises, they all carry the letters of the alphabet on the various components. And – most importantly – each device is fully dimensioned, not on the illustrations themselves, as in modern practice, but by referring to the identifying letters. Of the earlier writers Philo gives only a few dimensions, Hero and the Banū Mūsà scarcely any. Riḍwān does give some measurements but in a haphazard manner. Indeed, few European treatises before modern times give such accurate specifications and dimensions for the benefit of constructors.

It would not be very rewarding to attempt to fit the four major writers into a progressive sequence. The treatises of all the writers have merits peculiar to themselves: Philo's pioneering pneumatic devices, Hero's experiments with steam and the remarkable pulley systems of the *Automata*, the Banū Mūsà's use of conical valves in switching mechanisms. Al-Jazarī, in some aspects of his work, was clearly drawing upon the work of his predecessors, known and unknown, but his whole book carries the imprint of a brilliant engineering mind. The greatest tribute we can pay to these writers is to study their works and appreciate the wealth of engineering knowledge that they incorporate.

---

32 Cyril Stanley Smith, 'The Early History of Casting, Molds and the Science of Solidification', in *Metal Transformations*, ed. W. W. Mullins and M. C. Shaw, New York, 1968, pp. 23 and 27.

33 For example, al-Jazarī's first clock was reconstructed full-size by P. N. Haward in the Science Museum, London for the 1976 World of Islam Festival. It was built according to al-Jazarī's instructions and worked perfectly. Mr Haward also built a phlebotomy-measuring device for the same occasion and the Museum built a quarter-scale model of the piston pump. These worked equally well, as did a full-size reconstruction of the elephant clock built by Mr Haward in 1993 for the Musée de l'Horlogerie in Le Locle.

IV

Arabic Fine Technology and its Influence on European Mechanical Engineering

By fine technology we mean those types of machines or instruments that were designed to give pleasure to courtly circles, or for timekeeping or for the use of scientists (mainly astronomers). Fine technology is thus distinguished from utilitarian technology which is concerned with machines such as mills, water-raising devices and textile machinery. These were essential to the economic prosperity of society but were very much simpler technically than the constructions of fine technology. Also, the sources for the two categories are different. Information on utilitarian technology comes largely from archaeology (including the examination of existing machines), and references in the works of geographers, travellers and other non-technical writers. For fine technology we find most of our information in a few precious technical treatises. There is some overlap between the two categories, since the writers of the treatises borrowed some of their vocabulary from the craftsmen, such as millwrights and metalworkers. Sometimes they incorporated in their devices miniaturised versions of components from large machines. Occasionally they also described improvements they had made to the traditional machines. In general, however, the division into two separate categories is justifiable.

The traditions of fine technology began in the Hellenistic world in the third century BC. The most important works that have survived were those of Philo of Byzantium (*c.* 200 BC) and Hero of Alexandria (fl. mid-AD 100). Philo's *Pneumatics* exists only in an Arabic version;

The author acknowledges with gratitude the generosity of the Royal Society in providing him with a grant to assist him in his researches into Islamic technology.

it consists mainly of trick vessels operated by siphonic action, together with descriptions of various types of water-wheels. A number of Hero's works are extant, mainly in Greek, although the *Mechanics* has survived only in an Arabic version made by Qūstā b. Lūqā (d.*c.* 300/912–13) in Armenia towards the close of the third/ninth century. His *Pneumatics* is clearly derived from Philo's work of the same name, but includes more ingenious mechanisms, including a miniature steam turbine. His *Automata* describes two types of automatic theatre, one of which consists of a console that moves by means of a weight-driven motor.[1]

The works of Philo and Hero, as well as some shorter treatises of Greek origin by various authors, were certainly known to the Arabic writers. There is no space here, however, to make a detailed comparison among the various treatises — Greek and Arabic. Suffice it to say that the Arabs, although they took the Greek works as their starting-point, were considerably more advanced. In the following pages any ideas that were in existence in Hellenistic times will be mentioned where relevant, but the distinctive contributions of Arabic engineers will of course occupy most of our attention.

It is highly likely, from references we find in the works of Greek, Roman and Arab historians and travellers, that a tradition for the construction of ingenious devices had existed in the eastern Mediterranean from Hellenistic times and through the periods of the Roman and Byzantine empires into the world of Islam. Written works that actually describe machines in any detail are, however, very rare. The only two important Greek writers have already been mentioned. In Arabic also, although there are a number of short treatises, there are only two writers of major importance. The first of these was Aḥmad, one of three brothers known as the Banū (sons of) Mūsā from their father Mūsā b. Shākir who had been a close companion of Ma᾿mūn when the latter was residing in Khurāsān before he became caliph (he reigned from 198/813 to 218/833). The brothers were among the most prominent figures in the first flowering of Arabic culture that took place in Baghdad in the third/ninth century. They were patrons of scientists and translators, undertook public works and took part in the turbulent palace politics that characterised the ʿAbbāsid caliphate as the empire began to break up into regional units. Above all, they were erudite scientists and engineers in their own right. About twenty works are attributed to the brothers — Muḥammad, Aḥmad and al-Ḥasan — by the Arab biographers, but only three of these have survived. One is a mathematical treatise that exists only in a Latin translation, the second is the *Book of Ingenious Devices* by Aḥmad, and the third is the description of a musical automaton that we shall have occasion to refer to later. It is probable that this work is also largely attributable to Aḥmad. The most important of these works, however, is undoubtedly the *Book of Ingenious Devices.* About eighty of the devices described in it are trick vessels of various kinds, and the remainder include fountains, lamps, a "gas mask" for use in polluted wells and a clamshell grab. Despite the apparent triviality of many of the devices, Aḥmad's mastery of aerostatic and hydrostatic pressures and his use of automatic controls and switching systems places him well in advance of his Hellenistic predecessors. Indeed, his work in this field, though limited in scope, was not surpassed until modern times.[2]

The great machine book of Ibn al-Razzāz al-Jazarī (fl. seventh/thirteenth century) was completed in Diyār Bakr in 602/1206. This is certainly the most important of the Arabic treatises and probably the most important engineering document from any cultural area before the Renaissance. The work is divided into six categories: (1) water-clocks and candle-clocks, (2) wine dispensers, (3) phlebotomy measuring devices and water dispensers, (4) alternating fountains and musical automata, (5) water-raising devices and (6) miscellaneous. Several of the machines, mechanisms and techniques that appear for the first time in al-Jazarī's work were later to enter the vocabulary of European mechanical engineering. The work is also important because al-Jazarī described his methods of manufacture and construction with the avowed purpose of enabling later craftsmen to reconstruct his machines. Indeed, several have been successfully reconstructed by modern craftsmen using his instructions and illustrations. Detailed instructions and specifications are very rarely found in early mechanical treatises; in fact from this aspect al-Jazarī's work is almost unique before modern times. Other writers were usually unable to give clear instructions either because they had little practical experience or because they did not wish to reveal all the secrets of their profession.[3]

Reference to only two major treatises should not mislead readers into thinking that there were no other Arabic works of significance to the history of machine technology. Indeed, as will be made clear later, some significant ideas occur in some of the shorter treatises. It seems unlikely however, that any major works comparable to those of Aḥmad b. Mūsā and al-Jazarī await discovery. It may seem rash to make such an assertion, given the number of manuscripts in the libraries of universities, mosques and other institutions that have not been examined. The evidence suggests strongly that treatises such as the two mentioned above are of rare occurrence. All the major Arabic biographical dictionaries dealing with scientists have entries on the Banū Mūsā and on no other writers of comparable works on machines. Al-Jazarī lived too late to be included in the *Fihrist* of Ibn al-Nadīm (d. 385/955), and later biographers do not mention him, either because they usually copied from the *Fihrist* or because al-Jazarī's reputation did not spread beyond the limits of the Jazīra until some time after his death. He is, however, mentioned by the Egyptian scholar al-Qalqashandī (d. 821/1418) as pre-eminent in the science of machines.[4] The many manuscripts of al-Jazarī's work, dating from the seventh/thirteenth to the twelfth/eighteenth century, also testify to the esteem in which he was held in the world of Islam.

Al-Jazarī himself always acknowledges his predecessors. Thus he mentions the Banū Mūsā as the most illustrious workers in this field in earlier times.[5] He also mentions by name two Greek and two Arabic writers on machines. All these men are known to us. It seems likely that he knew of all the works of his predecessors that had a direct bearing on his work. From the available evidence, therefore, it appears that we possess the two major works of medieval Arabic machine technology, together with a number of minor works. Major works on machines were equally rare in Hellenistic times, for which we have only the works of Philo and Hero, whose writings were the only Greek works on this subject to reappear in medieval Europe. Clearly, the ability to describe machines accurately and fully has been a rare talent in earlier centuries.

For the remainder of this article we shall be considering some of the more important features of Arabic machine technology. Some of the ideas to be discussed are of Greek origin and such cases will be identified. Otherwise it may be assumed that ideas that occur for the first time in the writings of a given Arab engineer are inventions that were due to that engineer. (References to the Banū Mūsā and al-Jazarī, given in parentheses, are to the page numbers in Hill's translations. References to other writers are given in notes in the usual manner.)

It would be tedious to list all the components and ideas that were known in Islam before their appearance in Europe. Only some of the more significant will therefore be listed. Many of these come from al-Jazarī's works, partly because he was an innovative engineer and partly because his detailed descriptions permit us to understand the construction and purpose of most of the devices that he described. One of the most important components, however, the conical valve, was used to the greatest effect by the Banū Mūsā. It may have been of Greek origin, but nowhere is it used in more applications than by the Banū Mūsā throughout their work. (The normal practice among Arab writers of referring to the *Book of Ingenious Devices* as the work of all three brothers is followed from here onwards, even though its main author was Aḥmad.) The manufacture of these valves was not described until the latter part of the fourth/tenth century in *Mafātīḥ al-ʿUlūm* (The Keys of the Sciences), a dictionary of the sciences written by a certain Abū ʿAbd Allāh al-Khuwārizmī (d. 370/980–1).[6] Further details are given by al-Jazarī, who tells us that plug and seat, usually made of bronze, were cast together in the same mould, then ground together with emery powder until a watertight fit was obtained (20). Emery powder was also used in the calibration of orifices. Al-Jazarī tells us how he accurately calibrated an orifice for a monumental water-clock. He must have known the approximate bore needed and intentionally made the orifice, bored through a piece of onyx, too small. The orifice was then gradually enlarged, with continuous checks on its throughput, until the required rate of flow under a given head of water was achieved (24–5). Conical valves first appear in Europe in the notebooks of Leonardo da Vinci (d. 1519).

The crank is one of the most important components of many modern machines. Manual cranks were known in Europe by the ninth century but the incorporation of cranks in machines did not appear until the fourteenth. Rudimentary cranks appear in the Banū Mūsā; these did not make a complete rotation and were used for the automatic opening and closing of valves (204–9). In one of al-Jazarī's water-raising machines, however, a crank is an integral part of the machinery. It turns continuously to transmit power through a slot-rod to the water-raising apparatus (184–5). This section of al-Jazarī's work is the only one in which he describes large utilitarian machines, probably in order to record improvements that he had made to the traditional types. The fifth and last of these machines is very significant in the history of technology. It is a twin-cylinder water-driven pump with true suction pipes. The pumps of classical and Hellenistic times had vertical cylinders that stood directly in the water. Al-Jazarī tells us that this pump was similar to the "Byzantine siphon" used for discharging Greek fire, except that it was larger. This is the first evidence we have of the construction of this flame-thrower (186–9).

Among other techniques described by al-Jazarī were the static balancing of wheels on a mandrel using small lead weights (27); the use of paper models to assist in the design of intricate mechanisms (68); and the lamination of timber to prevent warping (80). In describing the casting of a large door made of brass and copper for the palace of Āmid (now Diyār Bakr) he mentions two methods of casting: one is the *cire-perdu* or lost-wax process, which had been known since antiquity; the other was using closed mould-boxes with green sand, a method that was not introduced into Europe until the end of the fifteenth century (192).

A very important feature of the thinking behind many of the devices described in the treatises is the preoccupation of the Arab engineers with controls, particularly with controls that would allow a given machine to continue working for long periods — hours, days or even longer — without human intervention. Many types of control, most of which are thought of as quite modern, were employed to achieve these results: feed-back control and closed loop systems, various types of automatic switching to close and open valves or change direction of flow, and precursors of fail-safe devices. In the space available it is possible to describe briefly only four such systems, two by the Banū Mūsā and two by al-Jazarī.

Figure 1 shows a constant-level vessel by the Banū Mūsā (196–7). The main container *DJHT* is divided into two chambers, the upper one, *DJCK*, being the reservoir for the water which is poured in at hole *X*. In the lower chamber is a tank *S* in which is a float *Z*, soldered to the top of which is a valve rod. The plugs of two conical valves are soldered to the upper part of this rod, their seats being fitted to a hole in plate *CK*; valve *L* opens upwards, valve *M* downwards. A pipe *BE* connects tank *S* to a tank *G* outside the main

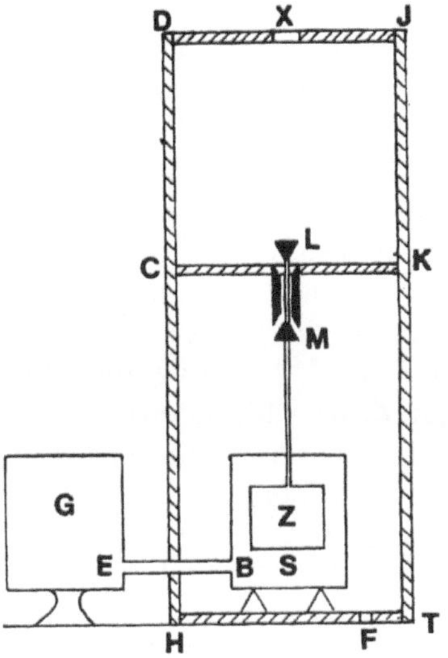

Figure 1 Banū Mūsā: Constant-level device

container. The water-levels in tanks *S* and *G* are the same. When a moderate amount of water is withdrawn from tank *G*, valve *M* opens to replenish the two tanks, then closes when the water reaches its original level. If however, a large amount is withdrawn, valve *L* closes, and no replenishment can take place. Although this device was intended only to mystify onlookers it embodies a principle that was to be important in fail-safe mechanisms. Also, for the first time, we have a double-acting valve.

A fluting machine attributed, probably correctly, to the Banū Mūsā, does not appear in the *Book of Ingenious Devices* but in a separate unique manuscript.[7] The flow of air to the flute is achieved by hydraulic means. Water is directed to one of two chambers alternately, the switching being done by means of a water-wheel, a worm-and-pinion gear, a semicircular cam and a system of valves. When one chamber is filled the air expelled from it goes through a non-return valve into the flute. The flute was provided with eight pallets that were operated by cams on the surface of a cylinder. The cylinder was rotated by a water-wheel on whose axle was a small toothed pinion meshing with a large gear-wheel co-axial with the cylinder. Unfortunately, the manuscript does not include any illustrations, but Figure 2, from a German work of the year 1650, shows essentially the same system. The air supply system shown at the right is cruder than that of the Banū Mūsā, and the instrument is an organ, not a flute, but the method of causing a tune to be played is very similar. The Banū Mūsā's machine, only a very sketchy outline of which is given here, was very sophisticated both hydraulically and mechanically.

The water-machinery shown in Figure 3 was used by al-Jazarī to provide motive power for the first of his monumental water-clocks (17–41). The bottom *A* of the large reservoir is shown. In the reservoir was a float *B*, to which cord *C* was attached; this cord operated the various time-recording mechanisms, and it was of course essential that float *B* descended at constant speed and hence that the discharge rate from the reservoir was also constant. This was achieved as follows: a tube of cast bronze was led out from the bottom of the reservoir, its end turned down at right angles and

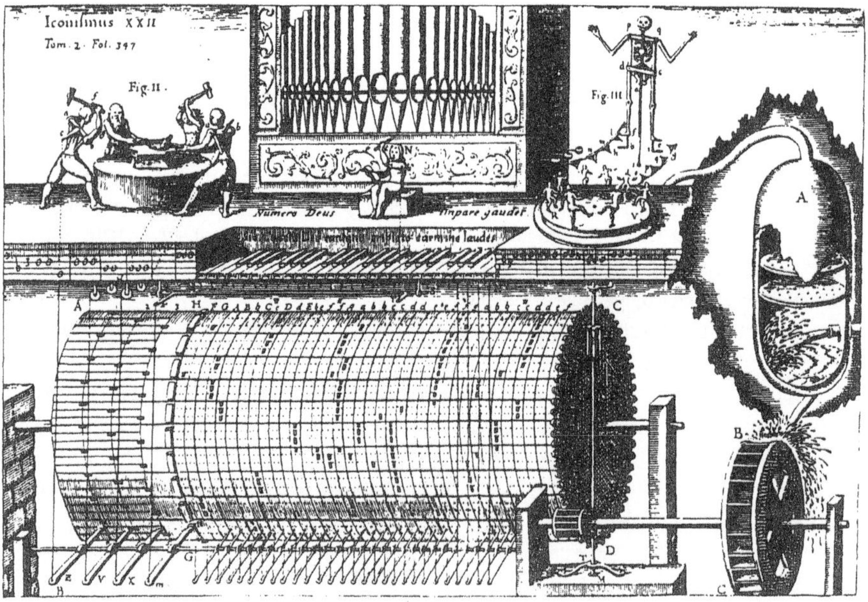

Figure 2 Hydraulic organ from a German treatise of AD 1650

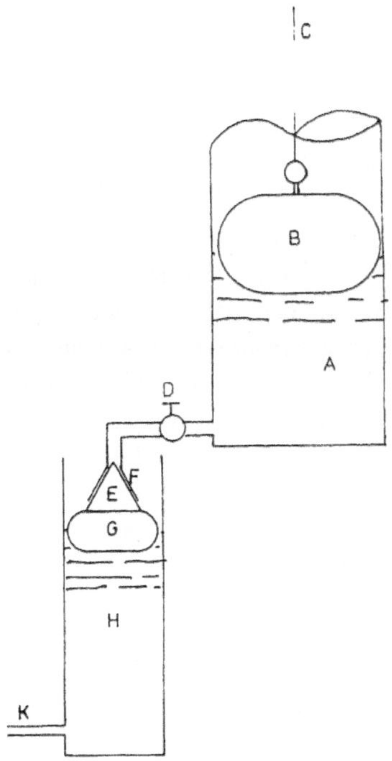

Figure 3 Al-Jazarī: Water-machinery of first clock

formed into the seat *E* of a conical valve. The pipe was provided with a tap *D*. Below the valve seat a small cylindrical float-chamber *H* was erected, having an orifice *K* at its lower end. In it was the small float *G* carrying the plug for the conical valve. To begin the operation the outlet *K* was temporarily closed and tap *D* was opened. The float-chamber *H* filled, float *G* rose and closed valve *E*. Now outlet *K* was opened, and water was discharged, allowing float *G* to fall. Water flowed through valve *F* which closed momentarily, then opened momentarily, and so on. The water-level in the float-chamber and hence the rate of discharge were therefore almost constant. This is a brilliant application of the principle of feed-back control. Al-Jazarī attributed its invention to Archimedes (d. 212 BC) and indeed there exists an Arabic manuscript attributed to

Archimedes that includes a description of this system. Even so, it is a little strange that no other Greek work mentions conical valves.[8]

A system in another of al-Jazarī's clocks is probably his own invention. Figure 4 shows the essential machinery on his third clock (51–7). Inside the model of a boat a water tank *n* is concealed. On the surface of the water is a submersible bowl, *tarjahār*, that has an orifice in its underside that allows it to submerge in one hour precisely. The bowl is attached to the side of the tank by the links *b*. Diametrically across the top of the bowl *a*, a rod *k* is soldered. This has a hole in its centre to which a wire *h* is attached. This wire goes vertically upwards and is connected inside the "castle" to a ball-release mechanism (not shown). A staple is soldered to the underside of the bowl, opposite the orifice, to which a light chain *d* is connected. Above the boat four tubular metal columns are erected and these support the "castle", a metal box, inside which, as already mentioned, is the ball-release mechanism. From this mechanism a channel leads to the head of a falcon *f*, soldered to the outside of the

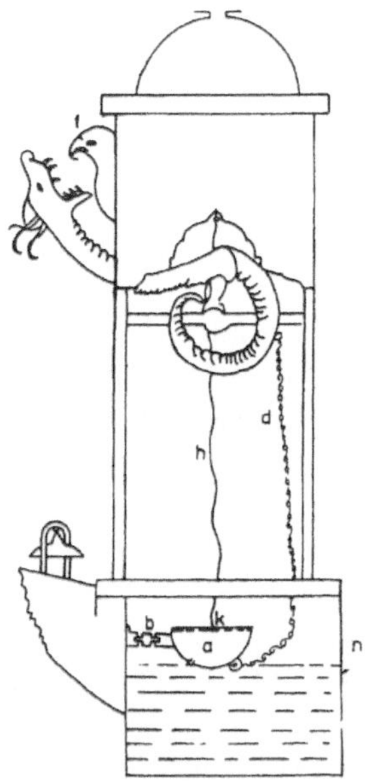

Figure 4 Al-Jazarī: Water-machinery of third clock

castle. Between the columns, below the castle, are two crosspieces between which is an axle to which the feet of a serpent are soldered. The tail of the serpent is bent into a circle which is in fact a pulley; its neck extends outside the castle, bringing its open jaws close to the beak of the falcon, this being the position at the beginning of the operation, when the empty bowl is on the surface of the water. Chain *d* is connected to a staple on the tail of the serpent, and the balls are loaded into the magazine of the release mechanism through a detachable dome at the top of the castle. The action is as follows: the bowl sinks slowly until at the end of an hour it submerges suddenly. Wire *b* then tautens and jerks the ball-release mechanism. A ball rolls into the head of the falcon and emerges from its beak into the jaws of the serpent whose head descends. At the bottom of its travel it drops the ball onto a cymbal, which rings, and the falcon, made lighter by the release of the ball, resumes its former position. When the serpent descended chain *d* tautened and the combined action of this and links *b* raised the bowl and tilted it. It therefore discharged its contents, settled back on the surface of the water and the whole sequence recommenced. A closed loop was therefore in operation: the clock would continue to work as long as balls were loaded into the magazine. This gives only the essential parts of the clock. There were several more automata on this clock and ever more on the following one, the clock of the elephant.

The foregoing examples of techniques, mechanisms and control systems described in the treatises will, it is hoped, give some idea of the level of sophistication achieved by Arab engineers. It has been indicated in several cases that Arabic ideas preceded their introduction into the west, but it is felt that the influence of Arab engineering on the development of European machine technology can best be demonstrated in a structured manner by examining the antecedents of a single machine. The mechanical clock was perhaps the most significant machine to have been introduced in the Middle Ages. The mechanical escapement, a mechanism for slowing down the descent of a heavy weight, was without doubt the invention of an anonymous western European clockmaker towards the close of the seventh/thirteenth century. It can be shown, however, that all the other elements of the mechanical clock, as it emerged in Europe in the fourteenth century, were present in Arabic horology many years before that time. The elements of the early European mechanical clocks were as follows: (1) weight drive, (2) escapement, (3) "going-train" of gears, (4) "striking-train" of gears, (5) assembly of clockwork into a limited space, (6) audible signals, e.g. striking bells, and (7) visual signals, e.g. dials, but more importantly moving jackwork figures of humans and animals; celestial automats. Let us see which of these elements were present in Arabic horology.

A. Weight drive and escapement, (1) and (2)

It can be argued that the large floats that were used in Greek and Arabic water-clocks were weight drives controlled by hydraulic escapements. Also, Hero of Alexandria, writing about the year AD 60, described automata activated by a heavy weight resting on a bed of millet seed in a container. As the seeds slowly escaped from a hole in the bottom of the container, so the weight descended. The idea of using a freely suspended weight, however, seems to have exercised the minds of medieval Arab engineers. There is a treatise, dating to the fifth/eleventh or sixth/twelfth century, that exists in a number of manuscript copies. The manuscripts all contain other material, but the section in question is entitled "Wheels that move by themselves".[9] These devices are usually dismissed as dead-end essays in perpetual motion, but there is more to them than that, although they are admittedly obscure. They were probably serious investigations into the possibility of using gravity as motive power in machines.

One of the devices shows two chain-of-pots pumps mounted on a horizontal axle, in the centre of which was a wheel fitted with curved tubes each partially filled with mercury. The wheel was supposed to turn by the movement of the mercury within the tubes and so operate the pumps. This is of course impossible. The second device shows two weight-driven gear-wheels which meshed with a third gear-wheel that could be connected to a machine. No method for controlling the speed of descent of the weights is given. It is often assumed that this treatise was written for the amusement of courtly circles and not for practical application. It is possible, however, that the writer knew quite well what it was about and that he omitted certain essential components in order to safeguard "trade secrets". Weight is lent to this supposition by the existence of a practical machine that combines the main principles of both machines — mercury escapement and weight drive.

The *Libros del Saber* were prepared in Spain in 1277 under the direction of Alfonso X of Castile (1226–84), with the express purpose of making Arabic knowledge available in Castilian. It is a compilation of translations and paraphrases of Arabic works, some with known authors, some anonymous. In the latter category come the five timepieces described in the last section of the work; only one of these is relevant to our present theme. This clock consisted of a large drum made of walnut or jujube wood and sealed with resin. The interior of the drum was divided into twelve compartments, each with small holes through which mercury flowed. Enough mercury was enclosed to fill just half the compartments. The drum was mounted on the same axle as a large wheel which had a rope wrapped around its perimeter, to the end of which the driving weight was attached. Also on the axle was a pinion with six teeth that meshed with thirty-six oaken teeth on the rim of an astrolabe dial. The drum and the pinion made a complete revolution every four hours and the astrolabe dial made a complete revolution in twenty-four hours.[10] The inclusion of astrolabe dials or similar celestial representations was not uncommon in Islamic monumental water-clocks. The remains of two such clocks can still be seen in Fez, Morocco, and one of them includes an astrolabe dial. Incidentally, the remains of these clocks, built in the eighth/fourteenth century are an important attestation of the tradition of building large water-clocks in Islam.[11]

The compartmental cylinder clock had been in Islam since the fifth/eleventh century — two hundred years before the first appearance of the mechanical clock in Europe. This type of clock, probably derived from the example in the *Libros del Saber*, became very popular in Europe. In the European variety the liquid was water and the drum itself descended. The clock consists essentially of an upright frame which is open and has hours marked down one or both sides. The drum is supported at the centre on two cords wound around the centre of the drum. The arbor is extended to cut the hour scale(s). Under its own weight the drum is free to revolve slowly down the cords, and if the quantity of water within and the sizes of the cords and the orifices are properly adjusted it will indicate the time on the hour scale at the side. Starting in the seventeenth century and throughout the eighteenth and nineteenth centuries, especially in agricultural regions, the falling drum clepsydra in its various forms was extremely popular as a cheap and reliable timekeeper.[12]

B. Gear trains, (3) and (4)

In a mechanical clock, there are two distinct systems of gearing. One of these, the "going-train" consists of the gears that are used to transmit power from the weight drive through the escapement and keep the movement of the clock in correct line. The "striking-train" is also connected indirectly to the drive; its purpose is to cause the bells or other audible signals to sound at regular intervals. One of the gear-wheels in the striking-train is of the segmental type, that is to say it has teeth on only part of its perimeter, thus permitting intermittent action.

The standard types of gearing had all been known since Hellenistic times: parallel meshing; meshing at right-angles; worm-and-pinion; rack-and-pinion. A very complex astronomical calculator, discovered in a shipwreck near the island of Antikythera, probably dates to about 80 BC. It contains many intermeshing gear-wheels and also a differential turntable. Although it could be used for a number of astronomical computations, it was manually operated and so did not have to transmit high torque.[13] Gears that did transmit high torque were the mill with a vertical water-wheel and the water-raising machine known as the *sāqiya*, both introduced in Hellenistic times. In these machines, however, the gears were simply a pair of right-angle meshing wheels.

The Arabs were of course familiar with all the normal types of gearing. Gears that begin to resemble the gears of clockwork occur in some geared calendars and astrolabes. Geared calendars were known in Byzantium by the late fifth century AD, and were also constructed by the Arabs, as shown in a treatise written by the famous scientist al-Bīrūnī (d. 440/1048–9) about the year 1040. The gear-wheels were fitted to the inside of a flat circular brass box, bearings being soldered to the base of the box to receive the arbors of the wheels. Three pairs of wheels were mounted concentrically. On the surface of the box, for three of the wheels, there were pointers mounted on extensions of the arbors. One of these moved over a scale divided into the days of the week, the other two, one for the sun and the other for the moon, moved over scales divided into the zodiac and its divisions. The gear ratios were carefully calculated so that the calendar remained synchronous. The pointer for the week was moved each day, causing the pointers for the sun and the moon to move to their correct degree of the zodiac for that day. There was another wheel which displayed the shape of the moon in its correct phase for each day.[14]

In Greek and Arabic instruments we therefore have fairly complex manually-operated gearing, and in the utilitarian machines we have simple gearing that transmitted high torque. Until recently, however, we had no evidence for the combination of the two, i.e. complex gears that transmitted high torque. In the mid-70s a treatise on automata and water-clocks was discovered. This was written in the fifth/eleventh century by a Spanish Muslim called Ibn Khalaf al-Murādī (fl. fifth/eleventh century). The unique manuscript is very badly damaged, the descriptions are too short and the illustrations are badly drawn. It is therefore impossible to discover exactly how any of the thirty devices worked. It can be determined unequivocally, however, first that the devices were large and robust, and secondly that complex gear-trains were used to transmit power from the prime movers to the mechanisms and automata. Sufficient text remains for us to be sure that these included segmental gears, which can also be clearly seen on the illustrations. We have already seen that such gearing was used in the striking-trains of medieval mechanical clocks (al-Murādī may also have used epicyclic gears, but this cannot be established with certainty).[15] Al-Jazarī also used segmental gears in two of his water-raising machines for the transmission of intermittent power (179–81). Taking the evidence as a whole, there is no doubt that the Arabs' knowledge of gearing was at least as sophisticated as that of the European makers of early mechanical clocks.

C. Assembly of clockwork into a limited space, (5)

There are a number of cases where automata and mechanisms were assembled in a limited space, foreshadowing the compact arrangement of the parts of mechanical clocks. Perhaps the most noteworthy example is the assembly of the mechanisms in the "machine-box" of al-Jazarī's fourth clock (63–8). The box was a cube, its sides approximately 30 cm. Into this volume the following mechanisms had to be assembled: ball-release mechanism and magazine for the balls; "balance" for directing the balls alternately to the left and right sides of the clock; a wheel for rotating the figure of a bird on top of the clock; a toothed wheel and a one-way hinge for slowly causing a set of roundels to turn from black to silver (63–9). Some of the Banū Mūsā's devices are masterpieces of intricate assembly, but the description of one such device would occupy too much space.

D. Audible and visual signals, (6) and (7)

Again we may profitably use one of al-Jazarī's machines as an example. This is his first clock. The description of the manufacture, assembly and operation of this clock is long and detailed, but, as is usual with al-Jazarī, he gives a summary of its outside appearance and operation at the start of the chapter. This summary is on pages 18–19 of Hill's translation. The clock consisted of a working face about one and a half metres across and three and a half metres high. At the top of the clock was a disc marked with the signs of the zodiac that turned with constant speed throughout the day. Below this, across the face of the clock were two rows of twelve doors each. In front of the lower row a small crescent moved with constant speed. Every hour, as it reached the sides of the doors, the upper door leaves opened to reveal a standing figure, and the lower door rotated to reveal a different colour. At either side of the clock was a falcon in a niche; every hour each falcon spread its wings, leant forward and dropped a ball onto a cymbal, which rang. Between the two falcons was a semicircle of twelve glass roundels. During the night one roundel became fully illuminated every hour, until at the end of the night all twelve were fully illuminated. Finally, on a platform across the bottom of the clock were the figures of two trumpeters, two drummers and a cymbalist. These performed at the sixth, ninth and twelfth hours. The motive power for all this array of automata was the water-machinery described in an earlier part of this article. Very similar displays were described in the clock built in the sixth/twelfth century at the Jayrūn Gate in Damascus, and another public clock built in Tlemcen, North Africa, in the eighth/

fourteenth.[16] Arrays of automata were to be seen on the early mechanical clocks of medieval Europe, in which dials were one of the less important features. To quote Lynn White:

> Something of the civic pride which earlier had expended itself in cathedral-building now was diverted to the construction of astronomical clocks of astounding intricacy and elaboration. No European community felt able to hold up its head unless in its midst the planets wheeled in cycles and epicycles, while angels trumpeted, cocks crew, and apostles, kings, and prophets marched and countermarched at the booming of the hours.[17]

It is never easy to trace the diffusion of technological ideas, since these ideas are usually transmitted by non-literary means: for example by contacts among craftsmen, by travellers' reports and by the inspection of earlier constructions. It is therefore not always possible to find absolutely firm evidence for the passage of a given idea from one cultural area to another. This is true of horological ideas. Even so, it is simply not credible that the mechanical clock suddenly appeared in Europe without any antecedents. Although our contention that many of the ideas upon which the mechanical clock was founded were Arabic ideas is partly conjectural, nevertheless it is felt that these conjectures are more credible than the genesis of the mechanical clock *ex vacuo*.

Writing in the year 1271, Robertus Anglicus indicated clearly that the invention of the mechanical escapement was imminent and that the invention was being sought among the ranks of the makers of water-clocks.[18] The earliest description we have of a European water-clock appears in MS Ripoll 225 from the Benedictine monastery of Santa María de Ripoll at the foot of the Pyrenees. The MS is tentatively dated to some time in the fifth/eleventh century. The description of the main water-machinery is missing but the section describing the striking-train is complete.[19] There are other, later references to water-clocks in Europe; indeed the existence of a considerable market for water-clocks in Europe in the twelfth century is implied by the establishment of a guild of clockmakers in Cologne in 1183.[20] There is also evidence to indicate considerable improvements in hydraulic timekeeping in the sixth/twelfth century.

Several routes for the transmission of Arabic ideas on fine technology can be suggested, but the most likely route was from

Muslim to Christian Spain and from there into northern Europe. We know for certain that the knowledge of the astrolabe entered Europe by that route. It has now been established that the first European treatises on the astrolabe were of Arabic inspiration and were produced in the monastery of Santa María de Ripoll in the fifth/tenth and sixth/eleventh centuries.[21] This new knowledge was probably disseminated into Europe by Gerbert d'Aurillac, later Pope Sylvester II (999–1003), after he visited Ripoll about the year 967. But we have already seen that the first known European description of a water-clock was written in the monastery of Ripoll. Since the church was keenly interested in methods of timekeeping, it seems likely that information about Arabic advances in horology was carried to the rest of Europe by churchmen.

All the ideas and components embodied in the mechanical clock, except the escapement, had long been present in Arabic fine technology. Apart from the Iberian peninsula, these ideas could also have entered at other points where the two cultures were in contact, e.g. Byzantium, Sicily, and Syria during the Crusades. The mechanical escapement, an invention of genius, undoubtedly originated in western Europe, but it was an invention that was grafted onto an array of ideas that were Arabic in origin.

Notes

1 A.G. Drachmann, "Ktesibios, Philon and Heron: a study in Ancient Pneumatics", *Acta Historica Scientiarum Naturalium et Medicalium* (Biblioteca Universitatis Hauiensis Copenhagen, 4, 1948), pp. 1–197.

2 Banū Mūsā, *The Book of Ingenious Devices*, trans. and annotated Donald R. Hill (Dordrecht: Reidel, 1979); Arabic text, ed. Aḥmad Y. al-Ḥasan (Aleppo: Institute for the History of Arabic Science, 1981).

3 Ibn al-Razzāz al-Jazarī, *The Book of Knowledge of Ingenious Mechanical Devices*, trans. and annotated Donald R. Hill (Dordrecht: Reidel, 1974); Arabic text, ed. Aḥmad al-Ḥasan (Aleppo: Institute for the History of Arabic Science, 1979).

4 Al-Qalqashandī, *Ṣubḥ al-Aʿshā*, ed. Muḥammad ʿAbd al-Rasūl Ibrāhīm, vol. 1–XIV (Cairo, 1913–30), p. 1,477.

5 Al-Jazarī, p. 157.

6 E. Wiedmann, *Aufsätze zur Arabischen Wissenschaftsgeschichte*, vol. 1 (Hildesheim: Georg Olms, 1970), pp. 209–10.

7 Banū Mūsā, "Al-Āla allatī tuzammar bi-nafsihā", Arabic text ed. L. Cheikho, *Al-Machriq*, 9 (1906), pp. 444–56; English trans. by H.G. Farmer, *The Organ of the Ancients*, (London: Reeves, 1931), pp. 88–118.

8 Pseudo-Archimedes, *On the Construction of Water-clocks*, trans. Donald R. Hill (London: Turner & Devereux, 1976).

9 H. Schmeller, "Beiträge zur Geschichte der Technik in der Antike and bei den Arabern", *Abhandlungen zur Geschichte der Naturwissenschaften und der Medizin*, 6 (Erlangen, 1922).

10 *Libros del Saber . . .*, ed. M. Rico y Sinobás (Madrid, 1863–7); see also Silvio A. Bedini, "The compartmented cylinder clepsydra", *Technology and Culture*, 3 (Spring 1962), pp. 115–41.

11 Derek de Solla Price, "Mechanical water clocks of the 14th century in Fez, Morocco", *ITHACA* (Paris, 1962), 26 VIII–2 IX.

12 Anthony Turner, *The Time Museum*, vol. 1, Part 3 (Rockford, 1984), pp. 40–3.

13 Derek de Solla Price, *Gears from the Greeks* (New York: Science History Publications, 1975).

14 Donald R. Hill, "Al-Biruni's mechanical calendar", *Annals of Science*, 42, (1985), pp. 139–63.

15 Donald R. Hill, *Arabic Water-clocks* (Aleppo: Institute for the History of Arabic Science, 1981), pp. 36–46.

16 For the Damascus clock, see Hill, *Arabic Water-clocks*, pp. 69–88. For the Tlemcen clock, see Abū Zakariyyā Yaḥyā ibn Khaldūn, *Bughyat al-Ruwwād*, ed. A. Bel, vol. 2 (Algiers, 1904–13), pp. 40–1.

17 Lynn White Jr., *Medieval Society and Social Change* (Oxford, 1964), p. 124.

18 Lynn Thorndike, "Invention of the mechanical clock about 1271 AD", *Speculum*, 16 (1941), pp. 242–3.

19 Francis Maddison, Bryan Scott and Alan Kent, "An early medieval water-clock", *Antiquarian Horology*, 3 (1962), pp. 348–53.

20 White, p. 120.

21 Emmanuel Poulle, "Les instruments astronomiques de l'Occident latin au XIe et XIIe siècles", *Cahier de civilisation médievale*, 15 (1972), pp. 27–40.

V

ARABIC MECHANICAL ENGINEERING: SURVEY OF THE HISTORICAL SOURCES

The intention of this article is to make an inventory of the source material of Arabic Mechanical Engineering, either published or as yet only available in manuscript form. The sources are listed in chronological order (insofar as this can be established) and for each source the relevant secondary works are given. It would clearly be irresponsible to claim that this inventory is comprehensive; omissions are inevitable. Nevertheless it is felt that all the known primary sources of significance have been included, together with the great bulk of important secondary works. It is hoped, of course, that more original material applying to mechanical technology will be discovered in the course of future researches. Meanwhile, the information given below should be of service to scholars working in this, and related, fields.

Two definitions should be borne in mind when considering the information given in this article. First, the term 'Arabic' is used in the title because the treatises were written in that language, although their authors were of various ethnic origins. The term 'mechanical engineering' is used with its modern connotation of machine technology. We are not, therefore, concerned with other branches of engineering such as civil or military, nor with theoretical mechanics, nor with instrumentation.

The article is divided into two sections, of which the first is the more important:

1 Technical treatises - Fine Technology
2 Utilitarian machines

The treatises provide us with by far the best information about the *technical* aspects of Arabic mechanical engineering. Of the

The author acknowledges with gratitude the generosity of the Royal Society in providing him with a grant to assist him in his researches into Islamic technology.

machines that they describe some – water-clocks, water-raising devices, specialized machines – served utilitarian purposes, while others – automata, trick vessels, fountains – were designed to give amusement or aesthetic pleasure to courtly circles. In general it is convenient to use the term Fine Technology for the machines described in the treatises. The range of mechanisms, components and techniques that they incorporate is astonishingly wide; much of this array was to enter the vocabulary of European engineering and to be of importance in the genesis of the Industrial Revolution. Moreover, the best of these treatises provide us with information about techniques of manufacture and construction that are to be found in no other pre-Renaissance sources.

The treatises do not describe everyday machines such as water-mills, wind-mills, traditional water-raising machines and so on. Almost certainly this is because they were such common objects that the writers assumed that their readers were familiar with them. Utilitarian machines appear in the treatises only when radical innovations had been made to the traditional designs, or when miniaturised components such as cams or water-wheels are incorporated in small machines.

We have stated that the technical aspects of Arabic mechanical engineering can best be studied in the treatises. The everyday machines were mechanically relatively simple. Nor, with the exception of the horizontal water-wheel, did they have a significant effect upon the development of machine technology. Nevertheless, the contribution made by the traditional machines to the economic and social life of the Arab world was enormous, and they can in no way be left out of account in any history of Arabic mechanical engineering. Because there is no detailed description of them in any technical treatise we must have recourse to other kinds of evidence in order to appreciate the rôle that they played in medieval Arabic life. This means combing through Arabic literature of various kinds in order to find references, however vague or tenuous some of them may be, to help us build up a picture of the diffusions and applications of the traditional machines. There are also a few scattered references to Fine Technology.

Archaeology, in the usual sense of the word, provides us with little information about Arabic mechanical engineering. On the other hand, the traditional water-raising machines, for example, have changed little over the centuries. Such machines, often still in use, have been examined by modern scholars and their published results contain valuable information about the design, construction and performance of these machines.

We will now consider the information available in these categories in some detail.

TECHNICAL TREATISES

Although all the treatises listed below are extant only in Arabic, not all of them are Arabic in origin. One – Hero's *Mechanics* – is known to be a faithful rendering of a lost Greek original. Others, such as the works of the Banū Mūsā and al-Jazarī, are by well-known Arabic writers. Between these clearly defined types, however, there are a number of treatises whose authorship is uncertain. They may be Hellenistic, Byzantine, Islamic or some combination of the three. In the following listing the origins of each treatise will be discussed under the entry for that treatise. It was felt that only by including all the known treatises in Arabic could the question of their origins, where these are conjectural, be presented to the readers.

Details of MSS, such as catalogue entries, locations, etc., are given only when the work in question has not been published. All the works that have been edited and published, in the original or in translation or both, give the relevant details about manuscripts.

(i) Pseudo-Archimedes, *On the Construction of Water-clocks*
The treatise exists only in Arabic, but there is no edited version in Arabic. The English version is *On the Construction of Water-clocks*, London, 1976. This is a fully annotated translation by Donald R. Hill, based upon the three known complete MSS, which are listed on pp. 5–7. A German version is E. Wiedemann and F. Hauser, 'Uhr des Archimedes und zwei andere Vorrichtungen', *Nova Acta. Abh. der Kaiserl. Leop.-Carol. Deutschen Akademie der Naturforscher*, 103,2 (1918), pp. 164–202.

The early chapters are probably by Archimedes, the later ones are Byzantine and Islamic additions. The water-machinery described in the early chapters was adopted by Riḍwān and al-Jazarī, both of whom attribute its invention to Archimedes.

(ii) Apollonius (of Perga? *c.* 210 BC)
Description of a musical automaton. No reliable attribution can be made. Extant only in Arabic but no Arabic edition has been made. MSS are given in Hill/Archimedes, pp. 5–7. There is a German translation by E. Wiedemann in *Aufsätze zur Arabischen Wissenschaftsgeschichte*, 2 vols., Hildesheim, 1970, vol. 2, pp. 50–6. The translation is unreliable. Several illustrations from the MSS are included, but there are no explanatory drawings. Altogether better is the English version of H.G. Farmer in *The Organ of the Ancients*, London, 1931, pp. 80–5. This includes explanatory drawings. See also: R. Hammerstein, *Macht und Klang, Tönende Automaten als Realität und Fiktion in der alten und mittelalterlichen Welt*, Bern, 1986, pp. 72–3.

(iii) Philo of Byzantium (*c.* 200 BC) *Pneumatics*
Arabic text and French translation published by Carra de Vaux, Paris, 1903; also includes Arabic text and French translation of nine water-raising machines from the Oxford Bodleian MS No. 954 (see below). There is also a version of the *Pneumatics* by F. D. Prager, *Philo of Byzantium – Pneumatics*, Wiesbaden, 1974. Prager gives the text of a Latin version of the *Pneumatics*, with translation. He also translates the Arabic version and reproduces the original drawings, whereas in Carra de Vaux they are redrawn in Renaissance style. Prager's translation, both of the text and the captions to the illustrations are, however, inaccurate and Carra de Vaux is to be preferred.

There has been considerable discussion about the authorship of some of the chapters in the *Pneumatics*. Although it is generally agreed that most of them are by Philo, the chapters that describe water-wheels may be Islamic additions. On the other hand, Philo's work was known to the Arabs, and we can assume that they had the text much as we know it today. 'Umar b. Muḥammad al-Kindī, historian of Egypt (283/897–350/961) said that Philo was renowned for the construction of water-wheels (*dawlāb*), mills and ingenious devices. See Wiedemann, *Aufsätze*, vol. 1, pp. 71–4. The most useful work on the Hellenistic engineers is A.G. Drachmann, 'Ktesibios, Philon and Heron, a study in ancient pneumatics', *Acta historica scientiarum naturalium et medicinalium*, 4, Copenhagen, (1948), pp. 1–197. Since the Arabs were known to be influenced by their predecessors, a knowledge of the Hellenistic treatises is essential to a full understanding of Arabic mechanical engineering.

(iv) The Murisṭus MSS (probably Hellenistic)
These three short treatises in Arabic, on hydraulic and pneumatic musical instruments, were published by L. Cheikho in *al-Machriq*, 9 (1906), pp. 18–28, from a MS in the Three Moons Orthodox College in Beirut. A certain Murisṭus is mentioned in the text, but nothing is known of this person. The instruments were discussed by E. Wiedemann and F. Hauser in 'Byzantinische und arabische akustische Instrumente', *Archiv für die Geschichte der Naturwissenschaft und der Technik*, 8 (1918). The instruments were also discussed by Farmer, pp. 63–72, 127–38. In addition to the MS used by Cheikho, Farmer used the British Library MS Or.9649. See also Hammerstein, pp. 60–2 and 74–9.

(v) Hero of Alexandria (*fl.* AD 62), *Mechanics*
The *Mechanics* was translated into Arabic by Qusṭā b. Lūqā (d. *c.* 300/912–13). An Arabic edition with French translation was published by Carra de Vaux, with the title 'Les Mécaniques ou l'Élévateur de Héron d'Alexandrie', Extrait du *Journal Asiatique* (1894). This was reprinted with an Introduction by D.R. Hill and Commentaries by A.G. Drachmann, Paris, 1988. This is certainly a genuine work of Hero, the well-known Alexandrian scientist and engineer. Much of the book is devoted to theoretical mechanics, but it also includes chapters on lifting machines and presses. The *Mechanics* is extant only in Arabic but the other works of Hero are in the original Greek, with the exception of the *Catoptrics* (burning mirrors) which has survived in a Latin version. All the works of Hero, except the *Belopoiica*, a treatise on war machines, were published, with German translations, under the title: *Heronis Alexandrini Opera quae supersunt omnia*, ed. W. Schmidt, 5 vols., Leipzig (1899–1914). For the *Belopiica* see E.W. Marsden, *Greek and Roman Artillery – Technical Treatises* (Oxford, 1971), pp. 17–60. From the point of view of the student of the history of mechanical engineering, the most important of Hero's works are the *Pneumatics* and the *Automata*, both of which are contained in vol. 1 of the Schmidt edition.

The remaining works are all Arabic in origin.

(vi) The Banū Mūsā b. Shākir (3rd/9th cent.), engineers and scientists

These three brothers, who worked in Baghdad during the most flourishing period of the Arab cultural and scientific awakening, were both patrons of scientists and translators, and eminent scientists and engineers in their own right. Their fame rests largely upon their book on machines, which is mentioned by all the Arabic biographical dictionaries. The most important versions of this work are as follows:

1 Arabic edition: *Kitāb al-ḥiyal*, edited by Ahmad Y. al-Hassan, Institute for the History of Arabic Science, Aleppo, 1981

2 English version: *The Book of Ingenious Devices*, annotated and translated Donald R. Hill, Dordrecht, 1979; reprinted, National Hijra Council of Pakistan, Islamabad, 1989

3 German version: F. Hauser, 'Über das Kitāb al ḥijal – das Werk über die sinnreichen Anordnungen der Benū Mūsà', *Abhandl. zur Geschichte der Naturwissenschaften und der Medizin*, 1, Erlangen (1922); Hauser's work is based upon a translation by E. Wiedemann

In order of seniority the brothers were named Muḥammad, Aḥmad and al-Ḥasan. Most of the *Kitāb al-ḥiyal* was the work of Aḥmad. About 80 of the 100 devices are trick vessels, and the remainder include fountains, lamps, a 'gas mask' for use in polluted wells and a clamshell grab. Despite the apparent triviality of many of the devices, the brothers' mastery of aerostatic and hydrostatic pressures and their use of automatic controls and switching systems places them well in advance of their Hellenistic predecessors. Indeed, their work in this field, though limited in scope, was not surpassed until modern times.

A second, shorter treatise entitled *The machine that plays by itself*, is also attributed to the Banū Mūsā, and we have no reason to doubt the attribution. The Arabic text was published by L. Cheikho, *al-Āla allatī tuzammir bi-nafsihā*, in *Al-Machriq*, 9 (1906), pp. 444–58. This was the MS from which the Murisṭus treatises discussed above were taken. Cheikho dates the MS to the 6th/12th century and lists its contents, which comprised 162 pages of treatises by various authors dealing with mathematical and mechanical subjects. Unfortunately, there are no illustrations to accompany the text, and no other MSS of this treatise are known.

A German version, part translation and part paraphrase, was published by E. Wiedemann, 'Über Musikautomaten', *Amari Festschrift* (1909), pp. 164–85. Some explanatory drawings have been added, but these are inadequate to elucidate all the difficulties of the text. It must, however, be admitted that the haphazard way in which the identifying letters have been provided for the non-existent drawings does not make for easy explanations. Nevertheless, a much better version is that of Farmer, pp. 88–118, who not only provides a full English translation, but also supplies adequate explanatory drawings. There is also a brief mention by Hammerstein, pp. 73–4.

This is a very advanced machine that embodies a number of delicate mechanisms and control systems. It should receive closer attention from historians of technology and musical instruments than has been the case up to the present time.

(vii) Abū ʻAbd Allāh al-Khwārizmī (*fl.* late 4th/10th century); *Mafātīḥ al-ʻulūm*, ed. G. van Vloten, Leiden, 1895
Not a treatise, but an encyclopaedia of the sciences. Our main concern is with the eighth treatise, which is devoted to the moving of heavy weights and to the hydraulic components used in Fine Technology. The work is valuable not only for the lexicographical information that it contains, but also because the manufacture of a number of components is described. A German translation of the eighth treatise is given in E. Wiedemann, *Aufsätze*, vol. 1, pp. 200–28.

(viii) Ibn Khalaf al-Murādī (Muslim Spain, 5th/11th century), a treatise on automata and water-clocks
The work did not come to light until the mid-1970s. Nothing is known of its author. Its importance can hardly be exaggerated. It describes water-clocks and a number of large automata-machines. The transmission systems incorporate complex gear-trains, which include segmental and possibly epicyclic gears. Some of the machines use mercury as part of hydraulic linkages. Unfortunately, the unique MS is badly defaced and the illustrations are very poor, so that it may never be possible to ascertain precisely how each machine was constructed and operated. Nevertheless, the treatise is so important as a precursor of mechanical horology that work on its elucidation should continue.

The work is discussed in Donald R. Hill, *Arabic Water-clocks*, Institute for the History of Arabic Science (IHAS), Aleppo (1981), pp. 36–46. The origins and authorship of the treatise were discussed by A.I. Sabra in ʻA Note on Codex Biblioteca Medicea-Laurenziana OR 152ʼ, *Journal for the History of Arabic*

Science (JHAS), 1,2 (1977), pp. 276–8. See also M.V. Villuendas, 'A further note on a mechanical treatise contained in Codex Medicea-Laurenziana OR 152', JHAS, 2,2 (1978), pp. 395–6. For an interpretation of the first chapter of the treatise, with reconstructed drawings, see J. Vernet, R. Casals and M.V. Villuendas, 'El Capitulo Primero del "Kitāb asrār fī natā'iŷ al-afkār"', *Awraq*, 5–6 (1982–3), pp. 7–18. The contents of the treatise and its relevance in the history of technology is considered by J. Vernet in 'Un texto arabe de la corte de Alfonso X el Sabio', *Al-Andalus*, 43 (1978), pp. 405–21. (The writings of Professor Vernet have now been collected in one volume: *De Abd al-Rahman I a Isabel II*, Barcelona, 1989. In this book the last two articles mentioned are on pp. 319–30 and 301–17 respectively.) There is no Arabic edition of this treatise.

(ix) Riḍwān b. al-Sa'ātī (Damascus), *On the Construction of Clocks and their Use* (completed 600/1203)
There is no Arabic edition of this work. The German version is E. Wiedemann and F. Hauser, 'Über die Uhren in Bereich der islamischen Kultur', *Nova Acta. Abh. der Kaiserl. Leop.-Carol. Deutschen Akad. der Naturforscher*, 100, 5 (Halle, 1915), pp. 166–266. There is also an English article on Riḍwān's treatise; see Donald R. Hill, in *Arabic Water-clocks*, pp. 67–88.

Neither of these works is an exhaustive treatment of the treatise, which is a lengthy and repetitive work describing the repairs that Riḍwān made to the clock at the Jayrūn gate in Damascus, built by his father about the middle of the 6th/12th century. The English work is a commentary on the construction and operation of the clock, with adequate explanatory drawings. The German work is partly a translation, partly a paraphrase; the explanatory drawings are unsatisfactory. The Damascus clock is inferior in its construction to al-Jazarī's first clock, which it closely resembles (see next entry). Moreover, because Riḍwān was a physician, not an engineer, his descriptions are often confused and his specifications imprecise. Nevertheless, the work has certain features that enhance its value. In the first place, because of Riḍwān's lack of technical training, he sometimes gives us details of manufacture that an engineer would regard as too mundane for comment. This applies, for example, to his meticulous instructions for the manufacture of copper pipes. Also, he gives us valuable information about the exchange of horological ideas between the Hellenistic world and Sasānid Iran.[1] The clock built by Riḍwān's father was commented upon by several travellers who inspected it, thus adding to our meagre store of evidence for the undoubted existence in Islam of large public water-clocks.[2]

(x) Ibn al-Razzāz al-Jazarī, engineer; book completed in Diyar Bakr in 602/1206

The Arabic edition of his great machine book is: *al-Jāmi' bain al-'ilm wal-'amal al-nāfi' fī ṣinā'at al-ḥiyal*, ed. Ahmad Y. al-Hassan, IHAS, Aleppo, 1979. An annotated English translation is: *The Book of Knowledge of Ingenious Mechanical Devices*, Donald R. Hill, Dordrecht, 1974; reprinted by the National Hijra Council of Pakistan, Islamabad, 1989.

Certainly the most important of all the Arabic treatises and probably the most important engineering document from any cultural area before the Renaissance. The work is divided into six categories:

1 Water-clocks and candle-clocks
2 Wine dispensers
3 Phlebotomy measuring devices and water dispensers
4 Alternating fountains and musical automata
5 Water-raising devices
6 Miscellaneous

Several of the machines, mechanisms and techniques that appear for the first time in al-Jazarī's work were later to enter the vocabulary of European mechanical engineering. The work is also important because al-Jazarī described his methods of manufacture and construction with the avowed purpose of enabling later craftsmen to reconstruct his machines. Indeed, several have been successfully reconstructed by modern craftsmen using his instructions and illustrations.[3] Detailed instructions and specifications are very rarely found in early mechani-

[1] Donald R. Hill, *Arabic Water-clocks*, pp. 12–13.

[2] Ibn Jubayr, *Riḥla*, Arabic text ed. W. Wright, 2nd edn revised by M.J. de Goeje (Leiden and London, 1907), pp. 270ff. See also Hill, *Arabic Water-clocks*, pp. 71–2

[3] The monumental water-clock described in Chapter 1, for example, was reconstructed full-size by P.N. Haward for the World of Islam Festival 1976 in the Science Museum, London. Mr Haward worked to al-Jazarī's own instructions and specifications, with some minor assistance from the present writer. The clock worked perfectly.

cal treatises; in fact from this aspect al-Jazarī is unique before the Renaissance. Other writers were usually unable to give clear instructions either because they had little practical experience or because they did not wish to reveal all the secrets of their profession.

(xi) The *Libros del Saber de Astronomia*, prepared in 1277 under the direction of Alfonso X of Castile (edited by M. Rico y Sinobas, 5 vols., Madrid, 1863–7)
The reason for including this work in a listing of Arabic treatises is that although it was written in Castilian, it was a collection of translations and paraphrases made from Arabic originals, with the declared intention of making Arabic knowledge available to the Christian world. The section that concerns us is at the end of the work and contains descriptions of five timepieces. Three of these have close connections with earlier Arabic constructions. One is a water-clock whose mechanisms and time recording systems are inferior in design to those of al-Jazarī. The system for maintaining a constant head is pneumatic, and similar to a method used by the Banū Mūsā. There is a candle-clock, but again this is much inferior to those described by al-Jazarī. The most significant of these clocks, since it incorporated a weight drive, was a clock consisting of a large drum made of walnut or jujube wood tightly assembled and sealed with wax or resin. The interior of the drum was divided into twelve compartments, with small holes between the compartments through which mercury flowed. Enough mercury was enclosed to fill just half the compartments. The drum was mounted on the same axle as a large wheel powered by a weight-drive wound around the axle. Also on the axle was a pinion with six teeth that meshed with 36 oaken teeth on the rim of an astrolabe dial. The mercury drum and pinion made a complete revolution every 4 hours and the astrolabe dial made a complete revolution every 24 hours. Clocks of this type had been known in Islam since the eleventh century – at least 200 years before the first appearance of weight driven clocks in the West.[4]

[4] See Hill, *Arabic Water-clocks*, pp. 126–30 for a detailed description of the water-clock. For the mercury clock see Donald R. Hill, 'Islamic Fine Technology and its Influence on the Development of European Horology', *Al-Abhath*, XXXV (Beirut, 1987), pp. 9–28 and also Silvio A. Bedini, 'The Compartmented Cylindrical Clepsydra',

(xii) Taqī al-Dīn b. Ma' rūf (mid-10th/16th century), *al-Ṭuruq al-saniyya fī al-ālāt al-rūḥāniyya (Sublime methods for pneumatic machines)*
This late work was discussed (in Arabic) by Ahmad Y. al-Hassan, in *Taqī al-Dīn and Arabic Mechanical Engineering*, IHAS, Aleppo, 1976. Together with an essay on Arabic mechanical engineering, the work describes several machines that occur in *al-Ṭuruq*, with some explanatory drawings. There is also a fascimile of the short treatise itself. The most interesting machine is a six-cylinder 'Monobloc' pump, driven by water-power. It may be significant that the work was written in 959/1522, thus contemporaneously with the machine book of Agricola and before that of Ramelli (1588).[5]

(xiii) Miscellaneous MSS
There are a number of manuscripts that contain a mixture of Greek and Arabic material. Parts of some of these manuscripts were used in the preparation of editions of Hero's *Mechanics* and Philo's *Pneumatics*. These cases are mentioned below. Otherwise the manuscripts, which are all in Arabic, remain unedited. A treatise that occurs in several of the manuscripts is entitled 'Wheels that move by themselves'. Although there are some variations among the various manuscripts, the treatise obviously comes from a single original. It has usually been assumed that it is a discussion of perpetual motion and as such dismissed as of no practical significance. The descriptions are difficult to understand and the accompanying illustrations are obscure. Nevertheless, the machines appear to embody important mechanisms and they would certainly repay further detailed study.

The most important of these Greco/Arabic MSS are given below. The list is not comprehensive and, moreover, it is almost

Technology & Culture 3, (Spring 1962), pp. 115–41. In later European versions of this type of clock the mercury was replaced by water and the cylinders themselves rotated and descended. A.P. Mills in 'The Mercury Clock of the *Libros del Saber*', *Annals of Science*, 45 (1988), pp. 329–44 doubted, on the evidence of a reconstruction he had made of the *Libros del Saber* clock, whether such clocks would be accurate timekeepers. This view is, however, contradicted by the fact that the falling drum clepsydra in its various forms was extremely popular in Europe in the eighteenth and nineteenth centuries as a cheap and reliable timekeeper; see Anthony Turner, *The Time Museum* (Rockford, 1984), vol. 1, pt. 3, pp. 40–3.

5 Taqī al-Dīn also wrote a book on mechanical clocks, see Sevim Tekeli, *The Clocks in the Ottoman Empire in 16th century and Taqī al-Dīn's 'The Brightest Stars for the Construction of the Mechanical Clocks'* (Ankara, 1966).

certain that similar compilations await discovery in various libraries throughout the world.

1	Manchester 351	Location: John Rylands Library, Manchester Catalogue entry: Arabic cod. 351 Contains Philo, Hero, Apollonius and various anonymous authors
2	Istanbul 2755	Location: Süleymaniye U. Kütüphanesi, Istanbul Catalogue entry: Hagia Sophia 2755 Contents are (see Carra de Vaux, Philo, 30): (a) Hero's *Mechanics* (used in vol. II of *Opera ... omnia*) (b) Wheels that move by themselves (c) A treatise on organs dedicated to Muriṣtus (d) Another treatise on organs, with the same dedication (e) Philo's *Pneumatics*
3	Istanbul 3713	Location: As Istanbul 2755 Catalogue entry: Hagia Sophia 3713 Contains only Philo's *Pneumatics*. Was the main manuscript used by Carra de Vaux in preparing his edition of Philo
4	Oxford 954	Location: Bodleian Library, Oxford Catalogue entry: Oxford 954 Marsh 669 Includes Philo's *Pneumatics*; 'Wheels that move by themselves'; some anonymous descriptions of machines. A full list of contents is given in Carra de Vaux, Philo, 29. Seven devices from this MS are given in E. Wiedemann and F. Hauser: 'Über Vorrichtungen zum Heben von Wasser in der islamischen Welt', *Jahrbuch des Vereins Deutscher Ingenieure*, 8 (1918), pp. 121–54.

The following MSS all contain 'Wheels that move by themselves' on the pages indicated:

5	Leiden 499	Location: Bibliotheek der Rijksuniversiteit Leiden Catalogue entry: Codex Orientalis 499 folios 74v to 87r
6	Istanbul 1884	Location: As Istanbul 2755 Catalogue entry: Esad Efendi 1884 folios 71r to 85v
7	Gotha 1348	Location: Universitäts-Bibliothek, Gotha Catalogue entry: Codex Gotha 1348 folios 109v to 124r
8	Istanbul 3472	Location: Topkapu Serai Müzesi Kütüphanesi,

		Istanbul folios 127r to 184r
9	Florence 152	Location: Biblioteca Medicea Laurenziana, Florence Catalogue entry: Or 152 folios 82–90

For a discussion of these devices see: H. Schmeller, 'Beiträge zur Geschichte der Technik in der Antike und bei den Arabern', in *Abhandlungen zur Geschichte der Naturwissenschaften und der Medizin*, 6 (Erlangen, 1922) and Juan Vernet, 'Alfonso el Sabio y la Mecanica', *Boletin de la Real Academia de la Historia*, 185 (1988), pp. 28–38; also pp. 331–40 in Vernet's collected works. Also Thorkild Schiøler: *Roman and Islamic Water-Lifting Wheels*, Odense University Press, 1973, pp. 62–5.

To conclude this inventory we will mention a few of the more noteworthy references to Fine Technology that are scattered among the works of Arabic writers. When compared to the relatively frequent mentions of utilitarian machines, references to this kind of construction are quite rare in general writings. Many of the references to automata are of a fanciful nature, such as those that occur in the *Thousand and One Nights*. We shall confine ourselves to some of the more sober statements, which as it happens are all concerned with water-clocks.

In AD 807 a clock with falling brass balls and horsemen that came out of doors at each hour was presented to Charlemagne by vassals of Harūn al-Rashīd.[6] The earliest unequivocal reference in Arabic to a clock that is clearly not a sundial occurs in a poem of 'Abbās b. Firnās (d. 274/887), poet and engineer to the Andalusian amirs 'Abd al-Raḥmān II and Muḥammad I. No description of the clock is given, but it was almost certainly a water-clock, since it could be used in cloudy weather or at night.[7] According to al-Khāzinī, the famous scientist Ibn al-Haytham (354/965–430/1039) mentioned a water-clock.[8]

There is a well-known though brief description of two water-clocks built on the banks of the Tagus about the year 473/1080

[6] Eginard, *Annales Francorum: Chronicum Turonese*; quoted and trans. John Beckmann, *A History of Inventions, Discoveries and Origins*, 4th edn (trans. William Johnstone, rev. and enl. by William Francis and J.W. Griffith), 2 vols. (London, 1846), vol. 1, p. 343.

[7] *Al-Muqtabis* (Beirut, 1939/1973), pp. 81–84. See also Juan Vernet, 'Mármol, Obra de Zarquel', *Hommage à Georges Vajda* (Louvain, 1980), pp. 151–54, reprinted in Vernet's collected works, pp. 297–300.

[8] Hill, *Arabic Water-clocks*, p. 49.

by the famous astronomer al-Zarqallu. According to the Spanish Muslim geographer al-Zuhrī (*fl.* 532/1137) these clocks each consisted of a basin (probably cylindrical) that was housed in a chamber built into the banks of the river at Toledo. They recorded not only the passage of the hours but also the phases of the moon: for the first 14 days of the month the basins filled at constant speed, then emptied at constant speed for the second half of the month. According to the description, if water was removed or added – either deliberately or accidentally – the rate of flow was automatically adjusted to bring the water back to the correct level. Both clocks were in use until 528/1133, and one of them after that date.[9] (It might be possible to reconstruct such a clock, using methods described by the Banū Mūsā and al-Jazarī; the adjustments for compensating for the abstraction or addition of water, however, present difficulties.) It is possible that the constructor was not al-Zarqallu, but that the description applies to an earlier machine.[10]

In 633/1235–6 a monumental water-clock that announced the appointed hours of prayer and the time both by day and by night was completed in the entrance hall of the Mustanṣiriyya college in Baghdad.[11] The remains of two large water-clocks still exist in Fez, Morocco. One of these is in the street opposite the Bū'anāniyya Madrasa and it was built by Abū al-Ḥasan 'Alī b. Aḥmad al-Tilimsānī in 758/1357. According to the historian of Fez, al-Jazna'ī in *Zahrat al-Ās* (ed. Alfred Bel, Algiers (1923), p. 40), at each hour pellets fell on to a gong and a pair of door leaves opened. The door openings and the brackets for the gongs still exist. A second clock was reconstructed in 763/1361–2, after it had fallen into disrepair, by 'Abd Allah Muḥammad al-'Arabī. This clock had an astrolabe dial and it also released pellets and had doors that opened. It is in the minaret of the Qarawiyyin mosque. The doors and the dial survive, but in neither of the Fez clocks does any trace of the water-machinery still exist.[12] Another North African water-clock, in Tlemcen,

[9] J.M. Millás, *Estudios sobre Azarquiel* (Madrid, 1943–50), pp. 6–9, in which a Spanish translation of al-Zuhrī's text is given. The Arabic text, with German and English translation is given by Armin Wittstein, 'Ueber die Wasseruhr und das Astrolabium des Arzachel', *Zeitschrift für Mathematik und Physik, historisch-literarisch Abtheilung*, 39 (1894), pp. 41–55.

[10] Juan Vernet, 'Mármol, Obra de Zarquel'.

[11] G. Le Strange, *Baghdad during the Abbasid Caliphate* (Oxford, 1900), p. 267.

[12] Derek de Solla Price, 'Mechanical Water Clocks of the 14th Century in Fez, Morocco', ITHACA – 26 VIII–2 IX 1962.

was mentioned by Abū Zakariyyā Yaḥyā b. Khaldūn in *Bughyat al-ruwwād fī dhikr al-mulūk* (2 vols., ed. Bel, Algiers (1904–13), vol. 2, p. 40).

In the opinion of the writer it is unlikely that any other major treatises on Arabic machine technology await discovery. It may seem rash to make such an assertion, given the number of manuscripts in the libraries of universities, mosques, and other institutions that have not been examined. The evidence suggests strongly, however, that treatises such as those of the Banū Mūsā and al-Jazarī are of rare occurrence. All the major biographical dictionaries dealing with scientists have entries on the Banū Mūsā and of no other writers of comparable works on machines. Al-Jazarī lived too late to be included in the *Fihrist* of Ibn al-Nadīm, and the later biographers do not mention him, either because they usually copied from the *Fihrist*, or because al-Jazarī's reputation did not spread beyond the limits of the Jazīra until some time after his death. He is, however, mentioned by the Egyptian scholar al-Qalqashandī (756/1355–821/1418) as pre-eminent in the science of machines.[13] The many manuscripts of al-Jazarī's work, from the 7th/13th century until the 12th/18th, also testify to the esteem in which he was held in the world of Islam.

Al-Jazarī himself mentions the Banū Mūsā as the most illustrious of his predecessors (p. 157 – page nos. refer to al-Jazarī, ed. Hill). He also acknowledges the pseudo-Archimedes for water-machinery (p. 17); a candle-clock made by Yūnus al-Asṭurlābī (p. 87); the paper by Apollonius on a musical automaton (p. 170); and a paper by Hibat Allah b. al-Ḥusayn on a musical automaton (p. 170). The treatises of the pseudo-Archimedes, Apollonius and the Banū Mūsā are of course known to us and are mentioned in the first part of this paper. Hibat Allah (d. 534/1139–40) is also known to us, as an astronomer – H. Suter, 'Die Mathematiker und Astronomen der Araber und ihre Werke', *Abhandl. zur Gesch. mathematischen Wissenschaften* (1900), p. 117. Yūnus b. al-Ḥusayn al-Asṭurlābī is known to have constructed a celestial globe, which still exists – see Emilie Savage-Smith, *Islamicate Celestial Globes*,

[13] Al-Qalqashandī, *Ṣubḥ al-a'shā*..., ed. Muḥammad 'Abd al-Rasūl Ibrāhīm, 14 vols. (Cairo, 1913–30), vol. 1, p. 477.

Washington, 1985, pp. 24, 217–8, 285. It seems likely, therefore, that al-Jazarī knew all the works of his predecessors that had a direct bearing on his work. (It is true that he does not mention Riḍwān but their treatises were almost exactly contemporary, and it is doubtful whether the two men knew of each other's work.)

From the available evidence it appears that we possess the two major works of medieval Arabic technology, together with a number of minor works. Major works on machines were equally rare in Hellenistic times, for which we have only the writings of Philo and Hero, again together with some minor treatises, and a few chapters in Vitruvius. The Arabs knew the works of Philo and Hero, whose writings on engineering were the only Greek works on this subject to re-appear in medieval Europe. Clearly, the ability to describe machines accurately and fully has been a rare talent in earlier centuries.

Professor David King informed the writer recently that he has discovered many astronomical works that are not listed in the biographical dictionaries or other written sources. It cannot be regarded as impossible, therefore, that there are major undiscovered Arabic works on machines. The cases of machinery and astronomy are, however, rather different. Astronomical observations are pointless unless the results are recorded and tabulated, together with the relevant computations. Astronomical data thus assembled are used for a variety of purposes, for example, for making predictions and in the manufacture of instruments. Furthermore, information must be prepared for different times and places and be frequently updated. One only has to think of the annual production of modern ephemerides to realise how much information must be produced on a regular basis. None of the above applies to mechanical engineering. A good engineer must have had a thorough practical training, and in earlier days this training was provided by master craftsmen and passed on by example and by word of mouth, often from father to son. There was usually no incentive for committing such knowledge to writing, especially when one's livelihood depended upon guarding the 'secrets of the trade'. We are in fact fortunate that a few precious treatises have been transmitted to us. This is not to say, however, that there are no short treatises and works describing single machines that remain undiscovered. After all, the treatise of al-Murādī was not discovered until 1976.

UTILITARIAN MACHINES

Reference to mechanical technology may be found in various types of Arabic literature; indeed, it is rarely safe to assume, from its main subject, that a given book will contain no references to machines, even if they are used only in similes. It would of course be quite impossible, even for a team of researchers, to comb through thousands of tomes, many of which would have few or no references relevant to the subject. One can only indicate the types of works most likely to yield valuable information. The task is not made easier by the fact that very few books have indexes that have entries on technical matters.

Without doubt, the greatest amount of information on mechanical technology, in works that are not specifically concerned with technical matters, is to be found in the books of Arab geographers and travellers.[14] Very rarely does this information take the form of practical details about the design and construction of machines. Usually the references are of a general nature, giving the number of mills, for example, to be found in the neighbourhood of a given city.[15] Further comments may be added, particularly the positions of mills: for example, on the piers of bridges, below dams, on ships moored in midstream. In the case of the ship-mills moored on the Tigris at Mosul, Ibn Ḥawqal actually tells us how many donkey-loads of flour a single mill could grind in 24 hours,[16] although this kind of information is very rarely given. Al-Muqaddasī tells us that at Baṣra mills made use of the ebb-tide, about a century before the first recorded example of this technique in Europe.[17] The same author gives us one of the rare mentions of the use of water-power for purposes other than the grinding of corn,

[14] Donald R. Hill, 'Information on Engineering in the works of Muslim Geographers', *History of Technology*, 9 (London, 1984), pp. 127–42.

[15] Donald R. Hill, *Engineering in Classical and Medieval Times*, Croom Helm (London, 1984), pp. 164–65.

[16] Ibn Ḥawqal, *Kitāb ṣūrat al-'arḍ* (completed 378/988), ed. J.H. Kramers, 2nd. edn of vol. 2 of *Biblioteca Geographorum Arabicorum* (BGA) (Leiden, 1938), p. 222.

[17] Al-Muqaddasī, *Aḥsan al-taqāsīm* ... (completed *c.* 380/990), ed. M.J. de Goeje, vol. 3 of BGA (Leiden, 1906), pp. 124–25.

in this case the fulling of cloth.[18] There is also a certain amount of data about water-raising machines, usually without constructional details. To summarise, the geographers give a good deal of information about the use of machines to improve the prosperity of certain regions of the Islamic world, but hardly any details about the machines themselves.

There are a few treatises on agriculture and others on irrigation that contain some information on water-raising machines. For example, there are some valuable remarks in the agricultural manual of Ibn al-'Awwām (end of 6th/12th century, Muslim Spain) about the construction and performance of the animal-driven saqiya.[19] In an anonymous Iraqi treatise of the 5th/11th century the performances of the various types of water-raising machines are compared.[20]

The *ḥisba* manuals are concerned with the guidance of the *muḥtasib*, the person responsible for the supervision of moral behaviour in a town, particularly its markets.[21] Since this supervision included quality control and manufactured goods, we can learn a good deal from the *ḥisba* works about various trades, including metalworking and woodworking. Since the duties of the *muḥtasib*, however, were confined to the trades practised in the towns, the manuals tell us nothing about machines such as mills and water-raising devices.

Occasionally references to mechanical matters can be found in historical works. The first reference to the use of a noria in Islam occurs in al-Balādhurī's history of the Arab conquests.[22]

It is probable that there is something to be learned about machines from a study of iconography, but there are considerable difficulties in research of this kind. Many of the manuscripts in libraries and other depositories have not been

[18] *Ibid.*, p. 409. The industrial uses of mills is an instance where many disparate sources must be combed in order to show that water-power was used for a number of purposes in Islam. To give but one example: al-Bīrūnī (362/973–440/1048), in his work on minerology, says that gold-bearing ores were crushed by water-powered trip hammers 'as is the case in Samarqand with the pounding of flax for paper', *Kitāb al-Jamāhir*..., ed. F. Krenkow (Hyderabad, Deccan, 1936), pp. 233–34.

[19] Schiøler, *Roman and Islamic Water-Lifting Wheels*, pp. 79–83.

[20] Claude Cahen, 'Le service de l'irrigation en Iraq au début du XIe siècle', *Bulletin d'études orientales*, vol. 13 (1949–51), pp. 117–43.

[21] For example, Ibn al-Ukhuwwa (d. 729/1329), *Ma'ālim al-qurba*, ed. R. Levy, with notes and an abridged English trans., Gibb Memorial Series, (London, 1938).

[22] Al-Balādhurī (d. 279/892), *Kitāb futūḥ al-buldān*, ed. M.J. de Geoje (Leiden, 1866), p. 363.

properly catalogued. Even when there are adequate catalogues there are very seldom indexes to the subject matter of illustrations.[23] As with textual material, it would therefore be necessary to search through many manuscripts in order to discover some relevant drawings. Even so, great caution is required. Artists usually depicted objects that were familiar to them from their own observations rather than those that were in use when the treatises they were illustrating were originally composed. It is possible for a mechanism to be depicted that was not invented when the treatise was written. In much the same way European artists depicted biblical characters in contemporary, for example, Renaissance, dress.

With the exception of water-raising machines there is at present little archaeological evidence available for the use of machines in Islam. One of the rare examples is the discovery in the Jordan valley in a recent survey of the remains of 32 water-powered sugar mills, dating to Ayyūbid and Mamlūk times.[24]

For water-raising machines, however, there is abundant archaelogical evidence for their use in Islam, both from ruined sites and from the survival of existing machines. There are a number of ruined castles which show evidence of saqiyas having been incorporated in the buildings. These ruins date from the 2nd/8th to the 8th/14th centuries.[25] Apart from such sites, we are fortunate that so many traditional machines have survived and are in operation. One might have expected that by now these would have been superseded by modern pumps, but in fact the saqiya and the noria, in certain areas, are increasing in numbers. A pump depends upon the supply of fuel, spare parts and the services of skilled mechanics, all of which may have to come from outside the area. The old machines can be built from local materials and maintained by the local carpenter. Moreover, in certain conditions they are at least as efficient as modern pumps. These factors, when an interruption of the supply of irrigation water, even for short periods, is quite literally a matter of life or death, militate in favour of the

[23] There is an index for the illustrations in the Persian MSS in the British Library: *Miniatures from Persian MSS*, Nora M. Titley (London, 1977).

[24] Saleh Hamarneh, 'Sugar-Cane Plantation and Industry under the Arab Muslims during the Middle Ages', in *Proceedings of the First International Symposium for the History of Arabic Science*, IHAS, Aleppo (1976), p. 221.

[25] Schiøler, *Roman and Islamic Water-Lifting Wheels*, pp. 90–96.

survival of the traditional machines. Since working machines can be readily inspected, the historian of technology is in a fortunate position. It might be argued that the design of the machines may have changed radically over the years, but this appears not to be the case, since descriptions in the writings of Ibn al-ʻAwwām and others tally closely with the construction of modern machines. The best study of Islamic water-raising wheels is that of Thorkild Schiøler, although this is concerned mainly with the saqiya.[26]

It is to be hoped that there will be further archaelogical discoveries of other types of machines, but the signs are not propitious. Water-mills and wind-mills have been almost completely superseded by machine-powered milling installations. The main constructional features of the traditional mills are well known, but the diffusion and applications of mills in medieval times can only be established by a painstaking study of literary sources.

[26] *Ibid.*

VI

Information on Engineering in the Works of Muslim Geographers

Historians of engineering concerned with periods before the sixteenth century are acutely conscious of the scarcity of evidence. It is probably the case that there is more material available for medieval Islam than there is for the classical/Hellenistic period or for the early Middle Ages in Europe, which is not to imply that the documents for the Muslim world are abundant. It is the purpose of this article to indicate how our fund of information on Islamic engineering can be increased by studying the works of Muslim geographers and travellers. However, to begin with it will be useful to survey the material that is *specifically* concerned with engineering.

First, there are a number of treatises concerned with *hiyal*, an Arabic word usually translated as 'ingenious devices' or 'automata', although its meaning can be very much broader, to the extent that it can embrace all aspects of mechanical engineering.[1] The earliest surviving such treatise was written by three brothers known as the Banū (sons of) Mūsà about AD 850 in Baghdad.[2] Most of the devices described are trick vessels but there are also a number of fountains, self-feeding and self-trimming lamps, and even a clamshell grab. The Banū Mūsà's work, although derived from their Hellenistic predecessors, goes well beyond them, particularly in the masterly use of small variations in aerostatic and hydrostatic pressures and in the use of automatic control systems using conical valves. Late in the tenth century, Abu ʿAbd Allah al-Khwārizmī compiled a scientific encyclopaedia called *The Keys of the Sciences* (*Mafātīh al-*ʿUlūm) in which the components of *hiyal* are described, with instructions for their manufacture.[3] In Muslim Spain in the eleventh century, a certain al-Murādī wrote a treatise on water-clocks and automata machines. Unfortunately, the only surviving manuscript of this work is so badly defaced that it is impossible to determine precisely how any of the machines worked. It is, however, a document of the utmost importance since it includes descriptions and illustrations of multiple gear-trains incorporating segmental gears and epicycles. These are the first-known examples of complex gear-trains intended for the transmission of heavy torque.[4] *The Book of the Balance of*

I wish to thank the Royal Society for their generosity in awarding me a grant to further my research into the history of technology.

Wisdom (*Kitāb Mizān al-Hikma*) was written by the great physicist al-Khāzinī in 1121. The book deals with a number of subjects including the specific gravities of liquids and solids, theory of gravity, measurement of gravities and applications of the balance. In the Eighth Treatise, al-Khāzinī describes the construction and use of steelyard clepsydras for the measurement of time. This is mainly of interest for the close attention he pays to accurate dimensioning and the close specification of materials.[5]

In a book called *On the Construction of Clocks and their Use*, written in 1203, Ridwān ibn al-Saʿātī describes the monumental water-clock built by his father by the Jayrun gate in Damascus. Despite some obscurities and inaccuracies due to Ridwān's unfamiliarity with engineering, the book tells us a good deal about methods of manufacture and construction.[6] It cannot be compared, however, with al-Jazarī's magnificent book on machines, written in Diyar Bakr in 1206. This contains descriptions of water-clocks, candle-clocks, automata, fountains, water-raising machines and a few miscellaneous devices. It is perhaps the most remarkable engineering document to have survived from any cultural area in pre-Renaissance times. It contains descriptions of techniques and detailed instructions for the construction of machines, from the smallest components to the completed assembly. Some of these techniques, components and machines are already known to us from the works of Hellenistic and earlier Arabic engineers, but a number appear for the first time in al-Jazarī's book and some of them may indeed be his own inventions. Perhaps the most important aspect of the book, however, is that the devices were described with the declared—and successful—intention of enabling later craftsmen to reconstruct them.[7]

Still on the subject of *hiyal*, there are a number of manuscripts extant only in Arabic which are compilations of Greek, Hellenistic, Byzantine and Islamic sources. These separate elements are not identified, and the treatises are usually ascribed to Greek or Hellenistic writers. For example, there are a number of manuscripts in Arabic attributed to Philo of Byzantium (fl. *c.* 230 BC), but all of these contain Islamic additions. Carra de Vaux' scholarly Arabic edition of Philo with French translation does not utilize all the 'Philonic' material, nor does it always succeed in making a convincing distinction between Philo's original work and the Islamic additions.[8] Another document, a treatise entitled *On the Construction of Water-clocks*, is attributed to Archimedes. I have published an English translation of this work, based upon the three known manuscripts, but my conclusions about the origins of the various sections are tentative.[9] There are also several Arabic manuscripts of a treatise on a musical automaton attributed to Apollonius 'the geometer, the carpenter'. On present evidence, this could have originated almost anywhere in the eastern Mediterranean area at any time from 250 BC to AD 750! This treatise was translated into German by Eilhard Wiedemann but the translation, admittedly based upon a very obscure text, is far from satisfactory.[10] The remaining manuscripts in this category have yet to be published, or even studied in detail.[11] While the value of this type of material is diminished by the uncertainty about origins, they are nevertheless of value in helping us to trace cultural diffusions. And, of course, in those cases where the origins can be established with some credibility, our stock of knowledge about engineering attainments in a given cultural area is thereby increased.

Two valuable treatises on hydraulic engineering have come down to us, both written in Iraq in the eleventh century. The first of these, by an Iranian called Muhammad b. al-Hasan al-Hāsib al-Karajī, is called *The Bringing to the Surface of Hidden Waters* (*Inbāt al-Miyāh al-Khafiya*). It contains a section on surveying instruments and their use, methods for detecting underground sources of water and instructions for the excavation of *qanats*—the underground conduits still in use today in Iran and parts of the Arab world.[12] The other treatise is concerned with the construction and maintenance of canals in the Iraqi irrigation system: this also contains descriptions of surveying instruments and their use. The various types of water-raising machines are listed, and although very little is said about their construction, there is a very interesting passage in which the output of each machine is given, together with the area of crops each could irrigate in winter and in summer. The final section deals with quantity surveying, i.e. how to calculate the quantities of excavation for canals and hence derive the cost of the works in labour and in money.[13]

One category of writing that deserves more attention from historians of technology is *hisba*, a word that is not easy to translate. On the one hand it means the duty of every Muslim to 'promote good and forbid evil' and, on the other, the function of the person who is effectively entrusted in a town with the application of this rule in the supervision of moral behaviour and more particularly in the markets; this person entrusted with the *hisba* was called the *muhtasib*. There are two kinds of *hisba* treatise, the first being general dissertations on the virtue of *hisba* and the second manuals for the instruction of the *muhtasib*. Examples of the second type, which are our present concern, are known from the eleventh century onwards.[14] The *muhtasib* had many duties, which included the supervision of every trade in the market in order to exercise quality control and to guard against malpractices. In describing this function the *hisba* manuals tell us a good deal about the materials and methods of manufacture in the various trades. A second duty of the *muhtasib* was analogous to that of a modern clerk of works, in that he had to supervise the building of houses and the materials used in their construction. We can therefore learn from the manuals about the manufacture of bricks—burnt and unburnt—tiles, timber, etc. and how these materials were used in building construction. Information from these sources, however, although very valuable, is limited by the fact that the duties of the *muhtasib* were confined to the trades practised in the town. They tell us nothing about machines such as mills and water-raising devices, nor about large structures and public works.

The evidence from non-literary sources is very uneven. It is fortunate that we are fairly well-informed about fine technology from the *hiyal* treatises, because none of the small automata devices has survived. There are, however, many surviving examples of Islamic astrolabes and a few other types of astronomical instruments. In Fez, Morocco, there are the remains of two monumental water-clocks, built of masonry, which appear to have had water machinery very similar to that described in the clocks of Riḍwān and al-Jazarī.[15] These remains are important, if only to demonstrate that the clocks described in the treatises were designs for actual clocks, rather than unrealized concepts. For despite the unmistakably practical intention of al-Jazarī's specifications, some writers have tried to show that his designs were simply paper ideas.

Many masonry structures built in medieval Islam still exist, and a number are still in use. The great mosques and palaces come immediately to mind, and there is plenty of literature available from which to arrive at a good understanding of the structural properties of these buildings. Some Muslim dams and bridges are still standing and substantial traces of others remain; a number of these can be examined in the works of modern writers.[16] Moreover, many of these structures can be visited without difficulty, although proper technical surveys would clearly be of more value than casual visits. In the field of hydraulic engineering, the construction of qanats using traditional techniques has been thoroughly studied and documented.[17] Medieval irrigation works have in many cases either been overlaid by new systems or have fallen into disrepair. Nevertheless, from the works that have survived, and from written records such as legal documents, histories and geographies, writers such as Thomas Glick have been able to present a coherent account of irrigation systems in certain communities.[18]

Our knowledge of heavy machinery comes mainly from archaeological sources, if this is taken to include the examination of traditional machines that are still in use. The *saqiya*, a chain-of-pots driven through a pair of gears by animal power, has been in continuous use for about two thousand years, and it is possible to verify, from scattered literary and iconographic evidence, that its design has not changed materially since the fifth century AD.[19] The noria, a water-driven wheel, has been in use for about the same period and, again, there is sufficient evidence to prove that its design has not been radically changed.[20] The case of water power, for grinding corn and for other industrial purposes, is rather different. Very few traditional water-mills are still in use, but in any case, as in Europe, the survival of an old mill does not necessarily tell us very much about its history. We cannot be sure, without further evidence, that there was a mill on the same site in medieval times, and in many cases milling installations were completely rebuilt on different lines in the eighteenth and nineteenth centuries. In the history of water power, certain important questions remain to be resolved. There are the questions of the origins, diffusion and relative importance of the three main types of wheel—overshot, undershot and horizontal. Of importance to economic and social historians is the extent to which water power was applied to industrial purposes. In an attempt to approach solutions to these problems we need to consult every type of evidence. Very little archaeological work has been done in this field; the remains of over thirty water-powered sugar-mills, dating to Ayyubid and Mamluk times, have been reported from excavations in Jordan.[21] Mills with horizontal wheels, some still working but most in disuse, are to be found in southern and eastern Europe, the Arab world and Iran. While the existence of these mills indicates a continuous tradition for their use, it does not tell us when the tradition began, nor does it permit us to locate the point of origin of this type of wheel in the geographical centre of the whole region.

The foregoing brief summary does not aim to be comprehensive, but it probably includes most of the important direct sources available to the student of Muslim engineering. At first glance it may seem that we are quite well supplied with information, but on closer inspection it will be seen that this is by no means the case. The best written sources are certainly those on fine technology—*ḥiyal*—but even in this field there is a barren period of about 350 years between the Banū Mūsà and al-Jazarī, and only a few brief works have survived from that time. In other fields, the deficiencies are more serious. We usually know what the finished constructions looked like and how they worked, from examinations of surviving machines and structures. On the other hand, none of the sources listed above tells us precisely what needs were served by particular constructions. In other words, how did technology serve the community? To attempt to answer this question, and to supplement our knowledge on purely technical matters, we must consult works of a general nature. Even so, we shall not find all the answers, partly because we cannot search everywhere and partly because some of the answers are nowhere to be found. One of the problems that bedevils the early history of water power, for example, both in Europe and in Asia, is that our sources very seldom tell us, in referring to a mill, the type of wheel which powered it. Indeed, they not infrequently fail to distinguish between the main sources of power—animals, water and wind. Even so, there is a good deal to learn by a careful reading of certain types of literature.

The surviving store of medieval Arabic literature, itself only part of the vast original production, is very large. Some, enough to fill the shelves of large libraries, has been edited and published, but even more is still in manuscript form, much of it uncatalogued. The task of cataloguing, editing, publishing and commenting upon the mass of manuscript material is an undertaking that would occupy the lifetimes of many scholars. Arabic writers of the Middle Ages were usually versatile, so that one expects to find references to many subjects in the works of a single scholar. From a technological point of view, for example, a close study of the works of the great scientist and polymath al-Bīrūnī would undoubtedly be very rewarding. His output, however, was voluminous and remains largely unpublished, so a study of this kind would take many years. One could suggest many other lines of research, all of them likely to be both rewarding and time-consuming. In the end, however, the best that the individual scholar can do is to dig into the strata which have a relatively high potential yield. For the historian of technology, the works of Muslim geographers and travellers have this potential.[22]

As mentioned at the beginning of this article, our main purpose is to indicate how the works of Muslim geographers may be used to increase our knowledge of Muslim engineering. Given that we are referring to a genre that comprises a large number of works, it would clearly be impossible in the space available to list all of them and review the merits and demerits of each one. I have therefore chosen to discuss the works of five writers, supplemented by citations from other works which I have had occasion to refer to in the course of my researches. The five are: al-Istakhrī, Ibn Hawqal and al-Muqaddasī—all geographers of the tenth century; Ibn Jubayr, the great Andalusian traveller of the twelfth century; and al-Idrīsī, a twelfth-century geographer, probably born in Morocco. The reasons behind this selection need not detain us long. It is not a 'representative sample' nor is it based entirely upon personal preference, although I must confess that I find all of these writers congenial. It is generally agreed that the tenth century witnessed the summit of achievement in Arabic geographical writing and that the three writers mentioned were the best of that school. Moreover, they wrote 'human geography' in the most flourishing period of Islamic civilization, and hence provide us with a mass of valuable information about economic and social conditions, not excluding technological matters, in the lands of the Muslim world. Ibn Jubayr was the first and one of the best of the Muslim writers of *Rihla* (*rihla* = travel, journey). His work has many similarities to the writings of the geographers, but devotes more space to personal experiences. Al-Idrīsī gives us more information than the other writers about life in North Africa and Spain, at a time of upheaval in the Iberian peninsula due to the successes of the Christian armies. All these works have been translated, in whole or in part, into modern European languages.

In his famous book *The Civilization of the Renaissance in Italy*, Jacob Burckhardt discusses how the Italians were the first Europeans to think of themselves as individuals:

> an *objective* treatment and consideration of the State and of all the things of this world became possible. The *subjective* side at the same time asserted itself with corresponding emphasis; man became a spiritual *individual*, and recognized himself as such. In the same way the Greek had once distinguished himself from the barbarian, and the Arab had felt himself an individual at a time when other Asiatics knew themselves only as members of a race.[23]

This way of thinking is very apparent in the works of the writers that we are discussing. They tell us about their own experiences, report conversations in the first person and, above all, consider nothing beneath their notice, provided it has some connection with human life. They can be scathing in their criticisms of governments, manners and morals, and living conditions. Only two factors detract somewhat from their value. All of them were members of the upper middle classes and their hosts and contacts were from the same class, so they seldom report directly on the lives of the rural or urban poor. Also, they were dependent upon patrons and tend, therefore, to paint a rosy picture of the lands under their control. The tenth-century geographers, for example, received benefits from the Samanid rulers of Khurasan, and they describe the province, especially the region of Bukhara and Samarkand, in glowing terms. This does not, however, lead to much distortion; it is generally accepted that the Samanids were enlightened rulers and that their lands prospered under their rule. In any case, although the writers may have exaggerated the wisdom of the rulers and the prosperity of the province somewhat, this did not prevent them from commenting objectively and sometimes adversely on the living conditions of the people.

For each province, the following information is given: first, a general description of the region, its extent, its boundaries and its principal towns and cities—in the manuscripts, maps were provided. There is a specialized vocabulary for each kind of locality—capital cities and seats of government, principal cities, military centres, towns, villages, suburbs, private and public lands, etc. Each city and its dependent towns, villages and countryside is then described in detail, including customs, religious divisions, dress and appearance of the inhabitants, and taxation. We are told about crops, manufacture, exports and imports,, methods and materials of building construction, streets, roads and bridges. And always, water runs through the pages. For the smallest village we are invariably told where their drinking water came from—from wells, springs or from a stream. On a larger scale we have descriptions of irrigation systems, canals, qanats, dams, reservoirs and mills. For each province, the concluding section is concerned with communications, because these books were for the guidance of travellers. We are given routes, sometimes more than one, between the various towns in the province and the main routes to other provinces. This is done by stages of a day's journey, each with details of the shelter, provisions and water available at each staging point. The above summary applies to the geographers' books. Much of the same information is found in Ibn Jubayr's work, but there is much less in the way of topographical information and, of course, more about his own experiences.

Very little is known about the life of al-Istakhrī, the earliest of our writers, whose full name was Abu Ishāq Ibrāhīm b. Muhammad. His name implies that he came from Istakhr, the Arabic name for Persepolis in the Iranian province of Fars, and he may have lived for some time in Baghdad. He almost certainly visited Arabia, Iraq, Khuzistan, Daylam (the province south-west of the Caspian) and Transoxiana. His work was written towards the end of the first half of the tenth century. The best edition is that of al-Hīnī (to which the page references apply) and there are partial translations in German and Italian.[24] The work is entitled *Book of the Routes and the Lands* (*Kitāb al-masālik wa'l-mamālik*).

Ibn Hawqal, Abu'l-Qāsim b. 'Alī al-Nasibī was a native of Nisibis in Upper Mesopotamia. Between 947 and 973 he travelled widely in the

Muslim world—to Spain, North Africa, Egypt, Iraq, Iran and Central Asia. His book, *Configuration of the Earth* (*Kitāb Sūrat al-Ard*), appeared in its final, definitive version about 988. As a young man, Ibn Hawqal had met al-Istakhrī and became in some ways his follower; many of the reports in his work are word-for-word copies of passages in al-Istakhrī's book. Ibn Hawqal, however, is very far from being a mere plagiarist. Not only was he able to make first-hand reports on places that al-Istakhrī had not visited, but his work has a breadth and a personality incomparably larger than that of his predecessor. Of all the Arab geographers, his work is probably of the greatest value for modern historians. To quote André Miquel:

> His constant care to depict a region precisely in the state and at the date that he himself had seen it, and occasional references to the distant or more recent past, give to his text, besides a vividness of description and even a depth of feeling which sometimes appear, undoubted value for the historian. This is particularly true of the notes on economic matters, which form a complete break with convention; for one thing, Ibn Hawkal is much less interested in rare or precious products than in the basic agricultural and manufactured products, and secondly he was able to study on the spot a given economic situation in relation to a particular period or with reference to an implicit norm. He was the only Arab geographer who really sketched a vivid picture of production.[25]

The Arabic text was edited by J.H. Kramers and there is a full French translation, based on Kramers' text, by G. Wiet (notes refer to pages in the Arabic text).[26]

Al-Muqaddasī, Abu ʿAbd Allah Muhammad b. Ahmad (d. 1000) was a contemporary of Ibn Hawqal. His book is entitled *The Best Division for the Understanding of the Provinces* (*Ahsan al-taqāsīm lī ma ʿarifat al-aqālīm*), completed between the years 985 and 990 after its author had spent many years travelling throughout the Muslim world.

> He rightly claims to have put Arab geography on a new foundation and given it a new meaning and a wider scope. Since he considered the subject useful to many sections of society, as also to the followers of many vocations, he widened its scope, including in it a variety of subjects ranging from physical features of the *iklīm* (region) under discussion to mines, languages and races of the peoples, customs and habits, religions and sects, character, weights and measures and the territorial divisions, routes and distances. He believed that it was not a science that was acquired through conjecture (*kiyās*), but through direct observation and first-hand information.[27]

Much of his work makes very lively reading. He expresses his opinions of people and places pungently, and his descriptions are animated and evocative. He was a native of Palestine, and often compares foreign towns to towns in Palestine, nearly always to the detriment of the former. The

Arabic text is edited by de Goeje and there is a partial French translation (references to the Arabic text).[28]

Ibn Jubayr, Abu'l-Husayn Muhammad b. Ahmad, was born in Valencia in 1145 and died in Alexandria in 1217. He made three journeys, but his fame rests on his day-by-day account of the first of these. He sailed from Tarifa in Morocco early in 1183 in a Genoese ship which took a month to reach Alexandria by way of Sardinia, Sicily and Crete. He then travelled to Mecca, where he spent nine months before continuing his journey to Medina, Iraq, Upper Mesopotamia, Aleppo, Damascus and Acre (then in the hands of the Crusaders). He took ship from there, again in a Genoese vessel, bound for Sicily and narrowly escaped with his life in a shipwreck in the straits of Messina. Re-embarking at Trapani, he arrived at Cartagena in March 1185.[29] He gives us an abundance of information about the peoples among whom he stayed and about navigation, buildings and communications. The standard edition is that of Wright, emended by de Goeje, and there are full translations in English, French and Italian.[30] (References to the Arabic text.)

Al-Idrīsī, Abu ʿAbd Allah Muhammad b. Muhammad completed his geography in 1154 at the court of Roger II of Sicily, on whose orders the book was written. For this reason it is usually called simply *The Book of Roger*. Biographical references to him are rare, probably because the Arab biographers considered him a renegade for having lived at the court of a Christian king and written in praise of him. He was born about 1100, perhaps in Ceuta, and may have studied at Cordoba before settling in Sicily, where he died about 1165. His work is a straightforward geography and contains a good deal of information about economic affairs in Spain and North Africa. Part of the book was edited by R. Dozy and M.J. de Goeje in 1866, with a French translation, and I have used this work in preparing this article (references are given to both the original and the translation, with the Arabic first, for example 187/228). A full critical edition of the *Book of Roger* is being undertaken by an international group of scholars, under the auspices of the Instituto Italiano per il Medio e l'Estremo Oriente.[31]

One of the main concerns of these writers is to depict human society in each part of a given province within its urban and rural environment. The second concern is with commerce—with the production of goods and materials and their interchange, inside the Muslim world and with foreign countries. They do not, in general, seem to have been very interested in the daily lives of the inhabitants. Descriptions of cities and towns are usually fairly brief and are confined to details of layouts, with the locations of the principal buildings, markets, main streets and gates. Building materials are mentioned in very broad terms, for example 'their houses are built of wood/burnt bricks/stone', and only important buildings such as large mosques are described in detail. On the other hand, there is a wealth of information about agriculture: the various types of crops are listed, we are told whether cultivation was by irrigation or rainfall and about methods of processing—grist milling, drying, pickling and salting.

Often this processing was done in order to prepare the foodstuffs for export. We are seldom told, however, precisely how this processing was done but simply that it *was* done. Textile materials, of course, also came from the fields, and our writers mention a bewildering variety of fabrics made from wool, cotton, flax and silk. These fabrics, which were woven in state-operated workshops called *tirāz*, were the most important commercial commodities in the Muslim world.[32]

For the most part, our information about technology is a by-product of our writers' concern with agricultural production and their pre-occupation with water. So there is a relative abundance of data about the hydraulic works necessary to impound water and conduct it to the fields and urban communities. At the other end of the cycle of production, we can learn a good deal about the scale of corn-milling operations. Frequently, the reports are of a general nature, describing a river and the various irrigation canals drawn from it and the lands watered by them. Such, for example, is al-Muqaddasī's description of the canal system in Transoxiana (p. 280ff.). The irrigation system in the region of Marw, on the River Murghāb in Khurasan, was extensive and highly organized. Ibn Hawqal tells us that the director of the system had more power than the prefect of the city and controlled a labour force of 10,000 paid employees who carried out maintenance work and supervised the allocation of the water (pp. 635-6). Al-Muqaddasī adds that included in the labour force was a team of 400 divers who kept the dams and dykes under 24-hour observation. Each man had to keep with him a supply of timber and brushwood for carrying out emergency repairs and if they had to enter the water when it was very cold they coated their bodies with wax (pp. 330-1). All the water in Nishapur, on the other hand, was supplied by qanats, both for the town supply and for irrigation (Istakhrī p. 145). The same was true for Tangier in Morocco, where the main feeder qanat was many miles in length (Ibn Hawqal p. 79). A number of new cities were founded by the Muslims, and sometimes completely new irrigation systems had to be constructed as part of the cities' infrastructure. The fullest description of the excavation of such a system, for Basra in Iraq, occurs in the work of the historian al-Balādhurī (d. *c.* 892). The city was founded as a temporary military encampment in 638, but the two main feeder canals from the Shatt al-ʿArab were not completed until about 660. Thereafter, the excavation of new canals, which al-Balādhurī lists in detail with the names of their constructors, kept pace with the growth of Basra from a settlement into a large city.[33] About 150 years later, even though Basra was somewhat in decline, al-Istakhrī was amazed to see the vast network of canals (p. 57).

Dams were used for various purposes and they are sometimes described in some detail. Al-Muqaddasī describes two dams in Egypt, made from earth and esparto, which were breached at the time of the Nile flood to allow the water to flow on to the fields (p. 206). Much more impressive are his descriptions of two large dams, both of which provided hydropower as well as irrigation water. One of these was built in pre-Islamic times at Ahwāz in Khuzistan (p. 411). The other was on the River Kūr, in the central Iranian province of Fars; it was built in the tenth century by the Buwayid Amir ʿAdud al-Dawla:

> ʿAdud al-Dawla closed the river between Shiraz and Istakhr by a great wall, strengthened with lead. And the water behind it rose and formed a lake. Upon it on the two sides were ten water-wheels, like those we mentioned in Khuzistan, and below each wheel was a mill, and it is today one of the wonders of Fars. Then he built a city. The water flowed through channels and irrigated 300 villages (p. 444).

This dam, known as the Band-i Amir, still exists today. Al-Idrīsī describes a dam at Cordoba in Spain: 'across the river, below the bridge, was a dam made of Quptiyya stone, with large marble pillars. In the dam were three millhouses and in each house were four mills' (p. 212/262). Until recently its three millhouses still functioned, but much changed from their original form.[34]

These writers convey a very positive attitude towards the use of water power. The Tigris at its source, says al-Muqaddasī (p. 136), would turn only one mill, and al-Istakhrī, looking at a fast-flowing stream in the Iranian province of Kirman, estimates that it would turn twenty mills (p. 166). There are frequent references to mills all over the Muslim world, from the Iberian peninsula to Central Asia, and near large cities there were often dozens of mills constituting, as it were, factory milling complexes. It could scarcely have been otherwise. At the height of its prosperity, Baghdad had a population of about 1·5 million and Fustāt (Old Cairo) and Cordoba, though smaller, were still very populous cities. The populations were agriculturally non-productive and had to be fed, clothed and housed, and their merchants supplied with the materials of commerce. One example of large-scale milling, from Ibn Hawqal, will have to suffice. It refers to Upper Mesopotamia, which was the granary for Baghdad:

> The ship-mills on the Tigris at Mosul have no equal anywhere, because they are in very fast current, moored to the bank by iron chains. Each has four stones and each pair of stones grinds in the day and night 50 donkey-loads. They are made of wood—sometimes of teak (p. 219).

If we take a donkey-load as 100 kg, then the output of one of these mills in 24 hours was 10 tonnes. Long after the time of Ibn Hawqal, Upper Mesopotamia continued as a large supplier of flour to Iraq. About 1184, Ibn Jubayr saw the ship-mills across the River Khabur, 'forming, as it were, a dam' (p. 243).

Ship-mills were one way of increasing the output of mills, by taking advantage of the high velocity of the water in midstream. For the same reason, water-wheels were installed on the piers of bridges (al-Muqaddasī p. 312). The use of dams to increase the head of water above the mills has already been mentioned. The power of the tides was also harnessed.

Al-Muqaddasī mentions tidal mills in the Basra area: 'The ebb-tide is also useful for operating the mills, because they are at the mouths of the rivers, and when the water comes out it turns them' (pp. 124–5). The quality of millstones has always been important. They must be hard but of homogeneous texture, so that pieces of grit do not become detached and get mixed with the flour. The stone from certain localities was therefore particularly prized for milling purposes and there are a number of references to suitable kinds of stone. In the area of Majjāna, in modern Tunisia, for example: 'They cut millstones from the nearby mountains, which are excellent and good for grinding; they can last a man's lifetime without dressing or other treatment, due to their solidity and the fineness of their grains' (al-Idrīsī pp. 118/138).[35]

There is therefore a good deal of information about mills and their importance in the economic life of the Muslim world. We do not learn much, however, about the actual construction of the machines—Ibn Hawqal's description of ship-mills, quoted above, is one of the few reports that gives any constructional details. We know from the *hiyal* treatises and elsewhere, however, that all the basic types of water-wheels were in use in the Muslim world, and we can infer that the type of wheel used in a given locality was that which best suited the local conditions. Nor do we find much information about industrial uses for water power, apart from corn milling. There is an indication in al-Muqaddasī that water power was utilized for the fulling of cloth (p. 409), but we must use other sources for information about industrial milling.[36]

The only references to windmills are to the machines with vertical axles which were built in the province of Sijistan or Seistan (the western part of modern Afghanistan). Al-Istakhrī was the first to mention them (p. 140), but for a full description of this type of machine, which were still in use until recently, we have nothing earlier than a passage by the Syrian geographer al-Dimashqī, in a book written about 1271. He tells us that the mills were supported on substructures built for the purpose, or on the towers of castles or on the top of hills. They consisted of an upper chamber in which the millstones were housed and a lower one for the rotor. The axle was vertical and it carried twelve or six arms covered with a double skin of fabric. The walls of the lower chamber were pierced with funnel-shaped ducts, with the narrower end towards the interior in order to increase the speed of the wind when it flowed on to the sails.[37]

There is abundant evidence for the widespread use of water-raising machines in the Muslim world and they are often mentioned by our writers, although most references simply state that saqiyas or norias were used in a given area. The first report for the construction of norias in Islam comes from al-Balādhurī, who tells us that in the second half of the seventh century (we can date it no closer) a certain Bilāl excavated a canal in the Basra area and erected norias along its bank.[38] Al-Muqaddasī was clearly impressed with the large norias on the river at Ahwāz in Khuzistan (p. 411) and al-Idrīsī describes a hydraulic installation at Toledo in which a noria lifted water through a height of 50 cubits to an aqueduct that led into the city (pp. 187/288). In the town of Ra's al-ʿAyn in Upper Mesopotamia, norias raised water from the River Khabur on to the gardens (Ibn Jubayr p. 243).

When we leave the field of hydraulic engineering, we find that reports are less frequent and generally briefer. An exception, as mentioned earlier, is the case of large buildings such as great mosques, for which the descriptions are often sufficiently detailed to be of value to students of architecture and building construction. On the other hand, many of these buildings still exist and can be examined *in situ* or through the works of architectural historians.

Another type of large structure that often invited comment was the arched bridge, built of masonry or occasionally of burnt bricks. Many of those described were of pre-Islamic origin—Roman or Sasānid. Al-Idrīsī, for example, describes the many-arched Roman bridge over the Guadalquivir at Cordoba (pp. 212/262). One bridge, which still exists, impressed all the Muslim writers who saw it. This is the bridge at Sanja in Upper Mesopotamia, a single arch of dressed masonry with a span of 112 feet built by the Emperor Vespasian about AD 70 (see, for example, Ibn Hawqal p. 123). A remarkable Muslim bridge, built over the River Tab in central Iran, is described, among others, by al-Istakhrī. He says that it was built by an Iranian, physician to the Umayyad governor al-Hajjāj (661–714). It was a single arch of 80 paces in span, and so high that a man on a camel with a long standard in his raised hand could pass beneath it (p. 91). Al-Qazwīnī, in his geography completed in 1276, has left us a graphic account of the rebuilding of a ruined Sasānid bridge at the town of Īdhaj in Khuzistan. It was built by a minister of the Buwayid Amir al-Hasan, who conscripted craftsmen from Īdhaj and Isfahan. The bridge was 150 cubits in height and consisted of a single arch supported on tapering masonry piers strengthened with lead dowels and iron clamps. The slag from iron workings was used to fill the space between the arch and the roadway.[39] Bridges of boats were common and are frequently mentioned, although seldom with any details of their construction. Ibn Jubayr, however, describing a bridge of boats over the Euphrates at Hilla, tells us that it had chains on either side to which the boats were attached. These chains, 'like twisted rods', were secured to wooden anchorages on the banks (p. 213).

References to fine technology are very rare and add little to the information in the *hiyal* treatises. Al-Muqaddasī remarked upon the beauty of the fountains in Damascus (p. 157) and about two centuries later, in the same city, Ibn Jubayr saw a fountain consisting of a single jet surrounded by small pipes that threw the water up like the branches of a tree (p. 269). Al-Muqaddasī refers to a portico by the Jayrun Gate in Damascus as the 'Gate of the Hours', indicating that a clock had traditionally been located at this spot, where Ridwān's father was to build his clock in the twelfth century (p. 158). This clock was seen and described by Ibn Jubayr; his description, though quite brief, tallies with the details given by Ridwān in his treatise (Ibn Jubayr p. 270 ff).

The foregoing discussion and extracts, it is hoped, indicate how the works of Arab geographers can assist the historian of technology. Quite apart from their value in this respect, however, writers such as Ibn Hawqal and al-Muqaddasī paint a vivid picture of the Muslim world at the peak of its prosperity. While not free from blemishes and prejudices, their works depict that society as they saw it, with its merits and its defects. One thing is certain: it was a society that made extensive use of technology to support large populations and a thriving commerce.

Notes

(*EI* is used as the abbreviation for *Encyclopaedia of Islam*, 2nd edn, Brill, Leiden, 4 vols. to date, continuing)

1. Donald R. Hill, 'Hiyal' in *EI*, suppl. to vols I to III, pp. 371-4.

2. Banū Mūsà, *The Book of Ingenious Devices*, trans. and annotated by Donald R. Hill, Reidel, Dordrecht, 1979. *Kitāb al-Hiyal*, Arabic edn of the Banū Mūsà's work, ed. A.Y. Hassan, Institute for the History of Arabic Science, Aleppo University, 1981. Donald R. Hill, 'The Banū Mūsà and their Book of Ingenious Devices' in *History of Technology*, Vol. 2, 1977, pp. 39-76.

3. Al-Khuwārizmī, Abu 'Abd Allah Muhammad, *Liber Mafātīh al-Olūm*, ed. G. van Vloten, Arabic text with Latin critical apparatus, Brill, Leiden, 1895.

4. Donald R. Hill, *Arabic Water-clocks*, Institute for the History of Arabic Science, Aleppo University, 1981, pp. 36-46.

5. *Ibid.*, pp. 47-68.

6. *Ibid.*, pp. 69-88.

7. Al-Jazarī, Ibn al-Razzāz, *The Book of Knowledge of Ingenious Mechanical Devices*, trans. and annotated by Donald R. Hill, Reidel, Dordrecht, 1974. *A Compendium on the Theory and Practice of the Mechanical Arts*, Arabic text of al-Jazarī's work, ed. A.Y. Hassan, Institute for the History of Arabic Science, Aleppo University, 1979.

8. Philo of Byzantium, 'Le Livre des Appareils Pneumatiques et des Machines Hydrauliques par Philon de Byzance', Arabic text edn with French trans. by Carra de Vaux, *Paris Académie des Inscriptions et Belles Lettres*, Vol. 38, Part 1, 1903.

9. *On the Construction of Water-clocks*, trans. and annotated by Donald R. Hill, Occasional Paper No. 4, Turner & Devereux, London, 1976.

10. E. Wiedemann, *Aufsätze sur Arabischen Wissenschaftsgeschichte*, 2 vols., Olms, Hildesheim, vol. 2, 1970, pp. 50-6.

11. There are, for example, a number of manuscripts containing a chapter called 'Wheels that move by themselves'. Despite the perpetual motion element in these devices, they would repay close study. See Thorkild Schiøler, *Roman and Islamic Water-Lifting Wheels*, Odense University Press, 1973, pp. 72-8.

12. al-Karajī, *Inbāt al-miyāh al-khafiyya*, Hyderabad, 1945.

13. C. Cahen, 'Le Service de l'irrigation en Iraq au début du XIe siècle' in *Bulletin d'études orientales*, Vol. 13, 1949-51, pp. 117-43.

14. C. Cahen and M. Talbi, 'Hisba' in *EI*, vol. III, 1971, pp. 485-8.

15. Derek de Solla Price, 'Mechanical Water-clocks in the 14th century in Fez, Morocco' in *Ithaca*, Hermann, Paris, 1962, Vol. 26, pp. VIII-IX.

16. For example, Norman Smith, *A History of Dams*, Peter Davies, London, 1971.

3

Tutia was available also in various mining areas in Spain.

Tin (*raṣāṣ qalʿī, qaṣdīr*) came from the Malaysian peninsula, which was known as Kala, hence the name *qalʿī* for the metal.

Iron and steel were in great demand in the Islamic empire. Hence iron ores were utilised whenever it was feasible. These ores were distributed in most Islamic lands. Five major iron mining areas were utilised in Spain. These included mines near Toledo, and near Murcia. In the Maghrib, about ten iron mining areas were exploited in Morocco, Algeria and Tunisia. These included mines in Jabal al-Ḥadīd in the Atlas mountains, the Rīf, Gawr al-Ḥadīd in Algeria and Majjanat al-Maʿdan in Tunisia. Iron ores were produced and exported from Sicily. Iron ore was mined in the Libyan desert and in the Fezzān. Egypt exploited those ores that were available, for example in Nubia and on the Red Sea coast. Syria was famous for its iron and steel metallurgy (Damascus steel), and the iron ores were obtained in the south and in the mountain ranges between Damascus and Beirut. The Islamic countries of the east (al-Mashriq) were better endowed with iron ores than Egypt, Syria and Iraq. The province of Fārs had at least four important iron mining centres. There were
also iron mines in Khurāsān, in Transoxania, in Ādharbāyjān and in p. 968
Armenia.

Mercury (*ziʾbaq*) came chiefly from Spain. Al-Idrīsī mentions the mine to the north of Cordova, where more than one thousand men worked in the various stages of mining the ores and extracting the mercury. Another source was Farghāna in Transoxania.

Salt (*milḥ*) was produced in numerous localities. It was an essential commodity, and production was undertaken in some areas on a very large scale for export purposes, for example in the Maghrib, where the salt mines were on the desert edges in the south. Salt was produced and carried by caravans south of the Sahara to be exchanged for gold. Thousands of men and camels were involved in these operations. Other important salt mines or production centres were in Khurāsān, Arabia and Armenia.

The alum (*shabb*) of Yemen was famous for its quality, but according to al-Idrīsī, the major source was in Chad. It was exported to Egypt and to all the countries of North Africa. Egypt was also a major production centre for both alum and natron. Among the many other minerals that were known and utilised was asbestos from Badakhshān, from which wicks and fire-resistant cloths were made. Coal was also known and used in some areas such as Farghāna in Transoxania, where it was mined and sold. It was used as a fuel for ovens (*tanānīr*, sing. *tannūr*), and its ashes were utilised as a cleansing agent. Petroleum deposits were exploited on a large scale [see NAFṬ].

Precious stones of various kinds were mined. Arabic works on lapidary such as al-Bīrūnī's *al-Jamāhīr* are celebrated and deserve detailed study. Rubies were mined in Badakhshān, and they were brought from Sarandīb (Sri Lanka). Diamonds came from Hindustān and Sarandīb, agate from Yemen, emeralds from Egypt, turquoise from Nīshāpūr, lapis lazuli from Egypt and onyx from Yemen. Corundum came from Nubia and Sarandīb. Crystal was mined in Arabia and Badakhshān. Diving for pearls (*lu'lu'* [*q.v.*] was a flourishing industry and corals were obtained from the coasts of North Africa and Sicily.

2. Mining technology

Although mining operations in Islamic civilisations were very extensive, yet like most aspects of Islamic technology no attempt has yet been made to study the technology of mining. This is another field that requires extensive research. There is therefore a noticeable gap in the history of mining in general, and whereas much information and archaeological research has been published on ancient and Roman mining, research is completely missing for the period between the 7th and 15th centuries. Not only should this gap be filled, but there are academic reasons for re-examining much of the material that has been published on ancient and Roman mining technology.

17. H. Goblot, *Les Qanats; une Technique d'Acquisition de l'Eau*, Mouton Éditeur, Paris, 1979; and A.K.S. Lambton, 'Kanāt' in *EI*, vol. IV, 1978, pp. 528–32.

18. Thomas F. Glick, *Irrigation and Society in Medieval Valencia*, Harvard University Press, 1970.

19. Thorkild Schiøler, as note 11.

20. Donald R. Hill, *A History of Engineering in Classical and Medieval Times*, Croom Helm, 1984, pp. 139–42.

21. S. Hamarneh, 'Sugar-cane Plantation and Industry under the Arab Muslims during the Middle Ages' in *Proceedings of the First International Symposium for the History of Arabic Science*, Aleppo University, 1976, p. 221.

22. Two modern works on the early Muslim geographers are worthy of mention. The first of these is by André Miquel, *La géographie humaine du monde musulman jusqu'au milieu du XIe siècle*, Paris, 1967. This is a detailed study of geographical writings in Islam, from the rise of Muslim literature until about 1050. The development of scientific geography is discussed and the achievements of individual geographers are assessed. The second work is by Guy le Strange, *The Lands of the Eastern Caliphate*, Frank Cass, London 1905, third impression 1966. From the works of a number of Muslim geographers, the author reconstructed the physical geography of medieval Islam, from Iraq eastwards. There are maps for each province and a good index.

23. Jacob Burckhardt, *The Civilization of the Renaissance in Italy*, Phaidon Press, Oxford and London, 1945, p. 81.

24. Arabic edition by M.G. ʿAbd al-ʿĀl al-Hīnī, Cairo, 1961; German trans, *Das Buch der Länder*, A.D. Mordtmann, Hamburg, 1845, partial Italian trans. by A. Madini, Milan, 1842; see A. Miquel, 'al-Istakhrī' in *EI*, Vol. IV, 1978, pp. 222–3.

25. A. Miquel, 'Ibn Hawkal' in *EI*, Vol. III, 1971, pp. 786–8.

26. Arabic edition by J.H. Kramers, 2nd edn of *Biblioteca Geographorum Arabicorum* (BGA), Brill, Leiden, 1938; French trans. *Configuration de la Terre*, by J.H. Kramers and G. Wiet, 2 vols., Beirut/Paris, Vol. 1, 1964, Vol. 2, 1965.

27. S. Maqbul Ahmad, 'Djugrafiyya' in *EI*, Vol. II, 1965, p. 582.

28. Arabic edn by M.J. de Goeje, vol. 3 of BGA, Brill, Leiden, 1906; partial French trans. by A. Miquel, *La meillure répartition pour la connaissance des provinces*, Damascus, 1963.

29. C. Pellat, 'Ibn Djubayr' in *EI*, Vol. III, 1971, p. 755.

30. Arabic edn by M.J. de Goeje, Leiden, 1907, from the 1852 edn by M. Wright; Italian trans. *Viaggio in Ispagna, Sicilia, etc.* by C. Schiaparelli, Rome, 1906; English, *The travels of Ibn Jubayr*, by C. Broadhurst, London, 1952; in French by M. Gaudefroy-Demombynes, 3 vols, Paris, 1949–56.

31. Arabic with French trans. of part of Idrīsī's work: *Description de l'Afrique et de l'Espagne*, by R. Dozy and M.J. de Goeje, Brill, Leiden, 1866; see also G. Oman, 'al-Idrīsī' in *EI*, Vol. III, 1971, pp. 1032–5.

32. R.B. Serjeant, *Islamic Textiles—Material for a History up to the Mongol Conquest*, Beirut, 1972.

33. Al-Balādhurī, *Kitāb futūh al-buldān* (*Conquests of the Lands*), ed. M.J. de Goeje, Brill, Leiden, 1866, pp. 345–71.

34. Norman Smith, *Man and Water*, Peter Davies, London, 1976, p. 143.

35. Donald R. Hill, as note 20, p. 166.

36. For example, al-Bīrūnī in his work on minerology, *Kitāb al-jamāhir etc.*, ed. F. Krenkow, Hyderabad, Deccan, 1936, pp. 233–4, tells us that water-driven trip-hammers were used for pounding gold ores 'as is the case in Samarkand with the pounding of flax for paper'.

37. Al-Dimashqī, *Cosmographie*, Arabic text edn A.F. Mehren, St Petersburg, 1866, p. 182.
38. Al-Balādhurī, as note 33, p. 363.
39. Al-Qazwīnī, *Athār al-bilād*, Beirut, 1960, p. 303.

Mining Technology*

(written jointly with A. Y. al-Hassan)

p. 967

1. Islamic mines

In the vast Islamic empire, minerals played an important part. There was a great need for gold, silver and copper for the minting of coins and for other uses. Iron ore was necessary for the manufacture of iron and steel for arms and implements. Other minerals such as mercury, salt and alum were indispensable for everyday life. The empire was richly endowed with the various mineral resources, and the mining industry was an extensive one. *Maʿdin* (pl. *maʿādin*) denoted "mine" in Arabic sources, and *muʿaddin* means "miner". In modern Arabic, the word *manjam* denotes "mine", while *maʿdin* or *maʿdan* is now used mostly for "metal" or "mineral".

Information on Islamic mines occurs in geographical works, in alchemical treatises, in books on mineralogy and in various other sources. But these, and archaeological discoveries, have not yet been searched for the purpose of writing a history of mining technology, although some studies have appeared on the distribution of Islamic mines. We shall mention here only some of the minerals and a few of the mining centres, since it is not possible here to list all the mines.

Gold (*dhahab* [*q.v.*]) mines were found in western Arabia, Egypt, Africa and in some eastern Islamic lands. But the first major gold mining area was at Wādī al-ʿAllāqī, which is a right-bank tribu-

*In this new reprint of the *Encyclopaedia of Islam* article a slightly simplified transliteration system for Arabic has been used and place names have been given in their anglicised form whenever possible. All references in square brackets are to other articles in the *Encyclopaedia of Islam*. References in the indexes of this book are to the original pagination, here given in the margins.

tary of the upper Nile. It lay in the Beja country, which was between Ethiopia and Nubia [see BEDJA]. The mines were in a desert area between the Nile and the Red Sea. The nearest towns were Aswān [see al-USWĀN] on the Nile and ʿAydhāb [*q.v.*] on the Red Sea. The second major gold mining area was called by al-Bīrūnī the Maghrib Sūdān. This is the area south of the Sahara in Senegal and on the Upper Niger in Mali. According to al-Idrīsī, Wangāra was the most important gold mining centre on the upper Niger. Salt, cloth and other commodities were exchanged for gold.

Silver (*fiḍḍa* [*q.v.*]) was mined either individually or in association with lead ores. The major silver mines were in the eastern Islamic provinces. Prominent among these were the mines of the Hindū Kūsh in the towns of Panjhīr and Jāriyāna, both in the neighbourhood of Balkh. According to one report, there were about 10,000 miners working at Panjhīr. Other important silver mines were in Spain, the Maghrib, Iran and Central Asia.

Lead (*usrub*, *raṣāṣ*) was obtained mostly from galena (lead sulphide), which was of very common occurrence. This lead ore is often associated with small quantities of silver. Only two other lead ores have any importance as raw materials. One is cerussite (lead carbonate) and the other, which is of minor importance, is anglesite (lead sulphate). Lead ores, especially galena, were exploited in Spain, Sicily, the Maghrib, Egypt, Iran, Upper Mesopotamia and Asia Minor.

Copper (*nuḥās*) ore deposits were exploited in various areas, including the important mines of Spain in the west, and several deposits in the east, such as those in Sijistān, Kirmān, Marw, Farghāna, Bukhārā, Ṭūs and Herat. The copper mines in Cyprus were always an important source.

The word calamine or tutia (Arabic *tūtiya*) was used to denote the natural zinc ores (especially zinc carbonate), or the white zinc oxide which was obtained during the treatment of the ores. The major mines for tutia were in the province of Kirmān in the east.

It is possible to give here an outline of Islamic mining technology by looking into the brief accounts of the geographers and some of the lapidary books. In Europe it was not until A.D. 1556 that the first book on mining, that of Agricola, made its appearance. Before that time, only a few books on lapidary techniques, mainly translated from Arabic, were available.

There was not just a single mining technology in Islamic lands. Methods differed from mineral to mineral, from country to country, and from one method of ownership to another. In any one country could be found a range of technologies, from the primitive to the highly developed.

As with modern mining, there were two major types of operation: the underground and the opencut. In brief words, al-Bīrūnī (*al-Jamāhīr fī maʿrifat al-jawāhir*, Hyderabad 1355 H. / 1936-7, p. 83) writes:

> The search for *laʿl* (a kind of ruby) is of two types: one is to dig the mine under the mountain, and the other is to search for it among the gravel and earth which result from the collapse of the mountains by earthquakes or their erosion by floods.

In underground mining, one method was to sink shafts vertically into the soil, and to drive horizontal passages when the veins were reached. In Syria, the shaft of the mines was called the *bīr*, *i.e.* the well, and the horizontal tunnel the *darb*, *i.e.* the road. In the Lebanese mountains, a typical shaft was only 6-7 metres deep, and the tunnels were "very long". Al-Idrīsī saw the mercury mines to the north of Cordova in Spain, and he was told that the depth from ground level to the bottom of the mine was 250 fathoms (*qāma*) (*Nuzhat al-mushtāq fī 'khtirāq al-āfāq*, section on al-Maghrib, Arḍ al-Sūdān, Egypt and al-Andalus, Leiden 1864, p. 214). Other mines of intermediate depth were reported. Thus in the silver mines in the Maghrib the depth was 20 cubits (*dhirāʿ*). The technique of drilling vertical shafts and horizontal tunnels was a familiar operation in the Islamic countries if we remember the great tradition of constructing the underground *qanāt* systems [see ḲANĀT], with the exacting technical experience which they require.

More often, however, in underground mining miners preferred to dig horizontal adits into the slopes of a mountain and follow the veins, rather than to sink shafts. This method was easier and less expensive for a miner who was working for himself. It is noticeable that the reports of mines with vertical shafts usually apply to mines owned by the state. A vivid description of the silver mining activities in Panjhīr in eastern Khurāsān, where 10,000 men were employed in the mining industry, is given by Abu 'l-Fidā'(*Taqwīm al-buldān*, ed. Reinaud and de Slane, Paris 1840, p. 465). He wrote:

> The people of Panjhīr made the mountain and the marketplace like a sieve because of the many pits. They only follow veins leading to silver, and if they find a vein they dig continuously until they reach silver. A man may spend huge sums of money in digging, and he may find silver to such an extent that he and his descendants become rich, or his work may fail because he is overpowered by water or for other reasons. A man may pick a vein, and it is possible that another man picks the same vein in another position. Both start digging. The custom is that the miner who arrives first and intercepts the passage of the other miner wins the vein and its results. Because of this competition, they execute a work that devils cannot achieve. When one arrives first, the expenses of the other are wasted. If they arrive together, they share the vein and then they continue digging as long as the lamps are burning. If the lamps are extinguished and cannot be relit, they stop their progress because anyone who reaches that position would die immediately. (In this business) you will see that a man starts his day owning one million,

and by nightfall he owns nothing. Or he may start poor in the morning and by evening he becomes the owner of uncountable wealth.

The main tool of the miner was the pickaxe (*minqār*, *ṣāqūr*). It had a sharp end to peck the stone and a flat end to hammer or to drive wedges. There were also various hammers, chisels or wedges, crow-bars, hoes and shovels.

Windlasses were used for hauling ores and materials out of shafts. An efficient and simple form of windlass was used in the iron mines in Syria, and is still used in constructing the *qanāt* system in Iran, in drawing water, and in the building industry. Here an assistant sits on a bench on one side of the mouth of the shaft or well, pulling the horizontal bars of the windlass towards him with his hands and pushing the opposite ones away with his feet at the same time. The ore is loaded into a small bucket about 30-35cm in diameter, which has two handles. The rope is attached to the bucket by hooks fastened at its end. More sophisticated capstans were used for haulage also, when loads were heavier.

Oil lamps were used for illumination. The lamps were useful for aligning the direction of the digging, and were also good indicators of the adequacy of fresh air supplies, as was the case in the silver mines of Panjhīr. In Arabic technical literature, there are different designs of ingenious lamps. One such design, suitable for outdoor use and protected against being extinguished by winds, was described by the Banū Mūsā.

Ventilation was an important problem. In Panjhīr, as we have seen, with thousands of small-scale miners working for themselves in a frantic search for silver, capital investment was kept to a minimum, and usually no provision for ventilation was made. The miner simply abandoned the digging if the lamps stopped burning. In the more organised mining work, especially in the state mines, a means of ventilation was always provided. This would be essential, particularly in the very deep mercury mines which we mentioned. Special ventilation shafts were provided. When installing a drainage system, several shafts were needed for this purpose, and these also served as ventilation shafts. Special ventilation shafts have been found in Iran, and some of these go back to pre-Islamic times. The problem of ventilating wells and mining shafts attracted the attention of Muslim engineers. They designed special ventilating machines. The Banū Mūsā (*Kitāb al-Ḥiyal*, Arabic text ed. A. Y. al-Hassan, Aleppo 1981, p. 374, Engl. transl. D. R. Hill, Dordrecht 1979, p. 240) describe a design for

> a machine for use in wells which kill those who descend in them. If a man uses this machine in any well, it will neither kill nor harm him. This machine is suitable for wells that kill and for dangerous pits. If a man has this machine, which we shall describe, with him, he can descend in any well immediately without fearing it and it will not harm him.

The other important problem in mining was drainage of the mines. Here also, small-scale miners of silver could not afford to
p. 969 solve the problem, as we have seen in Panjhīr. On the other hand, in the state silver mines in Zakandar in the Maghrib, drainage was carried out properly, as al-Qazwīnī (*Āthār al-bilād*, Beirut 1960-61, pp. 199-200) reports:

> Here are the silver mines. Anyone who wishes can undertake processing them. There are underground mines in which many people are always working. When they descend 20 *dhirā*'s, water appears. The sultan installs waterwheels and water is raised until the mud appears. Workers bring this mud up to the surface of the ground and wash it. (The sultan) does this in order to take the fifth ... Water is raised in three stages, since it is 20 *dhirā*'s from the ground level to the surface of the water. He installs

> a wheel down in the mine on the water surface. Water is lifted and it is discharged into a large tank. Another wheel is installed on this tank. It lifts the water and pours it into another tank. On this tank a third wheel is installed. It lifts the water and discharges it on to the surface of the ground to irrigate the farms and gardens. This operation cannot be undertaken except by a very rich person possessing thousands. He sits at the mouth of the mine and employs artisans and workers, who bring out the mud and wash it in front of him. When the work is done, the fifth of the sultan is put aside and the rest is given to (the owner). It may come to be smaller than his expenditure and it may be [more]. This depends on the man's efforts.

The technique of water-raising by water-wheels was highly developed in Islamic civilisation, and it flourished in North Africa and Spain. The above passage by al-Qazwīnī is important, and should prompt a reconsideration of archaeological findings, since it is quite certain that other mines in Spain and North Africa were being drained in a similar manner. The Muslims carried out extensive mining operations in central and southern Spain in the areas of Cordova, Almáden, Ovejo, Huelva, Murcia, Ilbira and elsewhere. Water-raising devices of various types were used. These included the compartmented wheel, the chain-of-pots wheel, the Archimedean screw, the rag-and-chain pump and the piston pump. Archaeologists have found a screw pump in a mine near Cordova with an oak screw and a barrel of sheet lead, but R. J. Forbes (*Studies in Ancient Technology*, VII, Leiden 1966, p. 219) has expressed doubts as to whether this pump was really of Roman date.

We come now to a discussion of the dressing of ores. This includes the crushing, sifting and washing of ores. Such operations were usually conducted at the sites of the mines before the transport of the ores to the metallurgical centres. The ores were pounded or reduced in size, and in the case of gold ores they were crushed or milled to a finer degree. The milling operation was more important with the reef gold which occurs in quartz veins. Milling or crushing of these ores was done either by querns or by water driven trip-hammers. Al-Bīrūnī (*Jamāhīr*, pp. 233-234) discusses this operation:

> Gold may be united with stone as if it is cast with it, so that it needs pounding. Rotary mills (*ṭawāḥīn*) can pulverise it, but pounding it by *mashājin* is more correct and is a much more refined treatment. It is even said that this pounding makes it more red, which if it is true is rather strange and surprising. The *mashājin* are stones fitted to axles which are installed on running water for pounding, as is the case in the pounding of flax for paper in Samarqand.

This is an important text since it indicates that water-driven trip-hammers (*mashājin*) were already established for crushing ores before the 10th century.

3. Pearls and underwater resources

The exploitation of corals and pearls involved very extensive activity. Pearls of the Gulf area were highly valued; the Chinese considered them to be the best pearls. The operations were organised by entrepreneurs. One of these would hire divers for two months and pay them regularly; such enterprises were often highly profitable. However, we read detailed descriptions of the great dangers which faced the divers.

Al-Bīrūnī again gives a detailed description of a diving operation, including a description of new diving-gear (*Jamāhīr*, pp. 149-150):

> I was told by a man from Baghdad that divers had invented in these days a method for diving by which the difficulty of holding the breath is eliminated. This enables them to

> frequent the sea from morning to afternoon, as much as they wish and as much as the employer favours them. It is a leather gear which they fit down their chests and they tie it at the (edges?) very securely and then they dive. They breathe in it from the air inside it. This necessitates a very heavy weight to keep down the diver with this air. A more suitable arrangement would be to attach to the upper end of this gear opposite the forehead a leather tube similar to a sleeve sealed at its seams by wax and bitumen, its length equal to the depth of diving. The upper end of the tube will be fitted to a large dish at a hole in its bottom. To this dish are attached one or more inflated bags to keep it floating. The breath of the diver will flow in and out through the tube as long as he desires to stay in water, even for days.

An ingenious dredging device was designed by the Banū Mūsā in their *Book of Ingenious Devices* (*Kitāb al-Ḥiyal*, text p. 376, Engl. transl. p. 342). They wrote:

> We wish to explain how to make a machine by which a person can bring out jewels from the sea if he lowers it, and by which he can extract things which fall into wells or are submerged in the rivers and seas. p. 970

This worked on exactly the same principle as the modern clamshell grab.

4. Non-ferrous metallurgy

Much information about metallurgy can be gained from the alchemical and chemical treatises. There was a close relationship between the work of the metallurgist and the chemist, and the Arabic sources reflect the experience gained in the laboratory and in the metallurgical furnace.

Gold. Speaking about native gold which is collected from gold mines, al-Bīrūnī (*Jamāhīr*, p. 234) says that it is usually not free from impurities and therefore this gold can be refined by smelting only, or by other methods. He gives details of the amalgamation method that was used in the mines on a commercial scale:

> After pounding the gold ore or milling it, it is washed out of its stones, and the gold and mercury are combined and then squeezed in a piece of leather until mercury exudes from the pores of the leather. The rest of the mercury is driven off by fire.

He further describes (p. 236) how gold is mined from the deep waters of the Sind River:

> At its sources there are places in which they dig small pits under the water, which flows over them. They fill the pits with mercury and leave it for a while. Then they come back after the mercury has become gold. This is because at its start the water is rapid and it carries with it particles of sand and gold like mosquitoes' wings in thinness and fineness. Water carries these particles over the surface of the mercury, which picks up the gold, leaving the sand to pass away.

The cupellation process was used extensively both on the laboratory and the industrial scales. The gold is alloyed with lead in a special crucible and it is then oxidised by means of a strong current of air blown onto the surface of the molten metal. The base metals such as copper are drossed, while the gold and silver remain as a button of silver-gold alloy. The separation of the gold from the silver was done by what is known as the salt and sulphur process. The liquation process was also used in combination with the cupellation process. What is new in these processes — and this was an Islamic innovation — is the use of nitric acid in the separation of gold from silver. Jābir b. Ḥayyān [see D̲JĀBIR B. ḤAYYĀN] described this process, and it was also mentioned by other later writers. Gold was tested by various methods. These included the touchstone (*al-miḥakk*), measuring the specific gravities, and noting the speed of solidification of gold after it had been removed from the furnace.

Silver and lead. Unlike gold, native silver was not found in alluvial deposits or in the sands and gravels of rivers, but was to be sought in mountainous regions in embedded veins. In general, however, native silver was not abundant, and the main source was from galena (lead sulphide), which was usually associated with small amounts of silver. The first step would be to obtain the lead itself (which usually contains silver) from the earth and stones (*i.e.* the galena). This was done first by roasting, followed by smelting. The resulting lead could then be treated to extract silver. In Arabic literature, we find the results of some experiments indicating the amount of silver which could be recovered from an ingot of lead. This was usually in *dirhams* in one *maṣla* (one silver *maṣla* = 50 *raṭl*s). There were smelters (*sabbākūn*) specialising in gaining silver from lead ingots.

Sometimes silver was associated with gold in what is called electrum. Here also, as we have seen, methods were adopted to separate these two precious metals. Silver was also obtained from its ores, or from lead or copper ores.

These different sources of silver necessitated the application of various techniques which were mastered by Islamic smelters and chemists, such as roasting, smelting, oxidation, liquation, leaching, cupellation and amalgamation. Details cannot be given here, but these methods fascinated the alchemists and were largely responsible, in our opinion, for the vast alchemical literature that resulted from the alchemists' experiments with stones (*aḥjār*) and metals (*ajsād*).

Tin, zinc, antimony and arsenic. Tin was one of the seven malleable metals or bodies. It was brought mainly to the classical Islamic countries from the Malaysian peninsula and, to some extent, from Spain and the West.

Zinc was not known as a distinct metal by the early Muslim metallurgists and chemists. It was first known, and used extensively, through tutia (zinc oxide) which was one of the stones (*aḥjār*). Later, as we shall see, zinc (*rūḥ al-tūtiya*) was known as a distinct metal. Tutia is usually the pure zinc oxide which is obtained from natural zinc carbonate. Various authors described the method of producing the pure product from the natural one. The ore is placed in furnaces which contain long ceramic rods. Upon heating the ores, the smoke of tutia ascends and adheres in films to the ceramic rods. Al-Muqaddasī (*op. cit.*, p. 470) saw the "curious tall furnaces in the mountain villages" in Kirmān. They later also attracted the attention of Marco Polo when he visited the same area. Before the 16th century, *rūḥ al-tūtiya* (zinc) as a metal was known and was used with copper to form brass. It replaced *kharṣīnī* as the seventh metal. Abu 'l-Faḍl in the *Āʾīn-i Akbarī* gives several compositions employing *rūḥ al-tūtiya* (E. Wiedemann, *Aufsätze zur arabischen Wissenschaftsgeschichte*, I, Hildesheim 1970, p. 706).

Antimony was obtained from antimony sulphide (Sb_2S_3) and was one of the constituents of copper alloys. Arabic chemical books described this process.

Arsenic was unimportant as a metal. But we read a description of the "preparation of the mercury of arsenic sulphide". This was a preparation of metallic arsenic from its sulphides. We may note that these metals which were not among the original seven were given such names as *rūḥ al-tūtiya* or "mercury of arsenic sulphide". The same remark applies to mineral acids, which were not classified originally, and hence were attributed to their salts.

Copper and its alloys. Copper was usually obtained from the sulphide ores. It seldom occurred as oxides or carbonates. These latter ores required only the simple treatment of heating with charcoal, while the sulphides (*zājāt*) required roasting, smelting with fluxes and partial oxidation. An interesting discovery took place in Spain. The sulphide ores, on exposure to air in the presence of water, are oxidated to soluble sulphates (Singer *et al.*, *A History of Technology*, II, Oxford 1979, p. 11): "The Moors then found that if water containing copper sulphate is allowed to run over iron, pure copper is deposited and the iron dissolved. As iron was cheap and abundant in Spain, this discovery yielded an efficient method of recovering copper from sulphide ore, and direct mining of copper ore became less necessary".

Bronze (*ṣafr*, *isfīdrūy*) is an alloy of copper and tin. It was much used for plain kitchen wares and implements, and was the alloy upon which coppersmiths based most of their work. p. 971

Brass (*shabah*, *bīrinj*) is an alloy of copper and zinc. Zinc with copper forms an alloy which is stronger, harder and less malleable than pure copper. Various kinds of brasses are obtained by varying the zinc content. A 20 per cent brass simulates the colour of gold. When zinc was not known as a metal, copper was heated in a mixture of powdered zinc ore and charcoal. A proportion of the zinc formed in the vicinity of the copper was diffused into it by cementation. Later we read in the *Ā'īn-i Akbarī* about three qualities of brass with increasing amounts of zinc content: one is ductile in the cold state, the second is ductile when it is heated, and the third is not ductile but can be cast (Wiedemann, *Aufsätze*, I, pp. 706-707).

A cheaper quality of alloy was called by al-Bīrūnī *bitrūy* and by some authors *rūy*. This was a kind of bronze alloyed from copper and lead. It was also *shabah mufragh*, and was used for hardware.

Kharṣīnī was a metal (or alloy) which was listed as one of the seven metals of early Islamic alchemy. It was attributed to China. All Muslim writers said that it was not available and was extinct. A time came when it was replaced by zinc in the list of the seven metals. Some historians now think that *kharṣīnī* was zinc. Others believe that it was a cupro-nickel alloy. *Ṭaliqūn* was another alloy (or metal) of uncertain composition. This uncertainty existed since al-Bīrūnī's days, but the general opinion is that it was a kind of copper alloy.

5. Iron and steel

The importance of iron (*ḥadīd*) in Islamic civilisation is exemplified in the Sūrat al-Ḥadīd in the *Qur'ān* (LVII, 25):

> God sent iron down to earth, wherein is mighty power and many uses for mankind.

Indeed, iron was considered as essential as food and clothing and it was always a source of power. Even after the sword had ceased to be a major military weapon, the manufacturing of iron and steel in our days is a basis for industrial civilisation and hence for economic and military power.

We have seen that iron mines in the Islamic empire were spread from Spain in the west to Transoxania in the east and there were famous steel production centres like Herat, Bukhārā, Damascus, Yemen and Toledo. Iron and steel technology in the Islamic lands has had a long and flourishing history, associated with the sword until recent times, and the excellence of steel for these swords was represented by Damascus steel. As is the case with other major issues in the history of technology, the history of iron and steel, and the history of Damascus steel in particular, have been a source of controversy. Islamic iron technology was ignored and the important sources were not taken into consideration. At one extreme, some writers have alleged that Damascus steel was alien to Damascus and to Islamic lands.

In this vast subject, we can only give extracts from some of the major Arabic sources, the best authorities on this subject known until now being al-Kindī and al-Bīrūnī, both of whom had critical minds that rejected legends and subjected scientific knowledge to actual observation and testing. The second main group of sources is the alchemical and chemical treatises of Jābir, al-Rāzī, al-Jildakī and others. The third group is the military treatises which disussed the manufacture of steel for swords, its heat treatment and the care for its *firind*. But there are numerous other sources of information, all of which deserve attention.

Types of iron and steel. From studying some of these sources, we conclude that the following main kinds of iron and steel were utilised in Islamic metallurgical centres: (1) wrought iron (*narmāhin*); (2) cast iron (*dūṣ*); (3) meteoric steel (*shabūrqān*); and (4) manufactured steel (*fūlādh*).

Wrought iron (*narmāhin*) is soft and it is the "female". It could not be heat treated, but could be used where strength was not important, and it was used as a raw material for manufacturing steel (*fūlādh*).

Cast iron (*dūṣ*). It is very important to know that cast iron was produced, since this was not realised by historians of technology and was totally ignored. Al-Bīrūnī says in his *Jamāhīr* that *dūṣ* is the water of iron and that it is the liquid which flows during the melting and the extraction of metal from iron stones. Al-Rāzī defined *dūṣ* as water of iron. In a commentary by al-Jildakī on the *Kitāb al-Ḥadīd* by Jābir, we read the following description of the production of cast iron (Chester Beatty Library, MS 4121, fols 1a-2b):

> Chapter. Learn, brother, that it is your comrades who found (*yasbukūn*) iron in foundries (especially) made for that purpose after they have extracted it (*i.e.* the ore) from its mine as yellow earth intermingled with barely visible veins of iron. They place it in founding furnaces designed for smelting it. They install powerful bellows on all sides of them after having kneaded (*yaluttūn*) a little oil and alkali into the ore. Then fire is applied to it (*i.e.* the ore) together with cinders and wood. They blow upon it until it is molten and its entire substance (*jismuhu wa-jasaduhu*) is rid of that earth. Next, they cause it to drop through holes like (those of) strainers, (made in) the furnaces so that the molten iron is separated and is made into bars from that earth. Then they transport it to far lands and countries. People use it for making utilitarian things of which they have need.

The properties of this cast iron can be summarised from al-Bīrūnī's *Jamāhīr* as follows: (1) it is quick to flow like water when smelting iron ores; (2) it is hard and whitish-silvery in colour — its powder had sometimes a pinkish reflection; (3) it cannot be forged to make swords; (4) it does not resist blows — it shatters into pieces: "breakage and brittleness are characteristic of it"; and (5) it is mixed with wrought iron in crucibles for making steel. *Dūṣ* was sold as a raw material in the 15th century. We learn that there were at least two commercial brands, one from Iraq, the other from Iṣṭakhr.

Meteor steel (*shabūrqān*) is often mentioned in early Arabic literature, with the comment that this was a rare material.

Manufactured steel (*fūlādh*). *Fūlādh* was usually manufactured in the molten state. It was made, in some cases, from wrought iron bars by cementation. The iron bars were packed with charcoal and heated until they absorbed enough carbon. In the molten state, steel was made in Islamic lands by the following methods: (1) by decarbonisation of cast iron; (2) by carbonisation of wrought iron; and (3) by the fusion of a mixture of wrought iron and cast iron; here we obtain two qualities of steel, depending upon the degree of fusion.

We have given above a text from al-Jildakī commenting on Jābir describing the production of cast iron. Let us now give the rest
972 of the text to see how the rods of cast iron are utilised to produce steel by carbonisation:

> As for the steel-workers, they take the iron bars and put them into the founding-ovens (*masābik*) which they have, suited to their objectives, in the steel works. They install firing equipment (*akwār*) in them (*i.e.* the ovens) and blow fire upon it (*i.e.* the iron) for a long while until it becomes like gurgling water. They nourish it with glass, oil and alkali until light appears from it in the fire and it is purified of much of its blackness by intensive founding, night and day. They keep watching while it whirls for indications until they are sure of its suitability, and its lamps emit

> light. Thereupon, they pour it out through channels so that it comes out like running water. Then they allow it to solidify in the shape of bars or in holes made of clay fashioned like large crucibles. They take out of them refined steel in the shape of ostrich eggs, and they make swords from it, and helmets, lanceheads and all tools.

This refining of iron from its blackness is a decarbonisation process; already, the *Lisān al-ʿArab* states that steel (*fūlādh*) is refined iron.

The other method of producing molten steel in crucibles by carbonising wrought iron is described by al-Bīrūnī (*Jamāhīr*, p. 256):

> Mazyad b. ʿAlī, the Damascene blacksmith, (wrote) a book describing swords, specifications for which were included in al-Kindī's treatise. He commenced by dealing with the steel composition and the construction of the furnace (*kūr*) as well as with construction and design of crucibles, the description of (the varieties) of clay, and how to distinguish between them. Then he instructed that in each crucible five *raṭl*s of horseshoes should be placed, and their nails, which are made of *narmāhan* (Persian "soft iron"), as well as a weight of ten *dirham*s each of *rusukhtaj*, golden marcasite stone, and brittle magnesia. The crucibles are plastered with clay and placed inside the furnace (*kūr*). They are filled with charcoal and they (*i.e.* the crucibles) are blown upon with *rūmī* bellows, each having two operators, until it (*i.e.* the iron) melts and whirls. Bundles are added containing *ihlīlaj* (myrobalan), pomegranate rinds, salt (used in) dough and oyster shells (*aṣdāf al-luʾluʾ*, lit. "pearls shells"), in equal portions, and crushed, each bundle weighing forty *dirhams*. One (bundle) is thrown into each crucible; then it (*i.e.* the crucible) is blown upon violently for an hour. Next, they (*i.e.* the crucibles) are left to cool and the eggs are taken from the crucibles.

The third method of producing molten steel in crucibles from a mixture of cast iron and wrought iron was also described by al-Bīrūnī. This was the method of producing cast steel in Herat. Two qualities of steel can be obtained. One is obtained if the components are "melted equally so that they become united in the mixing operation and no component can be differentiated or seen independently". Al-Bīrūnī says that "such steel is suitable for files and similar tools". A second quality of steel is obtained if the degree of melting of wrought iron and cast iron is different for each substance "and thus the intermixture between both components is not complete, and their parts are shifted and thus each of their two colours can be seen by the naked eye and it is called *firind*".

Some of the above important texts were confirmed by observers and travellers who described the making of crucible steel in Bukhārā in the last century. In 1820 Anossoff, a Russian expert, was in Bukhārā and found that crucible steel was made by carburising wrought iron with charcoal and other organic matter. Later Massalski, another expert who was also in Bukhārā, wrote in 1841 that Damascus steel was made there from a mixture of wrought iron and cast iron, as al-Bīrūnī had reported 900 years earlier from Herat. Observers thought that these reports were conflicting, but it is now clear that there is no contradiction; steel was made from different materials, and the quality obtained varied with these materials and with other manufacturing conditions (P. Anossoff, in *Annuaire du Journal des Mines de Russie*, 1843, pp. 192-236; Massalski, in *ibid.*, pp. 297-308).

The Damascus sword and the *firind*. C. S. Smith (*A History of Metallography*, Chicago 1965, p. 14) noted that: "In comparison with the relative neglect of structure by the European metallurgist, the enjoyment and utilisation of it in the Orient is impressive. In the Orient, etching to display patterns depending on composition difference was in use contemporaneously with the European pattern-welded blade, and was thereafter continually developed to a high artistic level". The best achievement in this direction is the Damascus sword, which was made in all the Islamic centres and in India. Because of its excellence, its name was given later to all swords with a pattern. Islamic lands exported and imported steel and swords. They imported from India and they exported to it. Al-Bīrūnī mentions that steel eggs were cast in Herat and then sent to India. Al-Idrīsī says that iron was exported from the Maghrib to India. In general, we may say that Islamic lands and India formed one area of Damascus steel culture.

Patterned swords (with a *firind* or *jawhar*) were in use before Islam. Imru' al-Qays (d. *ca.* 540 A.D.) describes the *firind* of the sword as resembling the tracks of ants. Another poet, Aws b. Ḥajar, a contemporary of his, describes the blade of the sword as if it has a water whose wavy streaks are like a pond over whose surface the wind is gliding. In fact, in Arabic poetry the beauty of the sword with the *firind* was always a source of inspiration. Damascus steel was thus a speciality of the Islamic world and India for many centuries. As Smith (*op. cit.*, p. 14) notes: "The geographical distribution of these swords seems to have been practically coextensive with the Islamic faith, and they continued to be made well into the nineteenth century".

In Europe, steel was produced by hearth-carbonisation of wrought iron rods. To imitate the *firind* or pattern of the Damascus sword, they resorted to placing together strips of iron and steel and welding these together; but these imitations never matched the qualities of the true Damascus steel.

Cast iron was first produced in Europe in the 15th century A.D. and crucible steel in the 18th. But a steel comparable in quality to Damascus steel was still required. For more than 150 years, a large number of metallurgists in European countries carried out extensive research on Damascus steel, including eminent scientists like Faraday. Smith, in a chapter "European attempts to duplicate Damascus steel", says that these attempts failed to reproduce true Damascus steel; what happened later was that the "interest in the duplication of the blade declined as European steelmakers developed their own techniques and the introduction of Bessemer and Siemens processes gave homogeneous steel more adaptable to large-scale production".

p. 973 There were, however, some advantages behind this research, in that we came nearer to an understanding of the structure of Damascus steel. According to Smith, "Damascus blades are made of a very high carbon steel (about 1.5-2.0 per cent) and owe their beauty and their cutting qualities alike to the inherent structure of the cakes of steel from which they were forged". "The light portion contains numerous particles of iron carbide (cementite), while the dark areas are steel of normal carbon content (approximately eutectoid). The structure, of course, is clearly visible only after etching, which was done with a solution of some mineral sulphate." It is amazing how close this modern interpretation of the *firind* is to that of al-Bīrūnī.

Welded "Damascus" gun barrels and swords. Another beautiful technique which flourished in later centuries in Islamic lands and India was the welding technique for gun barrels and swords, which was quite different from that of cast Damascus steel, but which was also called "Damascus". According to Smith (*op. cit.*, p. 30), "the technique seems to have originated in the Near East in the sixteenth century, and such guns were a famous product of the Kashmir smiths in the early nineteenth century." The manufacture of barrels of this type in Europe started in the 18th century, and during the 18th and 19th centuries great efforts were exerted in Europe to use this technique for both gun barrels and blades.

In 1798 Nicholson made a Damascus-textured metal by compressing fillings of steel and wrought iron in a die, restriking the compact at a welding heat and forging it into a plate; Smith says that this is an interesting anticipation of modern power metallurgy. But according to al-Bīrūnī (*op. cit.*), a somewhat similar procedure was practised in the 10th or 11th century:

I was told by somebody who was in the land of Sind that

he sat by a smith who was making swords. He looked into them and found that their material was wrought iron (*narmāhin*), on which he was sprinkling a finely pulverised drug (*dawā'*) whose colour gave a reddish shade. He sprinkles and welds by driving the powder deep, then he takes it out and elongates it by forging and he sprinkles again and repeats the work several times. (That same person) said: Then I asked him what that was. He glanced at me derisively. Then I looked carefully into it and realised that it was cast iron (*dūṣ*) which he mixes with wrought iron (*narmāhin*) by forging to elongate and hammering (to drive in) to obtain a steel similar to that of the eggs that are obtained in Herat by melting.

6. Furnaces, crucibles and other equipment

We can learn much about Islamic metallurgical equipment from a study of the alchemical equipment. The *istinzāl* operation is mainly the smelting of ores to obtain metals. There was, according to al-Rāzī, "equipment for melting metals (*ajsād*) and stones" and there was "equipment for the further processing of these metals". Iron melting was given special attention in the alchemical apparatus. We can safely assume that the alchemical equipment was the pilot-plant size of the actual metallurgical equipment. Sometimes the details in construction were identical (see below). Iron production from ores was achieved by blast furnaces. Further research will reveal how the design of these furnaces had developed in Islamic lands. We know that cast iron was produced before the 10th century. Al-Jildakī (see above) gave a description of the melting process; historians have reported experiments on the casting of large field cannon from cast iron at the turn of the 16th century in Egypt; and at the end of the 18th century, a typical blast furnace in the Lebanese mountains was 16 spans (3.84 metres) high by 7 spans (1.68 metres) wide and was constructed from masonry. Twenty men worked on the site. Layers of fresh wood and iron ore were stacked in the furnace; two horizontal bellows gave the necessary blast of air, each being operated by one man. The air from the bellows was combined in one outlet, and iron that accumulated at the bottom was taken out in small amounts. This iron was decarbonised in several hearth furnaces and forges on the same site in order to produce wrought iron for making useful articles. About 450 kg a day were produced from this furnace; similar furnaces were in use in the Maghrib.

Crucible steel was melted in small crucibles. The text by al-Bīrūnī which was cited above on the making of crucible steel by a smith in Damascus shows that several crucibles were put in the furnace. Several bellows were used, each operated by two men. Such an installation was still used in 1840 in Bukhārā to make Damascus steel, and was of the same design as that described by Jābir ten centuries earlier.

Bibliography (in addition to works cited in the text): A. Y. al-Hassan, *Mining and Metallurgy*, in *idem* and D. R. Hill, *Islamic technology: An Illustrated History*, Cambridge 1986, pp. 233-260; A. Y. al-Hassan, "Iron and Steel Technology in Medieval Arabic Sources", in *Journal for the History of Arabic Science*, 2:1 (1978), pp. 31-43.

VIII

Hydraulic Machines*

There is ample evidence from written and archaeological sources for the widespread use in pre-Islamic times of all the main hydraulic machines, described below, in all the areas that were to form part of the Muslim world. The *shādūf* was known in ancient times. The *sāqiya*, although it did not become fully effective before the introduction of the pawl in the 4th or 5th century A.D., was known in Roman times. Both machines are still in use today. The noria (*nāʿūra*) and the vertical undershot mill-wheel are both described by Vitruvius, without any claim to originality (*On Architecture*, Loeb Classics, ed. F. Granger, 1970, I, pp. 303-307). Vertical mill-wheels were sometimes mounted on boats moored to the banks of rivers (N. Smith, *Man and Water*, London 1975, p. 140). The origins of the horizontal, vaned mill-wheel are still obscure: it may have been referred to by a Greek writer of the 1st century B.C., and it was in use in Ireland in the 7th century A.D. (Smith, *op. cit.*, p. 142). It is described in a Byzantine treatise, probably of the 7th century A.D., extant only in Arabic versions (Wiedemann, *Aufsätze*, II, pp. 50-56, see the Bibliography). Hand-operated force pumps were used by the Greeks and Romans; these had single vertical cylinders that were placed directly in the water without suction pipes (Vitruvius, *op. cit.*, pp. 311-312, quoting Ctesibius, who lived

p. 860 p. 861

*In this new reprint of the *Encyclopaedia of Islam* article a slightly simplified transliteration system for Arabic has been used and place names have been given in their anglicised form whenever possible. All references in square brackets are to other articles in the *Encyclopaedia of Islam*. References in the indexes of this book are to the original pagination, here given in the margins.

ca. 200 B.C.). The problem of the origins and diffusion of these machines is largely unresolved, but our chief concern here is that they were all in existence in the 7th century.

The *shādūf* is a simple machine consisting of a wooden beam pivoted on a raised fulcrum. At one end of the beam is a bucket, at the other end a counterweight. The bucket is dipped into the water, then the beam is rotated by means of the counterweight and the contents of the bucket are emptied into a cistern or supply channel. The flume-beam swape is a development of the *shādūf*. Instead of a solid beam, a channel is connected rigidly to the bucket; when this is raised the water runs through it into the outlet.

The *sāqiya* is more complex, and indeed has over two hundred component parts. It consists essentially of a large vertical wheel erected over the water supply on a horizontal axle. This wheel carries a chain-of-pots or a bucket chain. On the other end of its axle is a gear-wheel that engages a horizontal gear-wheel to which the driving bar is attached. The animal is harnessed to the free end of this bar, and as it walks in a circular path, the gears and the wheel carrying the chain-of-pots rotate. The pots dip in succession into the water and when they reach the top of their travel they empty into a channel. The noria (sometimes confused with the *sāqiya*) is a large wheel driven by water. It is mounted on a horizontal axle over a flowing stream so that the water strikes the paddles that are set around its perimeter. The water is raised in pots attached to its rim or in bucket-like compartments set into the rim. The large norias at Hama in Syria can still be seen today; the first known mention of them is by Aḥmad b. al-Ṭayyib in 271 H (884/5) (Suter, *Die Mathematiker und Astronomen der Araber und ihre Werke*, Leipzig 1900, p. 33).

The Vitruvian mill-wheel turns a vertical gear-wheel that meshes with a horizontal gear-wheel to which the driving shaft is attached. The horizontal vaned mill-wheel is fixed directly to the driving shaft; there are no gears. It cannot be mounted directly in the stream since the water must be directed on to its vanes from a pipe or channel.

There can be no doubt that all these machines were in continuous use in Islam from the early conquests until the introduction of modern technology. (As mentioned above, the *shādūf* and the *sāqiya* are still in use; they are cheaper and more easily maintained than motor-driven pumps.) The evidence comes from treatises on machines, references in the works of historians and geographers, and archaeological investigations. For a selection of the available evidence, the reader is referred to the Bibliography under Wiedemann, *Aufsätze*; Schiøler; Hill, *al-Jazarī* and *Banū Mūsā*. The remaining discussion will be confined to developments of particular importance in the history of technology.

Mills were used in Islam for other purposes beside the grinding of corn and other seeds, *e.g.* for crushing sugar cane and for sawing timber (A. Y. al-Hassan, *Taqī al-Dīn and Arabic Mechanical Engineering* (in Arabic), Aleppo 1976, p. 51, quoting al-Nuwayrī and Ibn ʿAsākir). This suggests that rotary motion was converted to reciprocating, probably by means of trip-hammers. More examples of similar applications may be discovered when a systematic study of the historical and geographical works is undertaken. Another area of interest is the use of the overshot mill-wheel, in which the water is conducted through a channel to the top of the wheel, which has bucket-like compartments around its rim. The overshot wheel works mainly by the weight of the water, whereas the Vitruvian one is operated by its force. In many conditions, the former is the more efficient of the two. Its use is recommended by a certain al-Murādī in a treatise composed in Andalusia in the 5th/11th century (D. R. Hill, "A Treatise on Machines", *Journal for the History of Arabic Science*, 1 (1977), pp. 33-46. In this paper the treatise was wrongly attributed to the well-known astronomer Ibn Muʿādh). Shams al-Dīn al-Dimashqī, d. 727 H (1327), describes a similar wheel in operation near Tabrīz (*Nukhbat al-dahr fi ajaib al-barr wa-al-bahr*, ed. M. A. F. Mehren, Leipzig 1923, p. 188; French transl. by *idem*, *Manuel de Cosmographie du Moyen Age*, Copenhagen 1874, pp. 254-255). The overshot wheel did not come into general use in the West until about the 14th century. Al-Jazarī [see AL-DJAZARĪ in Suppl.] often uses small overshot wheels in his devices, but these are usually scoop-wheels, a kind of primitive Pelton wheel, the scoops being fixed to the ends of spokes that radiate from a solid disc.

It is reasonable to infer that the scoop-wheels used by al-Jazarī were miniature versions of wheels in full-size machines, an inference that is strengthened by the fact that he uses such a wheel in one of his water-raising machines (Category V, Ch. 3). The visible part of this is a *sāqiya*, which is provided with a model cow to give the impression that this is the source of motive power. The actual power, however, is provided in a lower, concealed chamber and consists of a scoop-wheel and two gear-wheels. This system drives the vertical axle that passes up into the main chamber, where two further gear-wheels transmit the power to the chain-of-pots wheel. Such devices (without the model cow) were in everyday use. A similar machine was in continuous use on the River Yazīd above Damascus from the 13th century until about 1960 for water supply and irrigation. It was restored to working order by the staff and students of Aleppo University (al-Hassan, *op. cit.*, pp. 58-59).

Al-Jazarī describes four other water-raising machines in Category V. Chapters 1, 2 and 4 deal with flume-beam swapes operated by animal pwer through gear trains. The first two of these incorporate segmental gears. The earliest known occurrence of these gears in Europe is in the astronomical clock completed by Giovanni de' Dondi in 1365. The fifth machine is a slot-rod pump, driven through gears from a paddle wheel. It is remarkable for having two horizontally opposed cylinders and true suction pipes. Taqī al-Dīn (16th century) describes a similar pump, but equally remarkable is his six-cylinder "Monobloc" pump (al-Hassan, *op. cit.*, pp. 47-50). The vertical cylinders are fitted into a single wooden block which rests in the water. Delivery pipes lead out from the sides of the cylinders, near their tops, and are brought together into a single outlet. Each cylinder has a clack-valve at the bottom. The pistons are provided with weights at the top and lever-arms at the sides. The lever-arms are supported at fulcrums and their free ends extend inside the perimeter of a scoop-wheel. As the water strikes the scoop-wheel it rotates, the scoops bear down in succession on the lever-arms and the pistons rise and fall in continuous succession.

It should be apparent from the foregoing brief discussion that Islamic engineers were active in the construction and development of hydraulic machines for water-raising and power supply throughout the mediaeval period and beyond. Similar activity took place in p. 862
Europe, India and East Asia. Each region used the machines that were best suited to its needs, to the local hydraulic conditions, and to the available constructional materials.

Bibliography (in addition to works mentioned in the text): Two treatises are available in annotated English translations by D. R. Hill, *The Book of Knowledge of Ingenious Mechanical Devices by al-Jazarī*, Dordrecht 1974, see esp. pp. 32, 95-96, 179-198, 275; and *The Book of Ingenious Devices by the Banū Mūsà*, Dordrecht 1979, see pp. 222, 226, for horizontal water-wheels; there is an Arabic-English glossary that includes many terms used in hydraulic engineering; an Arabic edition of al-Jazarī, also with Arabic-English glossary, is now available, ed. A. Y. al-Hassan, Aleppo 1979; Abu ʿAbd Allāh al-Khwārizmī, *Mafātiḥ al-ʿulūm*, ed. van Vloten, pp. 246-255 (this section contains definitions and brief descriptions of parts used in hydraulic and other machines); E. Wiedemann, *Aufsätze zur arabischen Wissenschaftsgeschichte*, Hildesheim & New York 1970, see Index under *raḥā* and *nāʿūra*; Wiedemann and F. Hauser, "Über Vorrichtungen zum Heben von Wasser in der islamischen Welt", *Jahrbuch des Vereins Deutscher*

Ingenieurs, 8 (1918), pp. 121-154; T. Schiøler, *Roman and Islamic Water-Lifting Wheels*, Odense 1973 (deals mainly with the *sāqiya*); useful discussions on hydraulic machines in various regions, including Islam, occur in J. Needham, *Science and civilisation in China*, IV:2, Cambridge 1965 (see Index under "mills", "water-raising devices", "water-wheel", *etc.*); Lynn White Jr., *Medieval Technology and Social Change*, Oxford 1962, ch. 3; R. J. Forbes, *Studies in ancient technology*, Leiden 1955-64, II, pp. 1-79; N. Smith, *op. cit.*, ch. 11.

IX

Les œuvres de Héron et leur contexte historique.

En dehors de ses œuvres, nous ne savons rien de Héron et, jusqu'à une période très récente, l'époque à laquelle il vécut était l'objet d'une controverse. Aucune source littéraire antérieure à Pappus (300 après J.-C.) ne mentionne son nom, Pappus cite ses *Mécaniques* et Héron lui-même cite Archimède (mort en 212 avant J.-C.), ce qui nous fournit l'autre borne dans le temps. Les spécialistes ont donné différentes dates, depuis 150 avant J.-C. jusqu'à 250 après J.-C. Cependant, la question a été tranchée par O. Neugebauer qui remarqua qu'une éclipse de lune décrite par Héron dans sa *Dioptre* (chap. 35), comme ayant eu lieu le dixième jour avant l'équinoxe de printemps et ayant débuté à Alexandrie la cinquième heure de veille de la nuit, ne peut correspondre qu'à une éclipse de l'an 62 après J.-C. et à aucune autre dans les 500 années en question. Une date astronomique est très fiable, c'est pourquoi, si l'on admet que Héron a bien écrit la *Dioptre* (ce qu'on n'a pas de raison de mettre en doute), la controverse sur les dates de Héron ne présente plus aujourd'hui qu'un intérêt historique [1].

Les ouvrages suivants nous sont parvenus sous le nom de Héron : les *Automates*, le *Baroulkos*, les *Belopoiica* (ou machines de guerre), la *Catoptrique*, la *Chirobaliste*, les *Definitions*, la *Dioptre*, la *Geométrie*, les *Mécaniques, des Mesures*, les *Métriques*, les *Pneumatiques* et la *Stéréométrie* [2]. Héron lui même nous dit qu'il a écrit quatre livres

1. O. Neugebauer, « Über eine Methode zur Distanzbestimmung Alexandria — Rom bei Heron, *Kongelige Dankske Videnskabernes Selskab Skifter*, 26, n° 2 (1938), 21-24.

2. Tous les ouvrages d'Héron, à l'exception des *Belopoiica*, ont été publiés, avec traduction en allemand, sous le titre : *Heronis Alexandrini Opera quae supersunt omnia*, 5 vol. ed. W. Schmidt (Leipzig, 1899-1914). Le volume 2 (1900) contient les *Mécaniques*, avec le texte arabe et une traduction par L. Nix. L'édition utilisée dans cette publication, de Carra de Vaux, est parue pour la première fois dans le *Journal asiatique*, neuvième série (1893) vol. 1, 286-472, vol. 2, 152-269, 420-514, sous le titre « Les Mécaniques ou l'Élévateur de Héron d'Alexandrie, sur la version arabe de Qostā Ibn Lūqā ».

Les *Pneumatiques* sont également disponibles en traduction anglaise : The *Pneuma-*

sur les horloges à eau, mais ceux-ci sont perdus [3]. Il semble que tous les livres techniques, à l'exception de la *Chirobaliste*, aient été écrits par Héron; parmi les ouvrages mathématiques, on ne peut lui attribuer que les *Définitions* et les *Métriques*. Les autres sont des livres scolaires byzantins comportant tellement d'additions qu'il est impossible d'identifier le contenu qui revient authentiquement à Héron [4]. Les *Mécaniques* ne sont disponibles qu'en arabe, la *Catoptrique* seulement en latin; les autres sont en grec.

Le plus long de ces livres est les *Pneumatiques* qui comporte 80 chapitres. Les sept premiers chapitres, tirés pour la plupart d'entre eux des *Pneumatiques* de Philon (voir plus bas), décrivent des expériences montrant que l'air est un corps, qu'il empêche l'eau de pénétrer dans un vase * à moins qu'il n'y ait une ouverture, et qu'il maintient l'eau dedans s'il ne peut y pénétrer. Les autres chapitres sont essentielement consacrés à divers types d'automates dont les effets sont obtenus en effectuant de petites variations dans les pressions hydrostatiques et aérostatiques. Certains d'entre eux sont tirés directement de Philon mais beaucoup sont inventés par Héron lui-même. On trouve quelques machines utiles, par exemple une pompe à feu et un orgue hydraulique, mais la plupart sont ce qu'on peut appeler des jouets : des vases truqués qui versent du vin ou de l'eau séparément ou dans des proportions constantes, des oiseaux qui chantent et des trompettes qui sonnent, des poupées qui s'animent quand un feu est allumé sur un autel, des animaux qui boivent quand on leur verse de l'eau et ainsi de suite. Certains utilisent l'air chaud ou la vapeur comme force motrice, mais on ne saurait en déduire que Héron ait anticipé en quelque façon l'invention de la machine à vapeur.

tics of Hero of Alexandria, facsimile de l'édition Woodcroft de 1851, avec une introduction de Marie Boas Hall (London and New York, 1971).

Pour les *Belopoiica* voir E. W. Marsden, *Greek and Roman Artillery – Technical Treatises* (Oxford, 1971), 17-60.

Les études pénétrantes et très complètes de A. G. Drachman demeurent encore les meilleurs travaux secondaires sur l'ancienne mécanique. La critique des *Mécaniques* de Héron qui accompagne cette édition est : A. G. Drachmann (1), *The Mechanical Technology of Greek and Roman Antiquity* (Copenhagen, 1963), 22-140. Nous avons également : A. G. Drachman (2), « Hero of Alexandria » dans le *Dictionary of Scientific Biography*, vol. 6 (1972), 310-314; et A. G. Drachman (3), *Ktesibios, Philon and Heron a Study in Ancient Pneumatics, Acta Historica Scientiarum Naturalism et Medicinalium* (Copenhagen, 1948).

Ces trois ouvrages sont la source de la plupart des informations données dans la première partie de cette introduction.

3. Schmidt, *Heronis Alexandrini Opera...*, vol. 1, 506-7.

4. Michael A. Mahoney, « Hero of Alexandria : mathematics », *Dictionary of Scientific Biography*, vol. 6 (1972), 314-5.

* N.d.T. : un vase renversé.

Par certains côtés, les *Automates* ou le *Théâtre d'automates* ressemble aux *Pneumatiques*. Il décrit deux genres de représentations scéniques, l'une mobile et l'autre fixe, toutes les deux fonctionnant sans aucune manipulation. La première s'anime toute seule devant les spectateurs et montre un temple dans lequel un feu est allumé sur un autel. Le dieu Dionysos verse une libation tandis que des bacchantes dansent en rond autour de lui au son des trompettes et des tambours. Après la représentation le théâtre se ferme. Le théâtre fixe ouvre et ferme ses portes sur la représentation du mythe de Nauplius. Les chantiers navals sont en activité, les bateaux sont lancés et traversent une mer dans laquelle évoluent des dauphins, Nauplius allume le signal trompeur pour les égarer, le navire fait naufrage et Athéna anéantit l'intraitable Ajax par le feu et le tonnerre. La force motrice, dans les deux cas, était un lourd poids de plomb reposant sur un lit de grains de millet qui s'échappaient par un trou. Le poids était attaché par une corde à un axe dont la rotation entraînait tous les mouvements au moyen de cordelettes et de tambours. Le système est significatif en ce qu'il met en pratique un contrôle du mouvement de descente du poids et qu'il constitue ainsi une sorte d'échappement, certes primitif.

La *Dioptre* contient la description d'un instrument topographique à deux usages – un niveau et un théodolite – avec des instructions pour son emploi dans un certains nombre de tâches. L'ouvrage contient également la description d'un odomètre actionné par la roue d'un char et utilisé pour mesurer des distances en roulant lentement le long d'une route horizontale; le système de comptage comporte des roues d'engrenage. Le *Baroulkos* est le chapitre 37 de cet ouvrage et figure aussi comme chapitre 1 du livre 1 des *Mécaniques* (voir le commentaire).

Le *Belopoiica* (traité des machines de guerre) contient la description des *gastraphètes*, une sorte d'arbalète, et de deux catapultes fonctionnant à l'aide de treuils; deux faisceaux de nerfs fournissent la force élastique pour propulser la flèche, le trait ou la pierre. Les descriptions sont assez détaillées et contiennent des formules permettant de calculer le diamètre des nerfs pour des poids donnés de projectiles. Héron nous dit que ces précisions ont été obtenues auprès des constructeurs après le perfectionnement de leurs dessins grâce à des tests expérimentaux répétés [5]. Ces catapultes ressemblent à celles décrites par Philon et Vitruve.

La *Catoptrique* traite des miroirs, aussi bien plans que courbes, et donne la théorie de la réflexion. Elle contient également des

5. Marsden, 38-9.

instructions sur la construction de miroirs pour différents usages et sur la manière de les disposer pour obtenir des illusions optiques.

Les *Métriques* constitue le plus important des ouvrages mathématiques de Héron; la découverte à Istanbul, en 1896, d'une copie manuscrite, unique, attira pour la première fois l'attention des spécialistes sur elles. Héron s'y occupe essentiellement de mathématiques appliquées, ce qui entraîne la réduction des résultats les plus remarquables de la géométrie classique grecque à de simples formules applicables. Dans le livre 1, proposition 1.8, on trouve la célèbre formule dite 'de Héron' pour déterminer l'aire d'un triangle à partir de ses trois côtés : $A = \sqrt{s(s\text{-}a)(s\text{-}b)(s\text{-}c)}$ (elle est en réalité déduite d'Archimède). Les autres ouvrages mathématiques furent réédités à plusieurs reprises après l'époque de Héron et adaptés pour être utilisés comme manuels scolaires. Ceci était probablement conforme aux intentions propres de Héron, ainsi que le montrent ses *Définitions* et ses *Commentaires des éléments d'Euclide* qui témoignent tous les deux d'une préoccupation pédagogique évidente [6].

Pendant au moins trois siècles avant l'époque de Héron, il y avait eu une tradition continue de pratique de la mécanique et de la technologie mécanique dans le monde hellénistique, particulièrement en Égypte. Les spécialistes modernes se sont donné beaucoup de mal pour identifier les sources de Héron sur les divers sujets auxquels il s'est intéressé. Nous allons résumer les travaux de ses prédécesseurs dans un ordre à peu près chronologique.

Aristote vécut de 384 à 322 avant J.-C. Bien que Héron ne le mentionne pas, ses 'Problèmes mécaniques' figurent dans le livre 2 des *Mécaniques* de Héron. On admet généralement que ces problèmes ne sont pas d'Aristote lui-même mais qu'ils furent composés au sein de l'école péripatéticienne, probablement du temps de Straton, qui fut à la tête de l'école de 287 à 268 avant J.-C. On n'a pas de preuve d'une influence directe d'Aristote sur le travail pratique de Héron.

Ctésibios (ou Ktésibios) fut probablement le plus important des premiers ingénieurs d'Alexandrie, mais aucun de ses travaux n'a survécu et nous ne le connaissons que par les références de Vitruve à ses inventions. On n'est même pas certain de ses dates, mais le plus probable est qu'il vécut de 300 à 230 avant J.-C. Selon Vitruve, il était le fils d'un barbier et naquit à Alexandrie. Vitruve rapporte qu'il fut le premier à étudier les effets de la compression de l'air (l. 9, chap. 8) et

6. Voir note 4.
7. Drachmann (3), 1-3.
8. Donald R. Hill, *On the Construction of Water-clocks*, Occasional Paper No. 4 (Turner & Devereux, London, 1976), 7-9.

lui attribue l'invention de l'horloge hydraulique monumentale (9.8), de la pompe aspirante et foulante et de toute une collection d'automates (10.7). Vitruve mentionne ses *Commentaires* (aujourd'hui perdus) et conseille au lecteur de s'y reporter pour de plus amples informations sur son travail (10.7).

Le grand savant et mathématicien Archimède mourut en 212 avant J.-C. à l'âge de 75 ans. Tous les ouvrages qui subsistent de lui sont seulement des ouvrages théoriques mais on sait qu'il a écrit un unique ouvrage de mécanique pratique sur la construction d'un planétarium sphérique. On lui attribue l'invention de la vis à eau qui porte son nom, de la poulie composée, de machines à soulever des poids et de machines de guerre. Il n'y a pas de raison de douter qu'il les ait réellement inventées. Il existe en outre un traité sur la construction des horloges hydrauliques, disponible seulement en manuscrits arabes. Bien que ce traité porte le nom d'Archimède, il contient certainement des additions byzantines et islamiques. Cependant, les auteurs arabes sont unanimes pour attribuer à Archimède l'invention mécanique hydraulique de base, un système de contrôle à feed-back perfectionné, ainsi que certains des automates associés. Ici encore, on n'a pas de preuve que cette attribution soit erronée.

La période féconde de Philon de Byzance se situe probablement dans le dernier tiers du troisième siècle avant J.-C, Il écrivit un manuel d'artillerie et les *Pneumatiques*[9]. La seule version plus ou moins complète de ce dernier ouvrage est en arabe ; il existe un fragment en latin qui est une copie postérieure, tirée du texte arabe. Il faut être prudent en évaluant l'héritage de Philon chez ses successeurs, car le traité tel qu'il nous est parvenu contient certainement des additions islamiques. Quoiqu'il en soit, l'essentiel de l'ouvrage est de la main de Philon et l'on peut omettre les chapitres douteux pour faire l'évaluation en question. Il est donc possible d'identifier un certain nombre d'idées que Héron a empruntées directement à Philon. Cependant, les *Pneumatiques* de Héron ne sont en aucune façon tirées pour l'essentiel de son prédécesseur et elles contiennent beaucoup d'innovations qui dénotent une grande ingéniosité.

Vitruve écrivit son *De Architectura* sous le règne d'Auguste, aux envitons de 25 avant J.-C.[10]. Il s'agit d'un manuel pratique à l'intention des architectes et des géomètres et il contient des

9. Voir Marsden 7-8 pour les dates de Héron, 105-184 pour l'ouvrage d'artillerie. Le texte arabe des *Pneumatiques* fut édité par Carra de Vaux avec une traduction française : « Le Livre des Appareils Pneumatiques et des Machines Hydrauliques par Philon de Byzance », *Paris Académie des Inscriptions et Belles Lettres*, 38 (1903), pt. 1.

10. Vitruve, *Les dix livres d'architectures*, trad. franç. de Claude Perrault, revue par A. Dalmas, Paris : Les Libraires Associés, 1965.

informations sur les méthodes de construction et les matériaux,les relevés topographiques, les horloges à eau et les automates, l'approvisionnement en eau, des machines pour élever de l'eau, des grues et des machines de siège. Vitruve dit clairement que son travail se base sur les résultats pratiques de ses prédécesseurs, principalement les Grecs. Ses données provenaient à la fois de sources littéraires et de l'examen des constructions existantes. Un grand nombre des auteurs auxquels il fait référence ne nous sont connus que par lui, mais il mentionne bien à la fois Archimède et Philon. On trouve très peu de discussion théorique dans son ouvrage qui résume les réalisations des architectes et des ingénieurs jusqu'à sa propre époque.

L'époque hellénistique et classique a connu une activité de haut niveau dans diverses branches des sciences de l'ingénieur. Les principaux progrès concernaient la construction de batiments, les routes, la topographie, l'ingénierie hydraulique, les élévateurs d'eau, les moulins, la fabrication de mécanismes à échelle réduite (horloges à eau, automates et instruments) et la technologie militaire. À l'évidence, tous les auteurs anciens énumérés plus haut, à l'exception d'Aristote, s'intéressaient dans une large mesure à l'invention, aux améliorations et à la diffusion de la connaissance de tout cet ensemble de sujets techniques. C'est davantage le besoin de s'appliquer à ces sujets pratiques pour répondre à la demande des gouvernants et de la société que les penchants propres des auteurs qui ont déterminé leurs sujets d'étude.

Au XIXe siècle, lorsque Héron n'était connu que par les *Pneumatiques*, on le considérait comme un 'simple artisan' n'ayant pas beaucoup de connaissance de la science ni des mathématiques [11]. La publication des *Mécaniques* et des *Métriques* entraîna la révision radicale de cette opinion bien que les *Peumatiques* eux-mêmes témoignent d'un degré élevé de compétence technique. Nous ne pouvons pas, bien entendu, savoir comment Héron lui-même considérait les divers aspects de son œuvre mais nous n'avons aucune raison de supposer qu'il jugeait les *Mécaniques* comme étant d'une espèce supérieure aux *Pneumatiques*. Nous devrions être prudents quand nous portons des jugements avec des critères modernes sur les œuvres de Héron (ou de ses prédécesseurs et de ses successeurs dans ce genre de travail); par exemple, en donnant une valeur supérieure aux travaux mathématiques ou scientifiques par rapport aux travaux de type pratique. Il est presque certain que de telles distinctions

11. Par exemple : H. Diels, « Über das physikalische System des Straton », in *Sitzungberichte der K. preuss. Akad. der Wissenschaften zu Berlin*, n° 9 (1893), 110, n. 3.

n'auraient eu aucune signification pour les anciens qui n'auraient pas compris (n'en déplaise à Platon) que l'on pût étiqueter et juger leur travail de cette manière.

Leur utilisation de la théorie était tout à fait différente de la nôtre. Ils suivaient rarement la séquence : hypothèse – expérience – loi scientifique. C'était plutôt la théorie qui surgissait de la pratique. Un de leurs objectifs, comme dans le cas des catapultes de Héron, était de déduire les formules à partir des machines mises au point sur le terrain et d'enregistrer ces formules dans des manuels pour aider les ingénieurs des chantiers. Le second objectif est moins évident, puisqu'il aboutissait à des analyses spéculatives des structures existantes. Par exemple, les chapitres 25 à 30 des *Mécaniques* de Héron, livre 1, composés à partir d'un ouvrage disparu d'Archimède, sont une discussion sur les réactions d'un alignement de colonnes sous le poids d'un mur, ou sur la répartition de poids d'une poutre de charpente sur les épaules d'un groupe d'hommes. Il se trouve que ces résultats sont erronés puisque des problèmes de ce type sont indéterminés en Statique, mais dans tous les cas, ils ne sont d'aucun intérêt pratique tels quels. Aucun entrepreneur ne songerait à faire des colonnes (et leurs fondations) plus légère à l'extrémité d'une rangée simplement parce qu'elles sont moins chargées que les autres – d'autant plus qu'elles ont souvent à supporter le poids d'un mur transversal. Les ouvriers ont généralement une assez bonne idée, d'après leur expérience antérieure, du nombre d'entre eux qui est nécessaire pour soulever un poids donné ; s'ils le trouvent trop lourd pour eux, ils font simplement appel à d'autres hommes pour les aider. De telles spéculations, bien sûr, ne sont aucunement sans valeur – elles représentent les premiers pas tentés vers le développement de la Théorie des Structures. Ce n'est qu'à une époque récente, cependant, que les ingénieurs ont pu utiliser cette théorie comme un outil effectif. La théorie et la pratique étaient donc des concepts très différents de ce qu'ils sont devenus à l'époque moderne, et, aujourd'hui encore, la part des 'sciences appliquées' dans l'ingénierie est souvent surestimée.

Un grand nombre des machines et des techniques de l'antiquité avaient été découvertes avant l'époque de Héron, en particulier les inventions de Ctésibios et celles d'Archimède. Cependant, Héron fut lui-même un innovateur comme le prouve, par exemple, la machine thermique dans les *Pneumatiques* et l'instrument topographique dans la *Dioptre*. La place manque ici pour évaluer ce qui, dans son œuvre, est travail original et ce qui est emprunté. En ce qui concerne les *Pneumatiques*, cette question a été traitée par Drachmann ; il est arrivé à la conclusion que Héron emprunta certains de ses appareils à des auteurs antérieurs (particulièrement à Philon), d'autres à des machines qui lui étaient familières, tandis que d'autres furent sa propre invention. Drachmann reconnaît cependant qu'il reste une grande marge d'incertitude au sujet des sources de Héron [12]. Son œuvre est particulièrement intéressante pour nous en raion de la part importante qui en a survécu. Il est ainsi, de loin, notre source la plus importante pour la théorie et la pratique de la technologie mécanique dans la période hellénistique et dans celle des anciens Romains.

Nous n'avons pas d'indication directe que quelqu'un ait connu et utilisé les travaux de Héron pendant les huit premiers siècles qui ont suivi son époque mais, comme il est probable que ses écrits mathématiques ont servi de base pour les manuels scolaires byzantins, nous pouvons supposer que ses écrits techniques ont été utilisés de la même façon. Il est certain que la tradition mécanique hellénistique s'est poursuivie dans la période byzantine. L'horloge hydraulique monumentale équipée du mécanisme hydraulique d'Archimède fut transmise sous les Sassanides à la Perse, où les automates furent modifiés, mais non le mécanisme de base. Le modèle transformé revint de nouveau à Byzance où, par exemple, une horloge de ce type fut construite, à Gaza, au XIe siècle, par un artisan inconnu [13]. La découverte récente d'un sablier byzantin portatif avec commande d'un calendrier datant d'environ 500 après J.-C., indique que la technique de la fabrication de l'instrument avait été préservée [14].

La résurgence des travaux de Héron dans la période islamique laisse clairement à penser qu'ils avaient continué de circuler après sa mort. Nous avons, bien sûr, la traduction des *Mécaniques* due à Qusṭā b. Lūqā vers la fin du IXe siècle, et il est également fait mention de Héron dans les ouvrages d'auteurs arabes. L'historien égyptien al-Kindī (897-961) rapporte que Héron fut un géomètre et un géomètre-expert, qu'il écrivit sur la géographie et la pneumatique et qu'il fabriqua des horloges et d'autres instruments pour mesurer le temps [15], Ibn al-Nadīm (env. 936-995) écrit que l'on trouve parmi ses ouvrages : un livre sur la résolution des incertitudes d'Euclide; un ouvrage sur l'astrolabe ; l'élévation des poids ; la puissance de la vapeur ; des objets qui se meuvent d'eux-mêmes [16]. Ibn al-Qifṭī (1172-1248) mentionne

12. Drachmann (3), 80-84.
13. Donald R. Hill, *Arabic Water-clocks*, Institute for the History of Arabic Science, (Aleppo, 1981), 13.
14. J. V. Field & M. T. Wright, « Gears from the Byzantines », *Annals of Science*, 42 (1985), 87-138.
15. E. Wiedemann, *Aufsätze zur Arabischen Wissenschaftsgeschichte*, 2 vol. (Olms, Hildesheim, 1970), vol. 1, 69-70.
16. *The Fihrist of Ibn al-Nadīm*, ed. et trad. Bayard Dodge, 2 vol. (New York and London, 1970), vol. 2, 642, 672.

ses *Pneumatiques* et son livre sur la résolution des difficultés rencontrées dans l'œuvre d'Euclide [17]. Il est donc presque certain qu'il y a eu des traductions arabes de ses ouvrages, bien que la seule à avoir survécu, à notre connaissance, est celle des *Mécaniques*. La meilleure preuve de l'existence des *Pneumatiques* en arabe est le *Livre des dispositifs ingénieux* par les Banū Mūsā, trois frères qui vécurent à Bagdad au IXe siècle [18]. Environ 25 sur les 100 dispositifs qu'ils décrivent sont tirés directement ou avec des modifications des *Pneumatiques* de Philon ou de Héron, ou des deux à la fois. Leur travail présente cependant des progrès nets par rapport à celui de leurs prédécesseurs, en particulier dans leur utilisation des contrôles automatiques. L'ouvrage des Banū Mūsā fut utilisé par des ingénieurs islamiques plus tardifs, y compris par le plus grand d'entre eux, al-Jazarī, qui acheva son livre des machines en 1206. Leur travail était bien connu en Islam tout au long du Moyen Âge [20]. Aux environs du Xe siècle, la science des ingénieurs mécaniciens islamiques se perpétuait elle-même ; les ingénieurs pouvaient prendre comme sources les ingénieurs musulmans antérieurs et n'avaient généralement pas besoin de se référer aux auteurs hellénistiques dont l'héritage fut intégré, malgré cela, à l'ingénierie islamique. Pour donner un exemple, al-Jazarī utilisa, en reconnaissant ses sources, le mécanisme hydraulique d'Archimède, dans deux de ses horloges hydrauliques [21].

Il est plus difficile de suivre la diffusion des travaux mathématiques de Héron et des parties théoriques des *Mécaniques* à l'intérieur du monde musulman. Au IXe siècle, pendant la période de traduction et des premiers travaux, il y eut des traductions des ouvrages d'Euclide, d'Apollonios, de Ménélaos, de Pappus, d'Archimède et des *Siddhantas* indiens [22]. Héron ne représente donc que l'une des diverses sources de l'épanouissement ultérieur de la géométrie et de la trigonométrie islamique et de leurs applications à l'astronomie et à la mécanique.

Quand on en vient à étudier l'influence éventuelle de l'œuvre de

17. Ibn al-Qitftī, *Akhbār al-ʿUlamā'*..., (Cairo A. H., 1326), 53.
18. Texte arabe : *Kitāb al Ḥiyal*, ed. A. Y. Hassan, Institute for the History of Arabic Science (Aleppo, 1981). Traduction anglaise annotée : *The Book of Ingenious Devices*, Donald R. Hill (Reidel, Dordrecht, 1979).
19. Texte arabe : *Al-Jāmiʿ bain al-ʿilm wal-ʿamal al-nāfiʿ fī ṣināʿat al-ḥiyal*, ed. A. Y. Hassan, Institute for the History of Arabic Science, (Aleppo, 1979). Traduction anglaise annotée : *The Book of Knowledge of Ingenious Mechanical Devices*, Donald R. Hill (Reidel, Dordrecht, 1974).
20. Les principaux dictionnaires biographiques arabes ont des articles sur les Banū Mūsā. C'est seulement dans le dernier quart du XIVe siècle que le grand historien Ibn Khaldūn a vanté le *Livre des appareils ingénieux* et dit qu'il était très largement connu (*Muqaddima*, trad. franç. de V. Monteil, Paris : Sindbad, 1967-1968, tome 3, p. 1062).
21. Al-Jazarī/Hill, 17.
22. M. Souissi, « 'Ilm al-Handasa » *Encyclopédie de l'Islam*, Suppl. pp. 411-415.

Héron sur la technologie mécanique en Europe, les problèmes se compliquent. Étant donné que l'on ne connaît qu'une version arabe des *Mécaniques* et un unique manuscrit grec des *Métriques*, on doit supposer que ces ouvrages n'étaient pas bien connus dans l'Europe médiévale. Le développement de la science européenne a reçu une influence prédominante des ouvrages d'Aristote et de ses commentateurs arabes, à laquelle s'ajoutent celle des textes grecs d'autres savants, qui ont survécu, et celle des travaux arabes originaux traduits en latin.

D'un autre côté, les *Pneumatiques* de Héron furent très lus en Europe pendant le Moyen Âge; on en a trouvé plus de 100 manuscrits [23]. La première histoire des automates en Europe a l'allure de contes légendaires qui reflètent probablement une vague connaissance des automates hellénistiques/arabes. Entrent dans cette catégorie les contes extraordinaires qui attribuent des pouvoirs magiques au poète Virgile, contes mis en circulation pour la première fois autour de 1159 par John de Salisbury, dans son *Policraticus*. Les histoires qui dotent les philosophes de pouvoirs magiques sont d'une veine semblable. On trouve des automates simples dans les *Carnets* de Villard de Honnecourt (env. 1254), mais ils sont très grossiers en comparaison des dispositifs perfectionnés de Héron ou des Banū Mūsā. Il existe peu d'indications de la fabrication d'automates compliqués avant le Moyen Âge tardif, lorsqu'apparaissent des personnages de Jaquemart sur les horloges hydrauliques puis sur les horloges mécaniques [24].

On ne peut mettre en doute que le développement de la technologie des machines en Europe, particulièrement dans le domaine des contrôles automatiques et des mécanismes délicats, ait reçu une forte impulsion des idées provenant de l'Europe du Sud et du Moyen Orient. En l'état actuel de nos connaissances, il n'est pas possible d'établir avec précision la manière dont ces idées ont pénétré l'Europe du Nord-Ouest, et il est même peu probable que cette transmission ait suivi un seul chemin. Il peut s'être agi d'une diffusion à partir de l'Islam à travers l'Espagne et ailleurs aussi bien que d'une diffusion à partir de Byzance, ou d'une conservation des idées répandues dans l'Empire romain qui se seraient disséminées plus tardivement. De toutes façons, que ce soit directement ou indirectement, les idées de Héron furent un élément important de ces diffusions.

Nous devrions nous intéresser à la panoplie complète des techniques

23. Drachmann (3), 312.
24. Donald R. Hill, *A History of Engineering in Classical and Medieval Times* (Croom Helm, London, 1984), 206.

que recèlent les machines hellénistiques et islamiques. Indéniablement beaucoup d'entre elles étaient seulement destinées à l'amusement, mais les jeux eux-mêmes, les jeux d'ordinateurs par exemple, peuvent nécessiter la mise en œuvre d'une technologie avancée. D'autre part, l'horloge mécanique est une des inventions les plus importantes de l'histoire de la technologie mécanique, et on peut montrer que tous ses composants, à l'exception de l'échappement à verge, se trouvent dans les machines de Héron et dans celles de l'Islam : trains d'engrenages ; poids descendant avec une vitesse contrôlée ; mécanismes de sonnerie aussi bien que les automates célestes et biologiques. Nous ne pouvons pas ignorer non plus l'importance des automates — seuls ou associés avec des pendules — en tant que symptomatiques d'un besoin, profondément enraciné dans l'homme, de représenter l'univers par des moyens mécaniques. Ce besoin mécaniste a conduit à la fabrication d'automates, et une grande part de notre technologie a résulté des automates, en particulier la partie comprenant les mécanismes fins et l'instrumentation scientifique [25]. Les ouvrages de Héron peuvent donc être considérés comme une étape importante dans le développement de la technologie mécanique.

Les *Mécaniques*

Dans l'édition de Carra de Vaux

Il faut ajouter quelques explications supplémentaires au commentaire du Professeur Drachmann :

Pour aider le lecteur, les références aux pages du texte arabe et de la traduction française sont données à l'intérieur du commentaire, par exemple 67, 4-9/112, 7-15 signifie : page 67 du texte arabe, ligne 4 à 9, correspondant à la page 112 de la traduction française, (dans la numérotation de l'édition originale, en haut de la page) lignes 7 à 15.

Alors que Carra de Vaux n'a utilisé qu'un seul manuscrit, celui de Leyde (L chez Drachmann), Nix en a utilisé trois autres : une copie à la main d'un manuscrit du Caire (C) ; un manuscrit du British Museum (B) recueilli pour lui par Carra de Vaux ; et un manuscrit de Constantinople (K) recueilli pour lui par Carra de Vaux et par Salih Zéhy Bey.

25. Derek de Solla Price, « Automata and the Origins of Mechanism and Mechanistic Philosophy », *Technology & Culture*, V, 1 (1964), 9-23.

Les figures des éditions de Carra de Vaux et de Nix ont été redessinées pour montrer à quoi elles ressemblaient – selon les éditeurs. Le lecteur n'a aucun moyen de savoir à quoi elles ressemblaient réellement, ni même s'il y avait vraiment une figure dans le texte, pour un chapitre donné. Dans les deux cas, on a utilisé sur les figures les lettres grecques, elles correspondent généralement, mais pas toujours, aux lettres équivalentes du texte. Cette méthode n'a rien de très recommandable. Drachmann a utilisé des photocopies de deux des manuscrits pour les figures, qu'il a recopiées à la main, en ajoutant des lettres romaines en correspondance avec les lettres arabes. Les transcriptions sont celle de la ' Draft ISO Recommendation 353 ', sauf pour la lettre ' ayn ', qu'il a traduit par O sur les dessins, mais normalement dans le texte.

Dans le chapitre 1 du livre 1 et dans le chapitre 29 du livre 2, le mot ' manivelle ' a été utilisé pour la poignée qui fait tourner un axe. La manivelle était inconnue en Asie de l'Ouest comme en Europe jusqu'au IXe siècle après J.-C. La traduction aurait dû être simplement ' poignée '.

Drachmann ne discute pas les chapitres 24 à 34, qui sont tirés de trois ouvrages perdus d'Archimède. Le chapitre 24 est une discussion générale sur les centres de gravité. Les chapitres 25 à 30 portent sur les réactions de colonnes sous le poids d'un mur ou d'une pièce de charpente transportée par un groupe d'hommes, le chapitre 31 sur les poids supportés par une seule poutre porteuse, et les chapitres 32 à 34 sur la balance. Le texte ne présente pas de difficultés mais, comme on l'a souligné plus haut, les chapitres 25 à 30 tentent de traiter de poutres continues par les méthodes de la Statique alors que de telles structures sont statiquement indéterminées.

BIBLIOGRAPHIE COMPLÉMENTAIRE

BEAUJEU J. : « La littérature technique des Grecs et des Romains », in : *Actes du Congrès Guillaume Budé 1948*, pp. 21-88.

GILLE B. : *Les mécaniciens grecs*, Paris : Seuil, 1980.

Histoire générale des techniques (sous la direction de Fr. Daumas), tome 1 : « Les origines de la civilisation technique », Paris : PUF, 1962.

REY A. : *L'apogée de la science technique grecque* (2 vol.), Paris : Albin Michel, 1946-1948.

X

CONSTRUCTION OF A FLUTING MACHINE BY APOLLONIUS THE CARPENTER

Kamal Shehadeh, Donald R. Hill and Richard Lorch

1. *Introduction*

The longer treatise presented in this article exists in four manuscripts. Three of these are virtually complete and appear in codices that also contain a treatise on the construction of water-clocks attributed to Archimedes[1]. These codices are:

1. British Library, London, Add. 23,391, folios 21r-25r. Referred to as 'London'. Siglum *N*.
2. Bibliothèque Nationale, Paris, Catalogue by G. de Slane, No. 2468, p. 437, folios 39v-42v. Referred to as 'Paris'. Siglum *S*.
3. Public Library, New York, Indo Persian, Spencer Collection, MS 2, *Risāla-i Ḥakīm Muḥammad*, folios 18r-20v. Referred to as 'New York'. Siglum *K*.

[1] See Hill (1976). This is an annotated translation of the 'Archimedes' treatise; the manuscripts are discussed on pp. 5-6.

These are the three manuscripts that were collated to prepare the Arabic text presented in the next section. There is, however, a further manuscript that we examined in the course of preparing this article: MS 223 of the Université St. Joseph, Beirut. According to the catalogue of L. Cheikho the Apollonius treatise occupied pp. 125-129 of a fifteenth-century manuscript consisting altogether of 162 pages.[2] The original manuscript seems to have vanished and we are left with negative photographs producing poor-quality prints. As far as can be ascertained from photographs of three pages of text and two illustrations, the manuscript was more or less an exact copy of the London manuscript, and it was considered that any attempt at deciphering the poor-quality photocopy would not have yielded results of much additional value.

In preparing the edition of the text we adhered to the following principles:

1. When the manuscripts agreed, we accepted the wording even if it involved grammatical mistakes or probably corrupt readings.
2. When the manuscripts differed, we took the reading that better fitted the meaning of the text or which was found in two of the three manuscripts. Rejected readings are put into the apparatus.
3. Passages not appearing in all three manuscripts are given in brackets and the details are given in the apparatus.

No attempt has been made to restore the original wording (whatever that may mean) by emendation, but have preferred to present the text as raw material for future scholars.

Also edited here is a fragment from the Ẓāhiriyya (Damascus) MS 4871, which is a collection of philosophical and scientific works. Only half of one page deals with the Apollonius machine: this is a brief and somewhat confused description of the device.[3] Whether it was extracted at some period from the original full-length text or whether the latter was compiled from sources such as the Damascus fragment is not clear. For ease of reference, numbers have been added in both text and translation. We note that the name of the constructor has been corrupted into Īliyūs and that, apart from the sentence we have labelled '3', the fragment is equivalent to the last three paragraphs of the main text (in the translation). Sometimes the

[2] See Muwafi and Philippou. The manuscript is referred to on p. 147 and Cheikho's catalogue on p. 165.

[3] Ragep and Kennedy, p. 97.

equivalence is close, but on the other hand not all the terminology is exactly the same.

The only secondary works of major importance on the treatise are the German translation by E. Wiedemann[4] and the English summary by Farmer[5]. Wiedemann's account is to be treated with caution. In particular, there is one passage that appears to have been deliberately misread in order to make the operation of the machine more convincing, and an important illustration has been omitted. These points are discussed in sections 7 and 8 below.

As mentioned above, the Arabic text and the English translation made from it have been based on three main manuscripts – London, Paris and New York. (The contents of the New York manuscript were not identified until 1975.) Since the illustrations in the New York manuscript closely follow those of London, only the illustrations from London and Paris are reproduced here. Additionally, Drawings A1 and A2 give in modern guise our interpretation of the construction of the machine. These drawings are themselves, in a sense, translations. We have, however, discussed the conjectural elements in our reconstruction, and consider these drawings assist the reader to understand how the machine was built and operated.

[4] Wiedemann, I pp. 50-56.

[5] Farmer, pp. 80-86.

2. *Arabic text and notes*

الرموز: مخطوط باريس: س

مخطوط نيويورك: ك

مخطوط لندن: ن

صنعة آلة الزامر[٦] لأبلنيوس النجار الهندسي

نتخذ صندوق مهندم مزفت أو مرصص من داخله كله، طوله[٧] أربعة أذرع في عرض أربعة أشبار في ارتفاع شبرين ونصف[٨] ويحجب من وسطه بحجاب جاف[٩] من جميع نواحيه، فيصير حينئذٍ خزانتين ويتّخذ في كل خزانة ثقب[١٠] مفتوح أربعة أصابع مضمومة مهندمة[١١] في مثلها ويتّخذ لكل ثقب منها باب مهندم[١٢] يقع في حومات تنهدم[١٣] حتى لا يخرج من الهواء مقدار حبة[١٤]، ويكون كل باب بنرمادجة[١٥] من داخل معلقة على كل باب مع[١٦] أرض الخزانة، ونتّخذ في ثلثي ارتفاع كل خزانة سهم على محورين شبيه بعمود الميزان، ويكون أحد رأسيه موازياً لطبق الباب الذي وصفناه. ويكون في هذا

[٦] صنعة آلة الزامر: صفة الزامر ن

[٧] طوله: غير واردة في ن

[٨] في ارتفاع شبرين ونصف: في ارتفاع شبر ونصف ك

[٩] ويحجب من وسطه بحجاب جاف: ويحجب في وسطه حجاباً جافاً س

[١٠] ثقب: بيت ن

[١١] مهندمة: غير واردة في ن

[١٢] ويتخذ لكل ثقب منها باب مهندم: ويتخذ لكل ثقب باب من هذه الأبواب من داخل مهندمة ن

[١٣] تنهدم: فتنهدم ن

[١٤] مقدار حبة: مقياس حبة ن

[١٥] بنرمادجة: نرمادجة س

[١٦] مع في س

الرأس الموازي لطبق الباب ثلاث سلاسل معلق بها طرجهارة و كأنها كفّة ميزان معلقة، ويكون في أسفل الطرجهارة رزّة وفيها عمود وفي طرف العمود رزة أخرى ثابتة في وسط الطبق ويكون في الطرف الآخر من السهم رمّانة[١٧] رصاص [ويكون وزنها أيضاً بمثل][١٨] حتى إذا مالت ارتفع الرأس الآخر من السهم، فترتفع الطرجهارة ويرتفع العمود، ويرتفع الطبق، فيكون الباب حينئذٍ قد انفتح وارتفع قدر ثلاث أصابع مضمومة[١٩] و أكثر قليلاً. هذا قدره وقدر العمود، ثم يعمل[٢٠] في الخزانة الأخرى كذلك سواء. وتكون الطرجهارة[٢١] إذا امتلت ماء تميل وتنجذب إلى أسفل فينطبق الباب، [وتتكي الطرجهارة على العمود ويتكي العمود على الطبق الذي هو الباب][٢٢] فيشتد انطباقه لكي لا يخرج من الهواء شيء، وترتفع الرمانة الرصاص من الجانب الآخر، هذا إذا كانت الطرجهارة ملأى ماء، فإذا لم يكن فيها شيء مالت الرمانة الرصاص فارتفعت الطرجهارة وانفتح الباب. ويكون في كل طرجهارة ثقب دقيق في أسفلها قريب من موضع الرزة. فاحكم هذا غاية الإحكام وزفت الخزانتين لئلاً يخرج منه شيء من الهواء ثم أطبق الخزانتين من فوق إطباقاً محكماً كهيئة ما عملت بباقيها فيكون حينئذٍ صندوق مصمت ليس يُعلم ما فيه من داخل، وتكون له قوائم قدر شبر ونصف فكأنه صندوق لايشك فيه، ثم اتّخذ فوق هذا الصندوق حوله كله بقدر أربع أصابع داخلة منه حجرة واتخذ في سقفي الخزانتين اللذين هما[٢٣] طبق الصندوق، بابين نافذين بإزاء الطرجهارتين سواء حتى إذا دخل في البابين شيء، وقع في جوف الطرجهارتين سواء وتكون هذه

[١٧] رمانة: زمانة، وردت في ن مراراً أي بنقطة على الراء

[١٨] الجملة بين الحاصرتين [] غير واردة في س

[١٩] مضمومة: مفتوحة س

[٢٠] يعمل: ويعمل ن

[٢١] وتكون الطرجهارة: ويكون وزن الطرجهارة ن

[٢٢] الجملة: ويحمل عليه العمود والطرجهارة فيشتد ... وردت في س

[٢٣] اللذين هما: الذي هو ن

الأبواب كأن بداخلها بيوت صغار في أواسطها[٢٤] أثقاب، أعني ثقب في سطح كل باب ويركب على كل ثقب أنبوب عدل ليدخل إلى الطرجهارة ولا يسرق شيئاً من الهواء. فليكن الصندوق اب والبابين اللذين ينطبقان بهندام وفيهما[٢٥] الرزتين كـ كـ، والعمودين اللذين في الرزتين ل ل، والطرجهارتين ه ه ، والمحورين وو، والرمانتين رر و بيتي العدل ط ط وأنبوبتي العدل ح ح، [وهذه صورة ذلك][٢٦].

ثم اتّخذ من فوق عند طبق الخزانة أبواب تنطبق ويرتفع على كل ثقب مما وصفناه الذي هو العدل باب نرمادجة[٢٧] قضيب طويل محدود ، فإذا كان الباب مطبق كان أصل القضيب المتّصل بالباب مرتفع عن وجه طبق الخزانة [والطرف الآخر البراني من القضيب ساقط على وجه الخزانة][٢٨] الذي هو طبق الصندوق فافهم ما وصفنا. فإذا فعلت هذا بالبابين وأحكمت ذلك إحكاماً جيداً فقد تم عملك لأن كل ما بعد هذا تبع له[٢٩]. وليكن البيتان اب والبابان دد والقضيبان ه ه والأنبوبان ح ح ، ثم اتّخذ في الخزانتين أنبوبين قائمين في أطرافها نافذين وعليها أنبوب آخر معارض. وفي طرفي الأنبوب المعترض للأنبوبين، بابين في داخل (يعملان) على عمل أبواب النضاحات[٣٠] التي يقال لها باب المدفع، ويكون في وسط الأنبوب المعترض للأنبوبين أنبوب طوله كطول الناي وهو الذي يخرج منه الصوت، ويثقب كما يثقب الناي وتركب على رأسه الحبّة[٣١] فإذا فرغت من هذا كلّه وأحكمته، فاتخذ على الحجرة دولاب برحاب

[٢٤] اواسطها: اوساطها س

[٢٥] وفيهما: اللذَيْن فيهما ن

[٢٦] الجملة بين الحاصرتين [] غير واردة في س

[٢٧] نرمادجة: بنامادجه ن

[٢٨] الجملة بين الحاصرتين [] غير واردة في س

[٢٩] له: غير واردة في س

[٣٠] أبواب النضاحات: أبواب الرصاصات س

[٣١] الحبة: هي القسم الذي يوضع في الفم من الناي

الرحا الرومي أو الرحا المولد القائم في سفود وهو الأبرار[٣٢] وفي هذا الأبرار الذي في الرحا[٣٣] دوارة مثل رحا الماء سواء، يديرها[٣٤] الماء. والدوارة تكون ذات أسنان قليلة لتبطئ حركتها، لكي يبطئ دخول الماء. وليكن السفود عدده مثل عدد أسنان الدوارة سواء، ويكون موضع نصب السفود فيما بين البابين سواء. ثم اتّخذ فوق هذه الحجرة، التي فيها الدولاب [ويكون الدولاب مرتفع من وجه الخزانة قدر ذراع فاتخذ حينئذٍ فوق الدولاب][٣٥] خزانة للماء تسع ما شئت ويكون لها ثقب إذا فُتح بشيونه انصب الماء على الرحاب. [وليكن الدولاب د وليكن الماء مرتفع عن وجه الخزانة قدر ذراع فإذا انصب الماء على الرحاب دار الدولاب ولا يكون دوره سريعاً جداً بل يكون][٣٦] بقدر ذلك حتى يكون معتدل الدوران. ثم اتّخذْ في أسفل السفود عند أسفله نصف دائرة صفيحة يكون وسطها عند ثخن البابين[٣٧] اللذَيْن فيهما الصفير وتكون هذه النصف دائرة إذا دارت دخل أولها الذي عند قطرها[٣٨] تحت القضيب الذي هو الباب. فإذا مرت النصف دائرة قليلاً، رفعت القضيب، فإذا ارتفع القضيب ارتفع البابِ ودخل الماء من تحت الباب من موضع ارتفاعه وصار إلى البيت الصغير الذي فيه أنبوب العدل وارتفع فدخل في أنبوب العدل فانصب في الطرجهارة والماء ينصب[٣٩] بسرعة والدولاب يتحرك بإبطاء ومالت الطرجهارة وانطبق الباب الأسفل واتكأ عليه العمود واتكت الطرجهارة على العمود واختنق الهواء وطلب الخروج[٤٠] فلم يجد مخرجاً إلاّ من الأنبوب، فدخل في

[٣٢] الأبرار: الأبراز ن

[٣٣] الرحا: البرجات ن

[٣٤] يديرها: التي يديرها ن

[٣٥] الجملة بين الحاصرتين [] غير واردة في س

[٣٦] الجملة بين الحاصرتين [] غير واردة في ن

[٣٧] عند ثخن البابين: عند نصف ثخن البابين ن ، ك

[٣٨] دخل أولها الذي عند قطرها: الذي دخل أولها عند قطرها س

[٣٩] ينصب: وهكذا ينصب س

[٤٠] الخروج: المخرج ن

الأنبوب حتى صار إلى عطفه ففتح باب المدفع المعلّق عليه من داخل، ومرّ مستوياً فوافى الباب الآخر فدفعه فأغلقه فرجع إلى المنتصف من الأنبوب فخرج من الأنبوب المعترض الذي فيه الزمّارة فزمّر.

فافهم ما وصفنا، فلا يزال يزمر والنصف دائرة تدور بإبطاء تحت القضيب رافعة له والباب مرفوع لارتقاع القضيب، ولا يزال كذلك والماء يجتمع في الخزانة حتى إذا تمّ دور النصف دائرة وخرجت جميعها من تحت القضيب، سقط القضيب على وجه أرض الخزانة كما كان أولاً. فانطبق الباب وانقطع الماء، ويبدأ طرف الدائرة يدخل تحت القضيب الآخر فارتفع وارتفع الباب ودخل (الماء) إلى أنبوب العدل الآخر في البيت الصغير المصنوع مثل الأول و صار الماء إلى الطرجهارة. فإذا امتلأ فعل مثل فعل الأول سواء وحدث الزمر مثل حدوث الأول متّصلاً به لا ينقطع، فلا يزال هكذا ما دام الماء يدخل ويخرج ليكون على الدائرة وعلى نايى موضع لمخرج الريح م ه وعلى بابي المدفع د د وعلى مدخل الهواء من الصندوقين ا ب وهذه صورة الدولاب ونصف الدائرة والمدفع[41].

ثم إن البيت الأول إذا انصب فيه شيء من الماء بدأ الماء يخرج من الثقب الذي فيه. فإذا خرج الماء خفّت الطرجهارة وجذبتها الرصاصة[42] التي في الطرف الآخر من السهم فارتفعت الطرجهارة وارتفع العمود وانفتح الباب فخرج الماء من الخزانة كلّه وتفرّغت ولا تزال الخزانة الأخرى تعمل حتى يبلغ وقتها وتبدأ الأولى التي كانت تفرّغت تعمل ثم تَفْرَغُ هذه، ولا تزال هذه الآلة تعمل ما كان لها ماء[43].

[41] والمدفع: غير واردة في ن

[42] الرصاصة: هي الرمانة الرصاصية

[43] في ن و ك فقط كان الختام جملة تتضمن الحمد لله والصلاة على النبي محمد وآله.

3. *Translation*

The Construction of the Fluting Machine by Apollonius, the carpenter, the geometer

We take a well-fitting box lined with asphalt or tin[44], in length four cubits and in breadth two and a half spans. In the centre of it we erect a partition, watertight at all its edges, so that it becomes two containers. We make a hole in each container, four closed fingers in length precisely, and for each of the holes we make a well-fitting valve that tightly fits into the hole so that not the smallest quantity of air escapes. Each valve has a hinge on the inside, connected on each valve to the floor of the container. At two thirds of the height of each container we take a beam on pivots, similar to the beam of the balance. One of its two ends is in line with the cover of the valve which we have described. In this end, which is in line with the cover of the valve, are three chains to which a *ṭarjahāra* (bowl) is suspended, as if it were the suspended pan of a balance. At the bottom of the *ṭarjahāra* is a staple, in which is a strut, and in the end of the strut there is another staple, fixed to the centre of the cover. At the other end of the beam is a lead counterweight, so that when it tilts downwards the other end of the bean rises and so the *ṭarjahāra* rises, the strut rises, and the cover rises. And so the valve at this juncture has opened and has risen by a distance of a little over three closed fingers. This is its distance and the distance of the strut. Then we do exactly the same thing with the other container. When the *ṭarjahāra* fills with water, it tilts downwards and is drawn downwards, and so the valve closes. The *ṭarjahāra* rests on the strut and the strut rests on the cover, which is the valve, so that it is tightly closed so that no air escapes. The lead counterweight rises on the other side – this is when the *ṭarjahāra* is full of water –, but when it contains nothing the counterweight tilts downwards, the *ṭarjahāra* rises and the valve opens. In each *ṭarjahāra* there is a small hole in the bottom near the position of the staple. Do this very carefully. Line the containers with asphalt so that no air escapes from them. Then cover the two containers at the top tightly in the manner that you did with the rest. It is then just

[44] *muraṣṣaṣ*. This can mean "leaded", but the usual meaning for water containers is "tinned".

a box – no-one knows what is inside it. It has feet one and a half spans in length [*qadr*], as if it were without doubt a box.

Then erect on top of this box, around its complete [perimeter], a rim[45] four [closed] fingers inside it [i.e. the perimeter]. Install in the roofs of the containers, i.e. the top of the box, two valves that penetrate opposite the *ṭarjahāra*s exactly, so that when anything enters through the valves it falls into the hollow of the *ṭarjahāra*s exactly. Inside these valves, as it were, [i.e. under them] are small chambers, in the centres of which are holes, i.e. a hole in the surface of each valve, and upon each hole is mounted a concentric siphon, to give entry to the *ṭarjahāra* without stealing any of the air. Let the box be *ab*, the two valves which close tightly and have staples *kk*, the two struts on the staples *ll*, the two *ṭarjahāra*s *ee*, the two axles *ww*, the two counterweights *rr*, the two chambers of the concentric [siphons] *ṭṭ*, and the two concentric siphons *ḥḥ*. This is the picture.

On top, on the cover of the container, fit valves – with hinges – that close and rise on each hole that we mentioned, which is [the entry to] the concentric [siphon]. [Each valve cover has] a long extended rod; and when the valve is closed, the end of the rod connected to the valve is raised from the surface of the cover of the container, and the other, outer, end of the rod lies on the surface of the cover of the container which is the cover of the box. So understand what we have described. When you have done this with the two valves and your workmanship is very accurate, you have completed your work, because the rest follows on from this. Let the two chambers be *ab*, the two valves *dd*, the two rods *ee*, the two pipes [i.e. siphons] *ḥḥ*.

Then install in the two containers two vertical pipes whose ends penetrate [the containers], and upon them is another transverse pipe. In the ends of the pipe which is across the two pipes are two valves on the inside [made] on the pattern of the sprinkler valves which are called *bāb al-midfaʿ* [clack-valves]. In the centre of the pipe [which is] across the two pipes is a pipe as long as the pipe of a flute; the sound issues from this. It is bored as a flute is bored and the mouthpiece is fitted to the end.

When you have completed all this and done it accurately, then install on the rim[46] a water-wheel like the *rūmī* mill or the newly-invented vertical wheel on the lantern-pinion. It is the *abrār* and in

[45] *ḥujra*. "Chamber" is another possible translation.

[46] See previous note.

this *abrār*, which is in the mill, is a rotator, exactly like the water-mill, turned by water. The rotator has few teeth, in order to slow down its motion and so slow down the entry of the water. Let the lantern-pinion have exactly the same number of teeth as the rotator. The position where the lantern-pinion is erected is between the two halves precisely. Then above this rim[47] that contains the wheel – the wheel is raised above the surface of the container a distance of a cubit – is a reservoir for the water of any capacity you desire and having a hole [from which], when the tap is opened, water discharges on to the paddles. Let the wheel be *d*, and let the water be raised from the [upper] surface of the [water-wheel's] container a distance of a cubit. When the water flows on to the paddles, the wheel rotates; its rotation is not very rapid, but the quantity is such that the rotation is moderate. [See *Construction and operation of the machine* and *Conclusion* for discussion of the foregoing paragraph].

Then install below the lantern-pinion, at the bottom of it, a semicircle, a plate whose centre is at the thickness [i.e. level?] of the two valves containing the whistle. When this semicircle rotates, its leading edge, which is at the diameter, enters below the rod for the valve. When the semicircle moves on a little, it raises the rod; and when the rod rises, the valve rises. The water enters under the valve, where it rises; it arrives at the small chamber containing the concentric siphon, rises, enters the concentric siphon and flows into the *ṭarjahāra*. The water flows rapidly and the wheel turns slowly. The *ṭarjahāra* tilts downwards and the lower valve closes; the strut rests on it [i.e. the valve] and the *ṭarjahāra* rests on the strut. The air is compressed and seeks an exit, but finds no outlet except through the pipe. It enters the pipe until it arrives at its bend and opens the clack-valve suspended on it from inside and passes straight on until it arrives at the other valve; it pushes it and shuts it. So it returns to the middle of the pipe and issues from the transverse pipe in which the flute [is placed]. So it makes a sound.

Understand what we have described. It continues to make a sound while the semicircle turns slowly below the rod, raising it, and the valve is raised through the rising of the rod. This continues, and the water collects in the container, until such time as the semicircle completes its [half-]rotation and all of it has come out from under the rod. The rod drops on to the floor-surface of the container, [so that it

[47] See note 46.

is] as it was originally and the valve closes. The water is cut off and the edge of the [semi-]circle starts to enter under the other rod: it rises, and the valve rises, and the water enters the other concentric siphon in the small chamber made like the first and arrives at the *ṭarjahāra*. And when it fills, it operates like the first one exactly. The flute works as it worked before, continuously and without interruption. It goes on in this way as long as the water enters and discharges. On the [semi-]circle and on the flute the place where the air comes out, let there be *m*, *e* on the two clack-valves *dd*, on the entry of air from the two boxes *ab*. This is the picture of the wheel, the semicircle and the clack[-valve].

Then when any water flows into the first chamber, the water begins to issue from the hole that is in it [i.e. the *ṭarjahāra*]. When the water comes out, the *ṭarjahāra* becomes lighter and the lead [weight] which is on the other end of the beam pulls it [down] and the *ṭarjahāra* rises, the strut rises, and the valve opens. All the water discharges from the container, which empties. The other container continues to operate until its time is up, and the first which had emptied begins to operate and then to empty. This machine continues to work as long as it has water.

4. *Arabic text of the Damascus fragment*

(١) بسم الله الرحمن الرحيم وبه أستعين

(٢) صنعة الآلة الزمرية لإيليوس الحكيم

(٣) الماء يسيل من خزانته من البزال ويدير الدولاب وذات الأسنان ويدور بدورانه البكرة والنصف الدائرة ويدخل طرفيها تحت القضيب الباب (٤) فإذا مرت قليلاً رفعت القضيب ويرتفع الباب (٥) ويفيض الماء إلى البيت الصغير ويرتفع فى العدل وينصب فى الطرجهارة (٦) وتثقل عند امتلائها (٧) واتكأ[٤٨] العمود المتصل بأسفلها على الباب (٨) فينطبق[٤٩] الباب الأسفل

[٤٨] واتكأ : MS واتكى || [٤٩] فينطبق: MS فينطق

ويفرغ الماء الذى فى الطرجهارة من ثقب[٥٠] في أسفلها (٩) فيختنق[٥١] الهواء الذى فى الخزانة ويطلب المخرج (١٠) فلم يجد مخرجاً إلاّ الأنبوبة فيدخل فيه (١١) وثم مستوياً فيفتح باب المدفع من داخل ويوافى الباب الآخر فيغلقه (١٢) ويخرج من الأنبوب المتصل بهما ويدخل الثاني فيخرج من ثقبها ويزمر (١٣) إلى أن يخرج النصف دائرة من تحت القضيب وتسقط الباب المتصل به ويدخل تحت القضيب الآخر، (١٤) فيحدث فى الجانب الآخر مثل ما حدث فى الجانب الأول سواء (١٥) ولا يزال هكذا ما دام الماء يدخل ويخرج (١٦) وعند فراغ الطرجهارة من الماء تشيلها[٥٢] المثقلة فينفتح الباب كما كان.

5. *Damascus fragment – Translation*

(1) In the name of God, the Merciful, the Compassionate. I begin asking for His help.

(2) The Construction of the fluting machine by Īliyūs the sage.

(3) The water flows from the orifice of its reservoir and turns the wheel and the gear-wheel and in its revolution the reel [*bakra*[53]] and the semicircle turn, of which the two edges enter under the rod [attached to] the valve. (4) When [this semicircle] moves a little, it raises the rod and the valve rises. (5) The water pours into the small chamber, rises in the siphon and flows into the *ṭarjahāra*, (6) which becomes heavier when it is full, (7) and the strut linked with its lower end rests on the valve. (8) The lower valve closes and the water in the *ṭarjahāra* empties from a hole in its base. (9) So the air that is in the reservoir is compressed, seeking an outlet. (10) It finds no outlet except the pipe, which it enters. (11) As [the air] goes on, it opens the valve from inside and arrives at the other valve, shutting it. (12) [The air] leaves the pipe connecting the two [valves] and enters the second to come

[٥٠] ثقب: MS ثقيب || [٥١] فيختنق: MS فستختنق || [٥٢] تشيلها: MS تشيل

[53] This must be the lantern-pinion.

out of its hole and makes a sound (13) until the semicircle goes from under the rod, drops the valve linked with it and enters under the other rod. (14) Exactly the same occurs on the other side as occurred on the first side. (15) It goes on in this way as long as the water enters and discharges. (16) When the *ṭarjahāra* is empty of water, the counterweight raises it: the valve opens as before.

6. *A note on technical terms*

The characteristic terms of Appolonius' text are here grouped under the names of the principal components of the fluting machine or other devices mentioned. Since the purpose is comparison with the usage in other texts, some common words appear as well as genuine *termini technici*. The English equivalents are those most often used in the translation.
D means that the word is found in the Damascus fragment only.

الأبرار *abrār* (type of gearing used in mills).

بشيون tap.

بكرة reel (*D*).

باب valve : طبق cover، مهندم well-fitting ، انفتح to open ، انطبق to close.

باب المدفع *bāb al-midfaʿ* clack-valve: دفع to push، غلق to shut.

باب النضاحة sprinkler valve.

بيت chamber.

حجاب partition : جافّ watertight.

حجرة rim.

محور axle, pivot.

خزانة reservoir, container : أرض floor، سقف roof.

دوارة rotator.

دولاب wheel (cf. الأبرار): رحبة paddle ، دولاب رحاب water-wheel.

ذراع cubit (measure).

رحى mill-wheel: رحى رومى rūmī mill.

رزة staple.

رمانة counterweight: مال to tilt downward.

ريح air (cf. هواء).

زمارة flute (cf. ناى، صفير): زمّر to make a sound، آلة زمرية fluting machine.

سفود	lantern-pinion.
سقط	to drop (on to), to rest (on).
سلسلة	chain.
سنّ	tooth : ذات الأسنان rotator.
سهم	beam.
شبر	span (measure).
إصبع	finger (measure): أصابع مضمومة closed fingers (measure).
صفير	whistle (cf. زمارة، ناى).
صندوق	box: جافّ watertight، مهندم well-fitting، زفّت to line with asphalt، رصّص to line with lead or tin، قدم foot.
طرجهارة	*ṭarjahāra* (pierced float): امتلأ to fill، ملآن full، جوف concavity، hollow، خفّ to become lighter، ثقل to become heavier، مال to tilt downward.
عمود	strut.
قضيب	rod.
ماء	water: انصبّ to flow, to discharge، انقطع to be cut off.
ميزان	balance : عمود beam، كفّة pan.
ناى	flute (cf. زمارة، صفير): حبّة mouthpiece، صوّت to sound.
أنبوب	pipe: عطف bend.
أنبوب عدل	concentric siphon (gen. constr.): بيت صغير small chamber.
نرمادجة	hinge.
نصف دائرة	semicircle (definite usu. النصف دائرة) (cam) : طرف edge، أول leading edge.
هواء	air (cf. ريح): اختنق to be compressed.

7. *The illustrations in the manuscripts*

The following manuscript pages are reproduced by kind permission of the British Library (London), the Bibliothèque Nationale (Paris) and the Ẓāhiriyya Library (Damascus). We are grateful to the Institute for the History of Arabic Science (Aleppo) for providing us with a photograph of the page from the Damascus manuscript.

As stated above, the illustrations in the New York and Beirut manuscripts appeared to be derived from those in the London manuscript. We therefore present only the drawings of British Library, Add. 23391, and Bibliothèque Nationale, ar. 2468. These are placed in the order found in the manuscripts and as far as possible in matching pairs on facing pages. In his German translation Wiedemann (pp. 52, 53, 54 and 55) presented modern redrawings of the manuscript illustrations of Plates 1a, 1b, 2, 4b (coupled with an illustration from the Beirut manuscript) and 5b.

From the similarity of subject and identity of order in the manuscripts, it is clear that the drawings are part of the transmitted text. The great differences must therefore be the result of deliberate revision.

Plate 2, from BL Add. 23391, has no equivalent in the Paris manuscript; but it should be noted that it complements Plate 1a, so that together they contain the same information as Plate 1b. Within the drawing is "the second diagram"; this is the only diagram in the two manuscripts to carry such a label. Plate 4a is equivalent to the upper part of Plate 4b, the lower part being equivalent to Plate 1b. It may be that the artist of the London manuscript (or the tradition he represented) had a policy of drawing the parts separately from a composite diagram; or, alternatively, the tradition represented by the Paris manuscript as far as possible made drawings of diverse parts into a composite diagram. Certainly the artist of the Damascus manuscript put everything into one drawing – including a poor representation of the contents of Plate 5a or 5b.

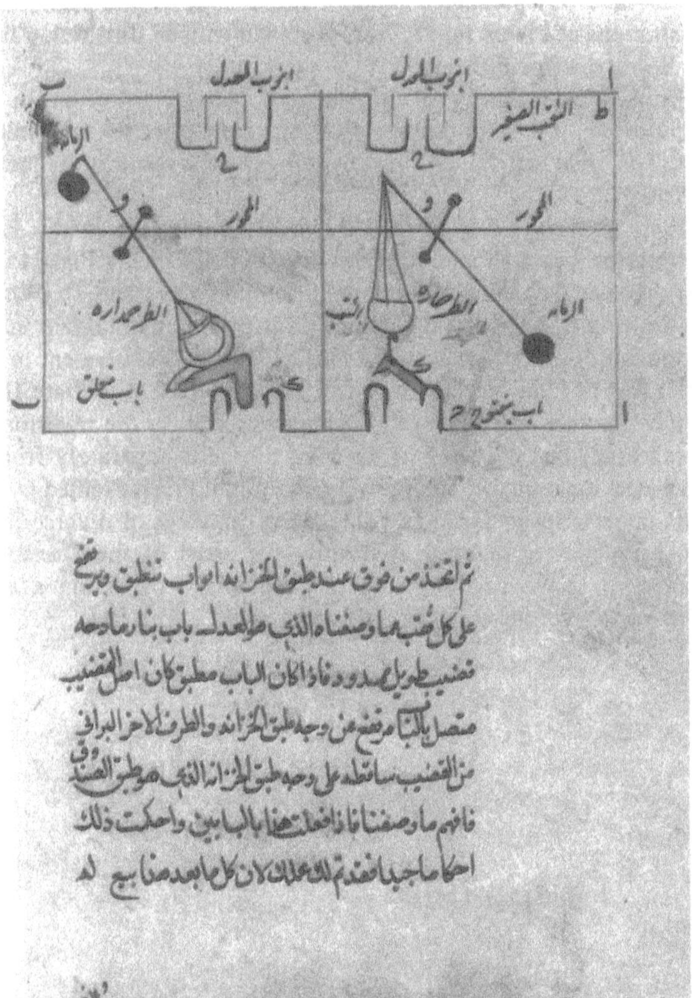

Plate 1a: the two *ṭarjahāra*s, the two concentric siphons and the two lower valves. BL Add. 23391, f. 22v

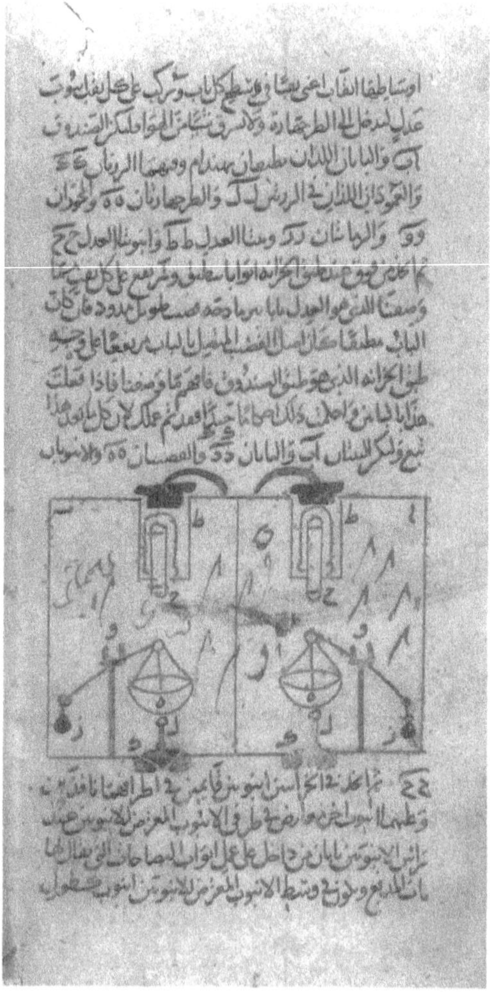

Plate 1b: the two *ṭarjahāras*, the two concentric siphons and the four valves. BN 2468, f. 40v

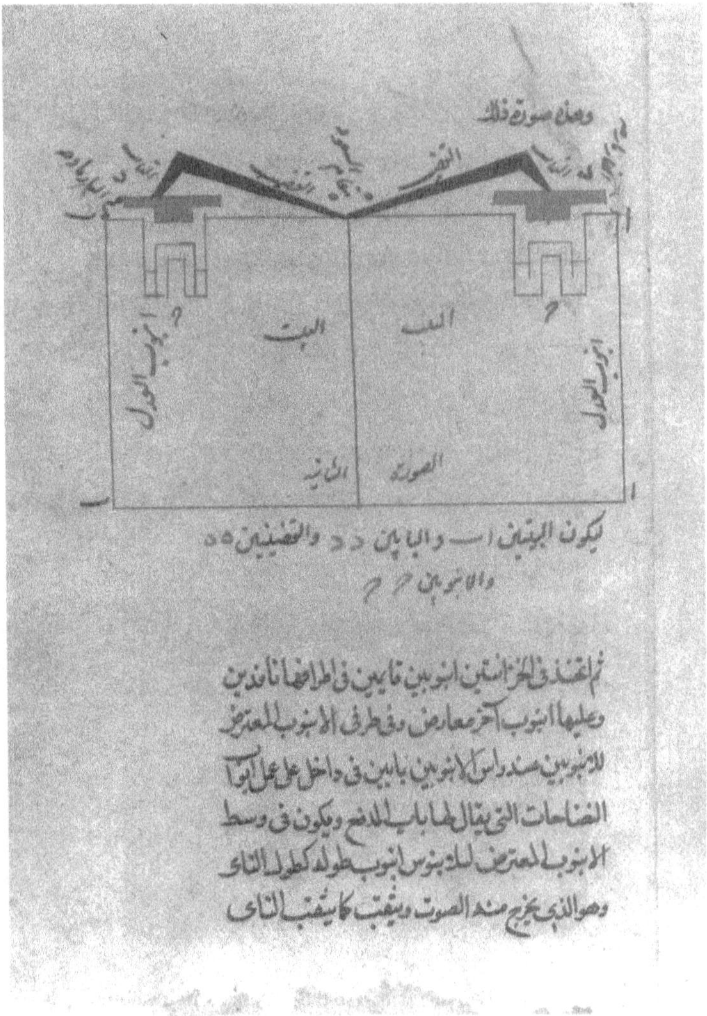

Plate 2: the two concentric siphons and corresponding valves with attached rods. BL Add. 23391, f. 23r

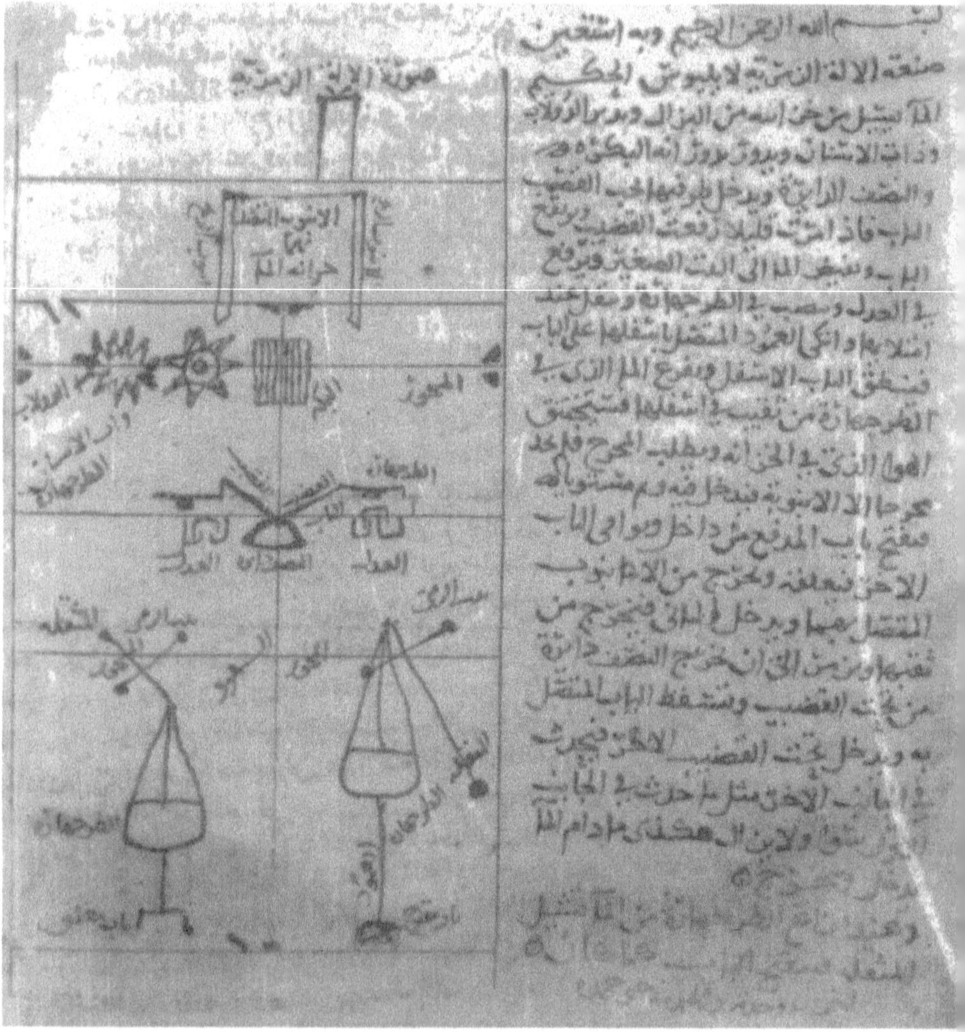

Plate 3: general view of the fluting machine.
Damascus, Ẓāhiriyya Library 4871, f. 75v

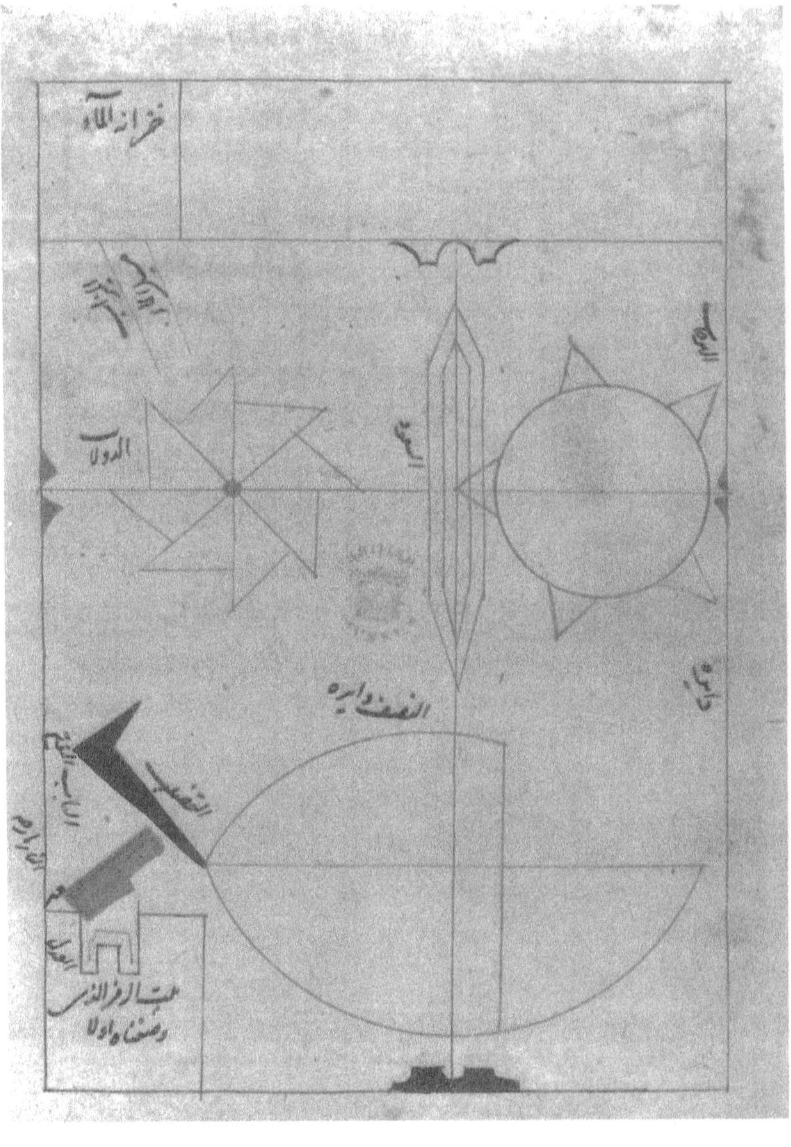

Plate 4a: the upper part of the apparatus, showing water-wheel, gearing and cam. BL Add. 23391, f. 25v

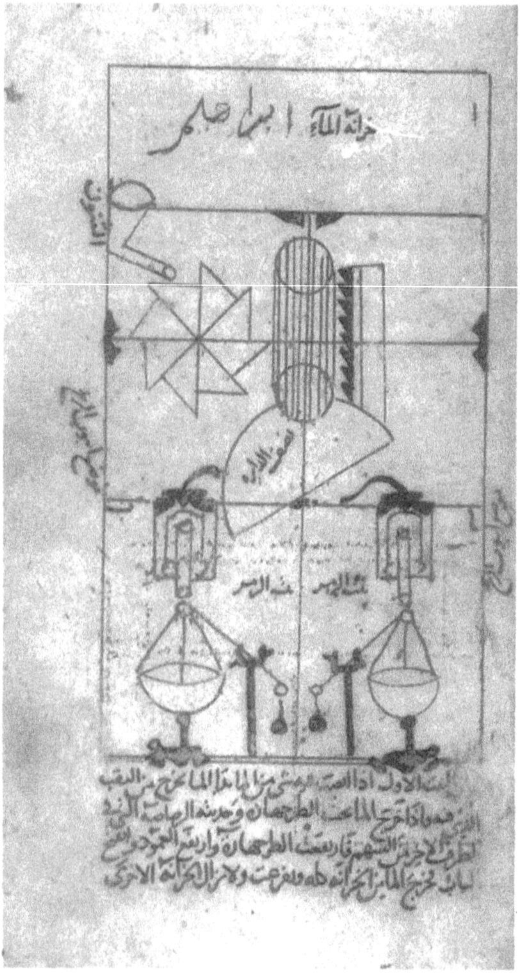

Plate 4b: general view of the fluting machine except the air pipes and flute. BN 2468, f. 42r

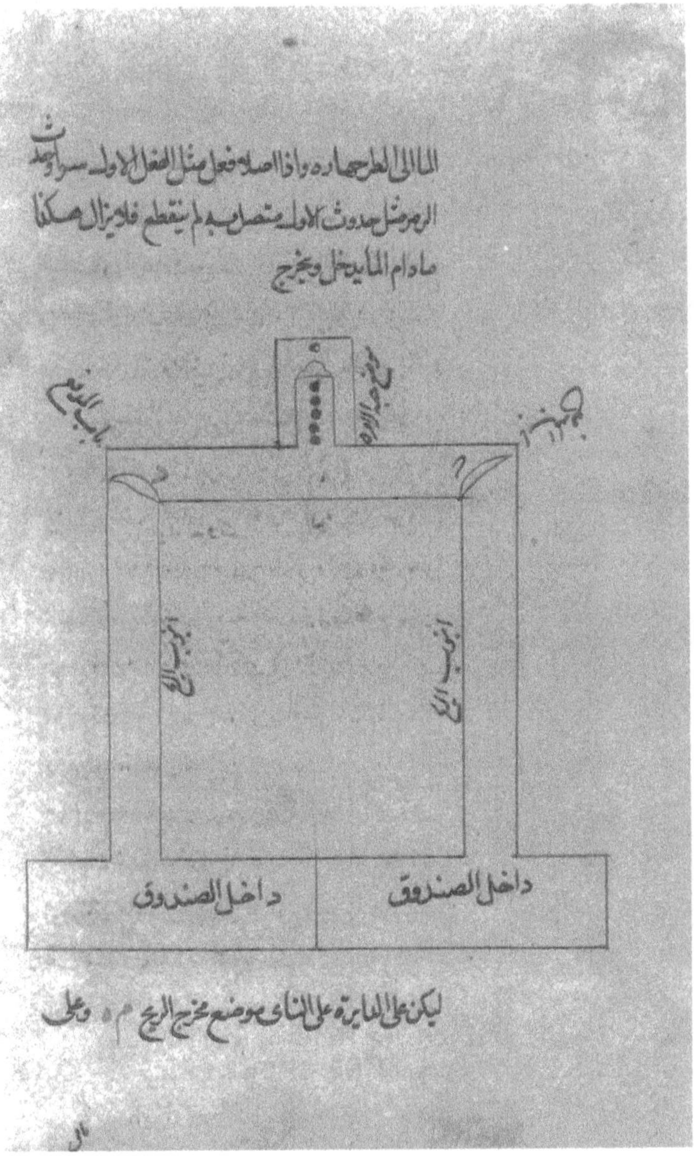

Plate 5a: the air pipes and flute. BL Add. 23391, f. 24v

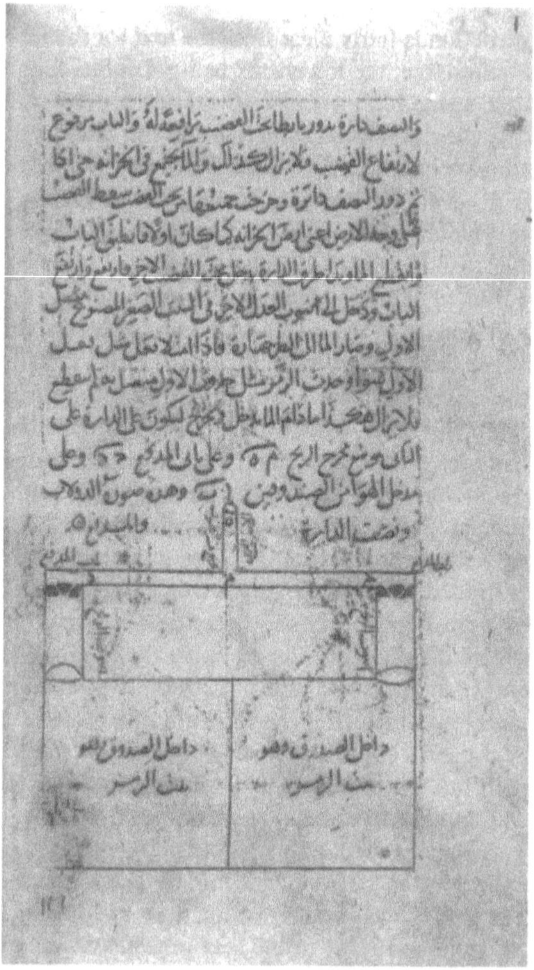

Plate 5b: the air pipes and flute. BN 2468, f. 41v

8. *Construction and operation of the machine*

The construction is fairly clear from the text for the most part. The machine is quite large, the lower box being 4 cubits long by 4 spans wide by 2 1/2 spans high – i.e. 2.40 metres x 0.96 metres x 0.60 metres, taking the cubit as 0.60 metres and the span as 0.24 metres. (These measures could, of course, fluctuate widely, but for our purposes their values need not be known precisely). The feet are about 0.36 metres long. The chamber above this is a distance of some 8 cm all round smaller than the lower chamber. This is to allow room for the pipes of the flute to be led out from the lower chamber – see drawings A1 and A2. The sizes of the water-wheel and gears are not specified – they and their chamber are simply sized in order to give a credible layout for the machinery. Nor is the size of the reservoir specified; it has been given the same dimensions as the chamber beneath it. It must have had a large hole for filling. It may also have had a continuous water supply and an overflow system, as is the case with al-Jazarī's musical automata[54], but the text is silent on this point.

The upper and lower valves in the lowest chamber are all airtight plate-valves hinged to the box. This is made quite clear in the text; these valves were erroneously shown as conical valves in the Paris manuscript – see Plate 1b. The valves in the flute's pipes are one-way clack-valves. Again these are shown incorrectly as conical valves in the Paris manuscript (Plate 5b), correctly in London (Plate 5a). Oddly, Wiedemann chose to omit this London illustration from his article. The flute itself has a mouthpiece and a number of holes drilled in it, but there is no indication of how the machine could have been made to produce any sound except a single note.

The arrangement of the water machinery is open to some doubt. The text mentions a *rūmī* wheel, which may have been horizontal, but the illustration and most of the text presuppose an arrangement with a vertical water-wheel. On the axle of this wheel is a vertical gear-wheel that meshes with a lantern-pinion, as in water-mills. Confusion arises because the text says that the water-wheel turns slowly, but does not indicate how this is achieved. A slow rotation of the central axle could have been produced by having more teeth on the gear-wheel than on the lantern-pinion, but the text expressly states that the gear ratio was 1:1. One cannot simply contradict the text, as

[54] See Hill's edition of al-Jazarī's treatise, p. 170-177.

Wiedemann did, by stating that the gear-wheel had more teeth than the lantern-pinion[55]. We may conjecture that the water-wheel had some form of escapement to slow it down, although it is significant that al-Jazarī thought that the speed of rotation of the Apollonius machine would have been too high[56]. That is to say that there may not have been time for the first half of the cycle to be completed before the second started. al-Jazarī's comments suggest that he may have had much the same text as we have. It does seem likely that somewhere along the line of transmission (translation, copying) some vital information about the rotating machinery was distorted or omitted – or both.

Theoretically, the operation was straightforward. The semicircular cam probably had a flat rim and a curved profile. Let us assume that its leading edge was just touching the rod of the left-hand valve V3 (Drawing A1). Both containers were empty. The water was turned on, the water-wheel rotated, and the cam began to lift the rod and hence opened valve V3. Water ran into the left-hand chamber through the valve and concentric siphon and fell into *ṭarjahāra* T1. The purpose of the concentric siphon was to prevent the air from bubbling back into the upper chamber, which would have occurred if the entry had been a wide pipe. As soon as the *ṭarjahāra* filled, it descended and closed valve V1. The only outlet for the air was through valve V5 (Drawing A2); it could not pass through valve V6 and so it entered the flute, which played.

When one semi-rotation of the cam was complete, valve V3 closed, valve V4 began to open, and the operation continued in the right-hand container. Perhaps the emptying of *ṭarjahāra* T1 and the opening of valve V1 and the subsequent opening of the left-hand chamber occurred simultaneously with the first part of the operation in the right-hand chamber. The operation continued in this manner, alternating from one side of the machine to the other, until the reservoir was empty.

[55] Wiedemann, I p. 54.

[56] Al-Jazarī, p. 170.

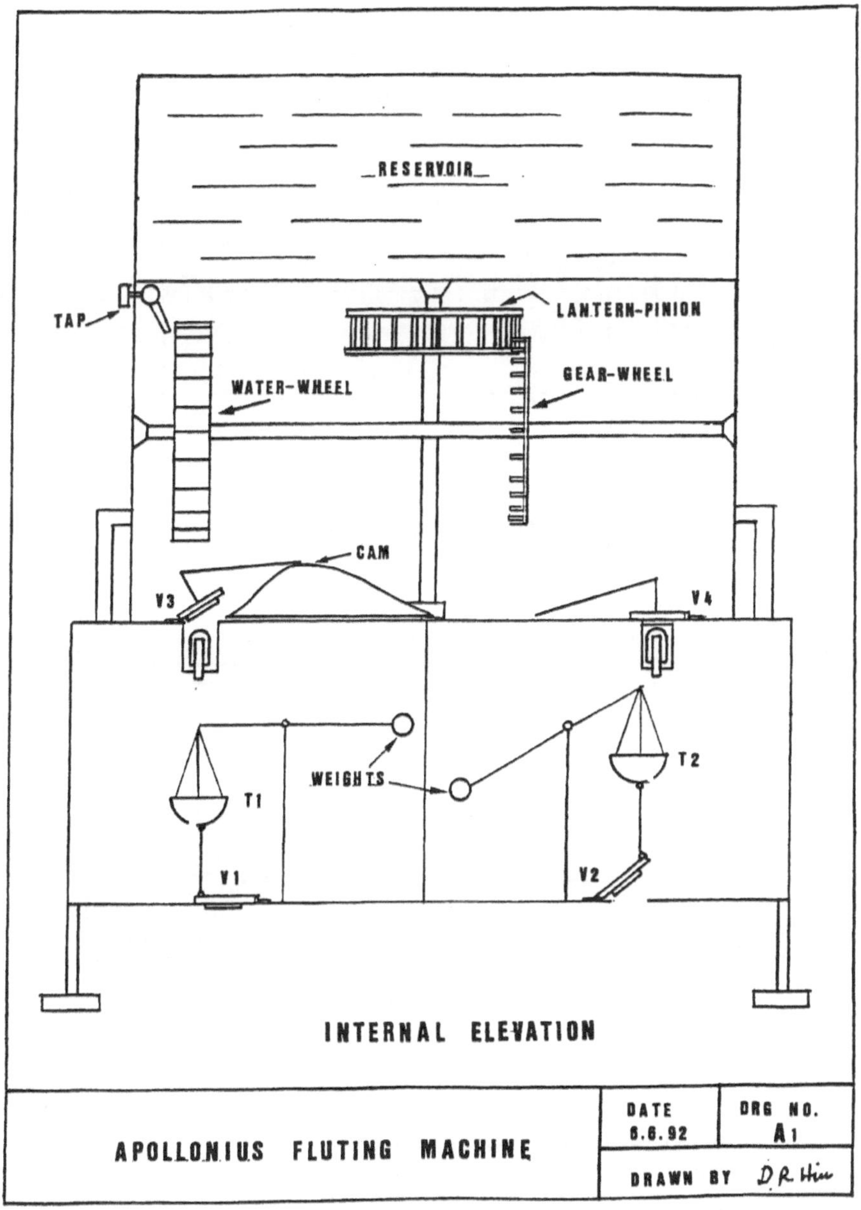
RESERVOIR
TAP
LANTERN-PINION
GEAR-WHEEL
WATER-WHEEL
CAM
V3
V4
WEIGHTS
T1
T2
V1
V2
INTERNAL ELEVATION
APOLLONIUS FLUTING MACHINE
DATE
6.6.92
DRG NO.
A1
DRAWN BY D.R. Hill

Drawing A1

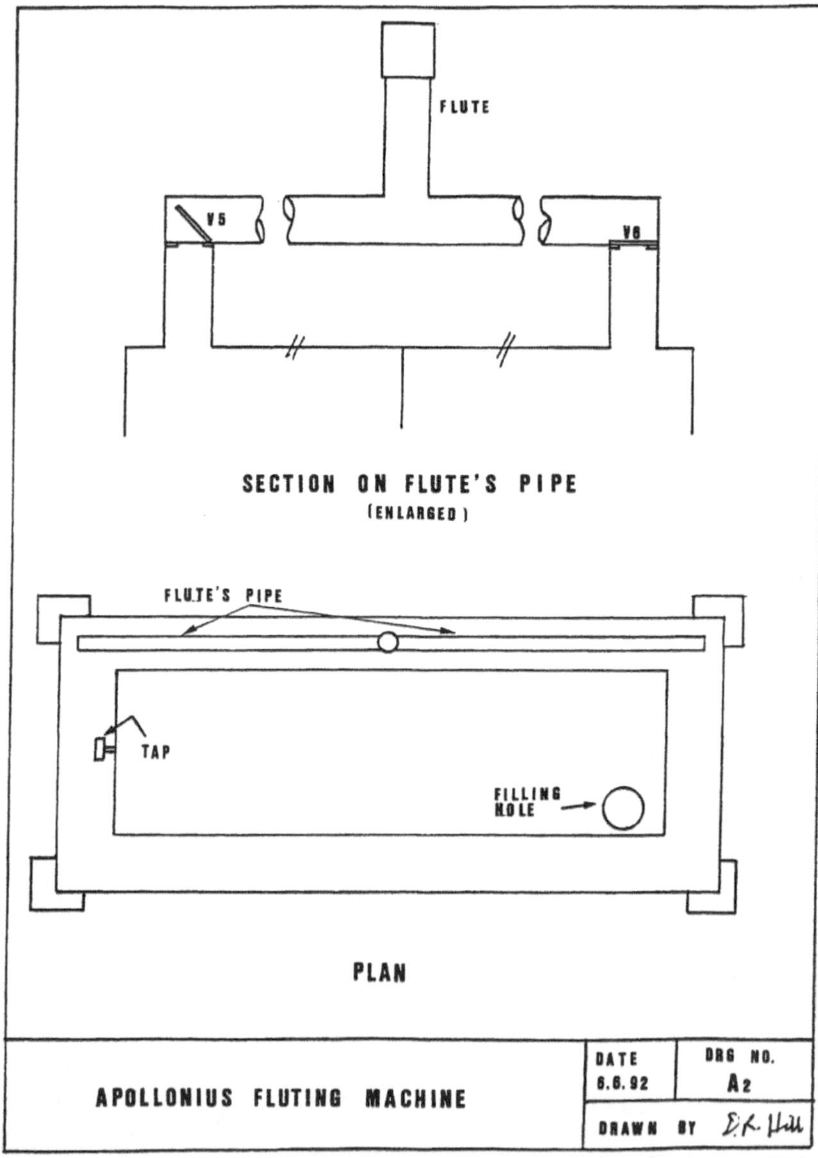
FLUTE
V5
V6
SECTION ON FLUTE'S PIPE
(ENLARGED)
FLUTE'S PIPE
TAP
FILLING HOLE
PLAN
APOLLONIUS FLUTING MACHINE
DATE
6.6.92
DRG NO.
A2
DRAWN BY

Drawing A2

9. *Conclusion – Origin of the treatise*

We cannot establish the date of origin of the treatise or its authorship. The best that we can do is to mention some of the features of its contents in the hope that we may thereby stimulate some useful lines of enquiry.

The three Arabic manuscripts that we have utilized here for the main text all contain, at least, the treatise attributed to Archimedes on a water-clock together with the treatise attributed to Apollonius. The combination of mathematical or scientific scholarship with mechanical inventiveness was not unusual in earlier times. Archimedes himself probably devised a celestial globe (Cicero, *Republic* I xiv, 21, 22) and was credited with the construction of siege machines for the defence of Syracuse. Hero of Alexandria (fl. mid-first century AD) wrote works on theoretical and practical mechanics as well as mathematical text books. In Islamic times the Banū Mūsā wrote works on mathematics, astronomy and machines; the great Andalusī astronomer al-Zarqāllu built two water-clocks at Toledo in the fifth/eleventh century[57]. One could adduce a number of other examples from both Hellenistic and Islamic times. There is therefore nothing intrinsically improbable about the attribution of mechanical devices to scientists.

Our knowledge of the interrelationships between patrons, scientists, engineers, architects and artisans in classical and medieval times is at present inadequate. In any case it is impossible to generalize; much depends upon the character and requirements of the patron and the inclinations and skills of his clients. There is little doubt, however, that the concept of demarcation was not understood in earlier times and that scholars might become proficient in a number of fields, as determined partly by their predilections and partly by the demands of their patrons. If a scientist concerned himself with mechanical matters, it is not necessary to assume that he made machines or instruments with his own hands. Manufacture could have been delegated to an artisan, supervised by the scientist[58].

In the light of these – admittedly very tentative – remarks about the activities of scholars in earlier times, there seems to be no reason

[57] Millás Vallicrosa, pp. 6-9.

[58] When al-Khāzinī (early 12c.) made his self-rotating sphere, the hand-work was done by a carpenter named ʿAlī al-Sarakhsī. See Lorch, pp. 304-307.

for the supposition that the treatise edited in this article could not have been written by a scientist or a mathematician. It is attributed to 'Apollonius the Carpenter, the geometer (or engineer)', and it is possible that this was Apollonius of Perga, who worked in the latter part of the third and the first part of the second century BC. Arabic writers referred to this scholar as 'the carpenter'[59] and it may be significant that our author is associated by implication with Archimedes, who lived perhaps a generation earlier.

There is one feature common to both treatises that merits attention. This is the vertical water-wheel with a vertical gear-wheel on its axis, meshing with a lantern-pinion. (The term *abrār* used by the author for the water-wheel has not been identified). This was the arrangement for a type of water-mill that was first described by Vitruvius (*De architectura*, X 5) in about 27 BC. Since Vitruvius was describing a well-established system, it is possible that it had been in use for two or three centuries before Vitruvius' time. 'Apollonius' refers to this as the 'newly invented wheel', as opposed to the *rūmī* type, which must therefore have been an earlier invention. It is probable that it was a horizontal wheel – it is difficult to imagine any other wheel that would be radically different from the Vitruvian wheel. The times and locations of the origins of the horizontal and vertical water-wheels are still unresolved and any indication, however vague, that contributes to a solution should not be neglected. We have here a suggestion, and no more than a suggestion, that the invention of the horizontal wheel predates that of the vertical type.

The only machine comparable to the Apollonius device before the Renaissance is a fluting automaton attributed to the Banū Mūsā[60]. The water machinery is very similar to that of Apollonius, but the musical apparatus is much more sophisticated.

The authorship of the present treatise may never be known. It has been suggested[61] that the early part of the 'Archimedes' treatise including the arrangement of the gears was indeed the work of Archimedes himself. If this is so, it lends weight to the supposition that the 'Apollonius' treatise was composed by Apollonius of Perga. There is no proof, however, that the two treatises had been embodied

[59] Wiedemann, II p. 86.

[60] The Arabic text was published by Cheikho in 1906. There is an English translation in Farmer, pp. 88-118.

[61] Hill (1976), p. 9.

in a single manuscript in Hellenistic times and the Arabic of the 'Apollonius' treatise is different from that of the 'Archimedes', a fact that militates against a single translator for both works. The two translations could have been made at different times and brought together subsequently. The 'Apollonius' treatise could therefore be of Hellenistic, Byzantine, or even Islamic, origin. The reference to the vertical water-wheel as a new invention strengthens the case for a Hellenistic origin. An attribution to Apollonius of Perga is by no means fanciful.

Bibliography

L. Cheikho, [Arabic edition of the Banū Mūsā's treatise on the fluting machine], *Al-Machriq*, 9 (1906) 444-458.

L. Cheikho, *Catalogue raisonné des manuscrits de la Bibliothèque Orientale de l'Université Saint Joseph de Beyrouth*, 2 vol., Beirut 1921 & 1929.

H. G. Farmer, *The Organ of the Ancients from Eastern Sources*, London 1931.

D. R. Hill, *On the Construction of Water-Clocks*, Occasional Paper No. 4, Turner and Devereux, London 1976.

Ibn al-Razzāz al-Jazarī, *The Book of Knowledge of Ingenious Mechanical Devices*, translated and annotated by D. R. Hill, Dordrecht 1974.

R. Lorch, "Al-Khāzinī's 'Sphere that Rotates by Itself'", *Journal for the History of Arabic Science*, 4 (1980) 287-329.

J. M. Millás Vallicrosa, *Estudios sobre Azarquiel*, Madrid 1950.

Amin Muwafi and A. N. Philippou, "An Arabic Version of Eratosthenes on Mean Proportionals", *Journal for the History of Arabic Science*, 5 (1981) 147-165.

Jamil Ragep and E. S. Kennedy, "A Description of the Ẓāhiriyya (Damascus) MS 4871: a Philosophical and Scientific Collection", *Journal for the History of Arabic Science*, 5 (1981) 85-108.

E. Wiedemann, "Beiträge zur Geschichte der Naturwissenschaften. III" reprinted in E. Wiedemann, *Aufsätze zur arabischen Wissenschafts-geschichte*, Hildesheim-New York 1970, vol. I, pp. 59-104. Originally published in 1905.

E. Wiedemann, "Beiträge zur Geschichte der Naturwissenschaften. XXXV. Über Musikautomaten", *Aufsätze*, vol. II, pp. 47-56. Originally published in 1914.

The Nilometer*

(an updated version of an earlier article by Julius Ruska)

MIQYĀS (Arabic), measurement, means of measuring, any simple measuring instrument; in Egypt the name of the Nilometer, *i.e.* the gauge in which the annual rise of the river can be measured. p. 39

Originally the rising of the Nile was measured by the gauge (*al-raṣāṣa*). According to the historians Ibn ʿAbd al-Ḥakam, al-Quḍāʿī and others, Joseph, the son of Jacob, built the first Nilometer at Memphis; at a later date, the "aged Dalūka" built Nilometers in Akhmīm and Anṣinā (Antinoë). These were the Nilometers in use throughout the Hellenistic period till the conquest of Egypt by ʿAmr b. al-ʿĀṣ. The latter erected a Nilometer at Aswan and a second at Dendera. Others were built in the reigns of Muʿāwiya and al-Walīd b. ʿAbd al-Malik. Finally, the caliph al-Mutawakkil had a large Nilometer built on the island of Rawḍa at Fusṭāṭ on the site of a Nilometer built in the reign of al-Walīd, which had subsequently been swept away by floods. This Nilometer, which still exists, was completed in 247 H (861–62) by Muḥammad al-Ḥāsib according to the inscription around the top of the pit. Ibn Abī Uṣaybiʿa (*ʿUyūn al-anbāʾ fī ṭabaqāt al-aṭibbāʾ*, ed. A. Müller, Cairo & Königsberg, 1882-84, I, pp. 207-208) says that the construction was supervised by Aḥmad b. Kathīr al-Farghānī but speaks disparagingly of his technical ability, so it may well be that al-Farghānī acted simply as

*In this new reprint of the *Encyclopaedia of Islam* article a slightly simplified transliteration system for Arabic has been used and place names have been given in their anglicised form whenever possible. All references in square brackets are to other articles in the *Encyclopaedia of Islam*. References in the indexes of this book are to the original pagination, here given in the margins.

the caliph's representative and had little to do with the actual con-
p. 40 struction of the instrument. When it was complete, al-Mutawakkil appointed Abu 'l-Raddād to look after it, and the office remained hereditary in his family down to the time of al-Maqrīzī who died in 846 H (1442) (*Kitāb al-Khiṭaṭ*, Būlāq, 1270 H., I, pp. 57 ff.).

The Nilometer was surveyed and described by K. A. C. Creswell (*A Short Account of Early Muslim Architecture*, Harmondsworth 1958, pp. 292-296, with a diagram at p. 293). It consists of a tall graduated octagonal column, which serves as a measuring gauge, standing in a stone-lined pit, roughly 6.20 metres square, with a staircase running down to the bottom. Below this is a cylindrical section with steps cut into its masonry walls. The four sides of the square pit are relieved by recesses, each covered by a pointed arch vault resting on a pair of engaged colonettes with clock-formed capitals and bases. This type of arch was, of course, an essential part of Gothic architecture, but the arches in the Nilometer are three centuries older than any Gothic example. The measuring column is a tall octagonal shaft measured into 16 cubits, averaging 54.05 cm, by transverse lines, and the ten uppermost are each subdivided into 24 "fingers" by 24 divisions, grouped four by four alternately on either side of a vertical line. Creswell then mentions that the column has been broken in two places: "(1) in the twelfth cubit, the length of which has been reduced, between 1798 and 1853, to 22.5 cm (9 in.), and (2) at the junction of the 16th and 17th (*sic*), the length of the cubits remaining unchanged." The column rests on a pedestal 83 cm square and 1.17 metres in height and this rests on a millstone 1.50 metres in diameter and 32 cm thick. The length of the column before it was broken, including these supports, was 19 cubits. The millstone rests on a floor of wooden planking supported by four heavy timber beams. Connection to the Nile is made by three tunnels, all opening into the east side.

Descriptions by Arab writers agree essentially with the foregoing details given by Creswell, except for the length of the column. According to Ibn Jubayr (*Riḥla*, ed. W. Wright and M. J. de Goeje, Leiden & London 1907, pp. 54-55), this was 22 cubits; a water-level of 17 cubits was most beneficial for agriculture but up to 19 was acceptable; the sultan was entitled to levy *kharāj* tax if the level reached 16 cubits and the actual amount of tax levied depended upon the measured level above that point. Al-Idrīsī, writing perhaps three decades earlier than Ibn Jubayr, has much the same information, including the fact that the sultan could levy *kharāj* once the level had reached 16 cubits. If the level exceeded 20 cubits it was detrimental because trees were uprooted and houses ruined, while a maximum level below 12 means a year of drought and sterility (*Description de l'Afrique et de l'Espagne par Edrîsî*, ed. Dozy & de Goeje, Leiden, 1866, Arabic text, p. 144, French transl. p. 173). There is therefore an indication that the column was longer in the 12th century than it is today.

Al-Muqaddasī (*Aḥsan al-taqāsim fī ma'rifat al-aqālīm*, ed. de Goeje, Leiden 1877, p. 206) gives little information about the construction of the Nilometer, saying simply that it was a pit in the centre of which was a long column divided into cubits and fingers. He tells us something, however, about the importance of the device to the people of the Nile Delta. When the water started to rise, the supervisor reported daily to the sultan the mark reached by the water on the measuring column. No public announcements were made until the level reached 12 cubits. Thereafter the crier made his rounds every day, announcing "Allah increased the blessed Nile today by such-and-such an amount". The people rejoiced when the water level reached 16 cubits, since they knew it would be a good year.

Bibliography (in addition to references given in the text): al-Qazwīnī, *Zakarija ben Muhammed ben Mahmud el-Cazwini's Kosmographie*, ed. Wüstenfeld, Göttingen 1849, I, p. 186; H. Ethé, *El-Kazwini's Kosmographie*, 1868, p. 376; U. Bouriant, "Description topographique et historique de l'Egypte", *Mémoire de la Mission Archéologique Française du Caire* 1 (1929), pp. 162 ff.; A. Mez, *Die Renaissance des Islâms*, Heidelberg 1922, p. 427; Ali Bey, *Travels*, London 1816, II, pp. 23ff.

XII

The Banū Mūsà and their *Book of Ingenious Devices*

The spread of Islam beyond the borders of Arabia began in A.D. 634 with the invasion of Syria and Iraq by large raiding parties. The success of these forces against the armies of the Byzantine and Persian empires was facilitated by the exhaustion of the two empires after a prolonged war between themselves and — in Syria, Egypt and Iraq — by the hostility of the indigenous populations to their imperial masters. Very soon the campaigns became wars for permanent conquest, fed partly by the desire for booty and partly by genuine religious fervour. By about A.D. 645 Syria, Egypt, Cyrenaica, Iraq and Mesopotamia had been effectively subjugated. Resistance was stiffer in the Iranian heartland and in the Berber country of north-west Africa but the first was conquered by about A.D. 656, the second by the end of the seventh century. The conquests were then extended into Spain and Central Asia, and by the middle of the eighth century the Arab empire had attained its greatest expansion, stretching from the Amu Darya river (Oxus) to the Pyrenees, and including the Mediterranean Islands of Cyprus and Sicily. At first the conquests were directed by the Caliphs from Medina, but in A.D. 660 the Umayyad dynasty assumed power and made their capital in Damascus. In A.D. 750 the Umayyads were in turn overthrown by the Abbasids and the centre of gravity of the empire shifted to Iraq, where the second Abbasid Caliph, al-Manṣūr, built the splendid new city of Baghdad as a fitting capital for his empire. The Abbasids were only able to maintain the empire as a cohesive unit for about one hundred and fifty years, after which it split up into a number of separate states, some nominally owing allegiance to the Caliphs but all virtually independent. In the second half of the ninth century the Caliphs themselves became the puppets of their Turkish military commanders and never regained temporal power. In the eleventh century Baghdad was incorporated in the Seljuk empire and was finally destroyed by the Mongols under Hulagu in A.D. 1258.

In the period from the early conquests to the end of the Umayyad dynasty the Arabs were concerned with conquest, with the pacification of the occupied lands, and with the setting up of administrative, fiscal and legal systems. The Arab conquerors formed a ruling élite, in receipt of financial and other privileges that were denied the conquered population, even when these became Muslims, although in this case discrimination was contrary to the teachings of the Koran. Islam slowly became the faith of the majority and Arabic superseded the native languages everywhere, except in Spain and Iran, although even in those countries Arabic became the main vehicle for written communication and remained so for several centuries. The Arabs were never more than a small minority of the population of the empire and as they intermarried with women from the native populations the numbers of those who could claim pure Arab descent diminished, while the status of the indigenous peoples gradually improved. The Empire of the Arabs became the World of Islam. It was by no means a homogenous world: racial and linguistic differences remained, there were sizeable minorities of Christians, Jews and other religious groups, and the Muslims themselves were split by religious schisms, the deepest and most enduring of which was that between the orthodox Sunni believers and the Shia, followers of ʿAlī the son-in-law of the Prophet. Nevertheless, and despite the political break-up, Islam was a real entity, held together by the bonds of religion and language.

We speak of Arabic science, engineering, medicine, philosophy and so on, because the books were written in Arabic and the beginnings of intellectual activity in Islam were due to the encouragement of Arab patrons. The scholars were of diverse origins, although in many cases their descendants would today be considered, and would consider themselves, Arabs. Arabs, Greeks, Persians, Jews and Sabeans from Mesopotamia, all made notable contributions to the origins and growth of Arabic science and technology, the foundations of which were laid in ninth-century Baghdad during the Abbasid Caliphate.[1] The Caliph al-Ma'mūn is generally regarded as the main founder of Arabic science, and there seems little doubt that without his encouragement the first steps would have been slower and less certain. Intensely interested in ancient learning and its 'verification', he founded an academy in Baghdad known as the House of Wisdom (*bayt al-ḥikma*),[2] and supervised the collection of manuscripts from Byzantium, Iran and India, which were then translated into Arabic, usually with high competence, in the House of Wisdom. Most emphasis was placed upon Greek works, sometimes received in Syriac versions, and most of the major Greek writings on mathematics, astronomy, mechanics, medicine and philosophy passed into Arabic during the course of the ninth century. Some of these works are now preserved only in Arabic versions. There were also significant transmissions from Iran and India and the importation from India of 'Arabic' numerals, the concept of zero being of the greatest importance for the future of mathematics and science. The translators wrote commentaries and criticism of many classical works, such as Euclid's *Elements*, the

Conics of *Apollonius* and Ptolemy's *Almagest*. Many of them were also scientists in their own right and composed original works of considerable value. By the end of the ninth century the great work of assimilation was complete, and learning had appeared in a new guise, in which Greek, Indian and Iranian elements were merged with Arabic additions to form a recognizably new science. In the same period the language itself had been transformed into a subtle and flexible instrument for the expression of philosophical and scientific thought. For the ensuing five centuries Arabic science flourished and developed, albeit sporadically, throughout the Islamic world, from Andalusia to Central Asia. In medieval Europe, notably in twelfth century Toledo, where there was a school of translators not unlike al-Ma'mūn's House of Wisdom, many Arabic works were rendered into Latin and became a potent force in the cultural revival known as the Renaissance.

Life and Works of the Banū Mūsà

The Banū (sons of) Mūsà were among the most important figures in the intellectual and political life of ninth century Baghdad and played a key rôle in the founding and development of Arabic science and technology. Fortunately we have a good deal of information about them in biographical and historical works, the most important of which are listed briefly below, with the abbreviations used in references given in parentheses after each source. (In other words each Arabic source is given a letter reference, e.g. F, Q, K, H, A and so on, and these are used in this paper in association with page numbers. Thus H.264 signifies page 264 of Bar Hebraeus', *History of the Dynasties*.) Separate articles are devoted to the Banū Mūsà in several works, of which the earliest is the *Fihrist* of Ibn al-Nadīm,[3] composed A.D. 988, pp. 378–9 (F). Ibn al-Qiftī,[4] A.D. 1172–1248, p. 208 draws some of his material from Ibn al-Nadīm and some from unknown sources (Q). Additional information is given by Ibn Khallikān, A.D. 1211–1282, pp. 161–3 in his *Biographies of Illustrious Men (K)*,[5] and by Bar Hebraeus, A.D. 1226–1289, *History of the Dynasties*,[6] pp. 264–7 (H). Abu' l-Fidā', A.D. 1273–1331, *Universal History*, Vol. I, Bk. 2, p. 49 provides what is virtually a copy of Ibn Khallikān's article (A).[7] There are other works which do not contain separate entries on the Banū Mūsà, but rather a number of scattered references to their activities. Perhaps the most important of these is the great history of al-Ṭabarī, A.D. 839–923, Series III (T).[8] Al-Ṭabarī was living in Baghdad while the brothers were active in the political and intellectual life of the city and his mentions of them, though brief, therefore have particular value. Of a similar nature are a number of references in the first volume of the biographical dictionary of Ibn Abi Uṣaybi ʿa, A.D. 1196–1270,[9] which are of interest for the light they throw upon the relationships of the brothers with the Caliphs and with other scholars (U). Finally, there are a number of references to the brothers' astronomical studies in Ibn Yūnus, d. A.D. 1009 (Y),[10] and in the *Chronology* of al-Bīrūnī, A.D. 937–*c*. 1050, (B).[11]

The three brothers were called, in order of seniority, Muḥammad, Aḥmad and al-Ḥasan. We do not know their dates of birth but Muḥammad died in January 873 (F.K.) and could hardly then have been less than seventy years old because the youngest brother al-Ḥasan was already a brilliant geometrician in the reign of al-Ma'mūn (813–33) (H.264). Their father, Mūsà bin Shākir, is usually described as al-Munajjim meaning either astronomer or astrologer, although in his youth he is said to have been a resourceful highwayman who made the roads in Khurāsān unsafe (H.264). By all accounts, however, he became a noted astronomer and a close companion of al-Ma'mūn when the latter was residing at Marw in Khurāsān before he became Caliph in 813. When Mūsà died he committed his sons to the care of al-Ma'mūn who entrusted Isḥāq b. Ibrahīm al-Musaʿbī with their guardianship. Isḥāq sent them to the House of Wisdom where their tutor was Yaḥyā b. Abī Manṣūr and they completed their education in that establishment. In general they were said to have been skilled in geometry, ingenious devices (*ḥiyal*), music and astronomy. According to Ibn al-Nadīm and Ibn Khallikān their weakest subject was astronomy, but this seems to conflict with the opinions of Ibn Yūnus and al-Bīrūnī, both good judges, who spoke highly of the accuracy of the Banū Mūsà's astronomical observations. Muḥammad, who was the most influential of the brothers, specialized in geometry and astronomy, and excelled Aḥmad in all the sciences except in the construction of ingenious devices. Al-Ḥasan was a brilliant geometrician with a retentive memory and great powers of deduction. A rival once tried to discredit him in front of al-Ma'mūn by saying that al-Ḥasan had read only six of the thirteen books of Euclid's *Elements*. Al-Ḥasan replied by saying that it was unnecessary for him to read the remainder because he could arrive at the answers to any of Euclid's problems by deduction. Al-Ma'mūn acknowledged al-Ḥasan's skill, but did not excuse him, saying 'laziness has prevented you from reading the whole of it — it is to geometry as the letters a, b, t, th[12] are to speech and writing.' (H.264). Al-Ḥasan is rarely mentioned by name elsewhere in the sources and may have preferred to devote his time to scholarship, whereas his brothers were involved in a variety of undertakings.

At the time of their entry into the House of Wisdom the Banū Mūsà were poor and needy (H.264) but under the successors of al-Ma'mūn they became wealthy and influential. They devoted much of their wealth and energy to the quest for the works of ancient writers, and sent missions to Byzantium to seek out such material and bring it to Baghdad (F.Q.). Muḥammad is said to have made a journey to Byzantium in person. The renowned translator Thābit b. Qurrà accompanied Muḥammad on his return to Baghdad and began his studies in Muḥammad's house (F.380. U.215). The brothers used to pay about five hundred dinars a month to a group of translators who worked in the House of Wisdom (U.187). The most outstanding of these translators were Thābit b. Qurrà and Ḥunayn b. Isḥāq, both of whom were proficient in Greek, Syriac and Arabic. They rendered numerous works into Arabic and made important original contributions.[13] Muḥammad was on friendly terms with these men, particularly with Ḥunayn, who translated and composed books at the request of his patron (U.102, 193, 199, 205). Several of these works were on medical subjects, which indicates that the brothers' interests were not confined to the physical sciences. Indeed, Thābit b. Qurrà wrote a report on atmospheric phenomena that had been observed by himself and the Banū Mūsà (U.219).

Apart from their rôle as scientists and patrons of scientists, the Banū Mūsà, particularly Muḥammad, were employed by the Caliphs on various undertakings, and became involved in the turbulent political life of Baghdad. One of their earliest tasks, carried out on the orders of al-Ma'mūn, was to check the circumference of the earth, given by the ancient writers as 24,000 miles. To do this they went to the level desert of al-Sinjar in northern Iraq, accompanied by a group of observers. They set up a base point from which they measured the altitude of the Pole Star, and then walked due north in a straight line, using a rope for measuring and marking their route by pegs, until the altitude of the Pole Star had increased by one degree. They then repeated this procedure towards the south, after which they went to the locality of al-Kūfa in southern Iraq and carried out similar measurements. The value they arrived at was 66⅔ miles for one degree of latitude, giving the circumference of the earth as 24,000 miles. (K.A.)

Muḥammad and Aḥmad later became concerned with public works. They were among a group of twenty men responsible for the construction of the new town of a Jaʿfarī for the Caliph al-Mutawakkil (T.1438) and are mentioned as having directed the excavation of a canal between Baṣra and Wāṣit (T. 1747). In this field, as in others, they did not take kindly to competitors and are said to have plotted against anyone who rivalled them in knowledge (U.207). They became enemies of the famous philosopher al-Kindī (T.1502) and caused him to lose the favour of al-Mutawakkil who ordered him to be beaten and allowed the brothers to confiscate his library (U.207). According to the account of Ibn Abī Usaybiʿa (207–208) the envious nature of the brothers almost led to their downfall. They had also alienated al-

Mutawakkil from Sanad b. Alī, a good engineer and a friend of al-Kindī. When al-Mutawakkil commissioned the brothers with the excavation of a canal known as al-Jaʿfariyya, they subcontracted the work to Aḥmad b. Kathīr al-Farghānī 'whose knowledge was greater than his achievements, because he never completed a work'. Al-Farghānī made a serious error in his levelling, such that the canal would never have filled to the required depth. When rumour of this reached al-Mutawakkil he threatened to crucify Muḥammad and Aḥmad on the banks of the canal if the report proved to be true. In desperation they turned to Sanad b. ʿAlī who would only give his assistance if they undertook to return al-Kindī's books to him. When this had been done, Sanad said that he would tell the Caliph that no mistake had been made; because the Tigris was at its highest level the error would not be detectable for four months, and the astrologers had predicted the death of al-Mutawakkil before this period had expired. Within two months the Caliph was assassinated and the conspirators escaped retribution.

In his last years Muḥammad was deeply involved with palace politics, in the period during which Turkish commanders were assuming effective control of the state. After the death of al-Muntaṣir he successfully campaigned for the nomination of al-Mustaʿīn in preference to Aḥmad b. al-Muʿtaṣim, because the latter was a friend of al-Kindī (T.1502). When Baghdad was besieged by the army of the Caliph's brother Abu Aḥmad, Muḥammad was sent by al-Mustaʿīn's commander ʿAbd Allah b. Ṭāhir to estimate the strength of the enemy forces (T.1557–8). He was at al-Mustaʿīn's side when the Caliph addressed the rebellious populace (T.1634) and he was one of a delegation sent by Ibn Ṭāhir to Abu Aḥmad's army to ascertain the terms for al-Mustaʿīn's abdication. (T.1641). These reports are interesting for the information they give us about Muḥammad's standing in the courtly circles, but his political activities were not of the first importance since it was of little moment who occupied the Caliphal throne when real power lay elsewhere.

The rôle of the Banū Mūsà in fostering the work of translators and scientists can hardly be overestimated, but their own scientific achievements were far from negligible. Al-Bīrūnī mentions their calculations for the length of the solar year (B.61) and Ibn Yūnus tells us that he made use of their astronomical observational records, which were numerous and accurate (Y.42, 62, 98, 132, 133, 142, 146, 148). A list of their works is given by Ibn al-Nadīm and Ibn al-Qiftī, as follows:[14]

1. Book on the steelyard (*qarasṭūn*). (F.Q.)

2. Book of Ingenious Devices (*ḥiyal*) — by Aḥmad. (F.Q.)

3. Book on the long, curved figure (i.e. the ellipse) — by al-Ḥasan. (F.Q.)

4. Book of the first movement of the sphere (spheres in Q) — by Muḥammad. (F.)

5. Book on conic sections — by Muḥammad. (F.Q.)

6. Book of the Three [?] — by Muḥammad. (F)

7. Book of the geometrical shape, as demonstrated by Galen — by Muḥammad.[15] (F.Q.)

8. Book of the Parts (proportions?) — by Muḥammad. (F.Q.)

9. Book in which it is demonstrated by didactic means and by geometrical method that there is no ninth sphere outside the sphere of the fixed stars — by Aḥmad. (F) In Q this reads: denial of a ninth sphere — by Aḥmad.

10. Book on the beginning of the world — by Aḥmad. (F) In Q this reads: on the first of causes.

11. Book on the question put to Sanad by ʿAlī by Aḥmad b. Mūsà. (F.Q.)

12. Book on the nature of speech (*Kalām*) — by Muḥammad. (F)

13. Book on the questions discussed between Sanad b. ʿAlī and Aḥmad b. Mūsà. (F)

14. Book on the measurement of the sphere, division of an angle into three equal parts, and calculation of the mean proportional between two quantities. (F.Q.)

15. Astronomical tables — as mentioned by al-Bīrūnī and Ibn Yūnus.

16. Two works on time, edited by Thābit b. Qurrà. (Q.117)

17. A work on war engines (Hajji Khalīfa I, 394).

18. On the sphere. (Hauser 10. Without exact references.)

19. On the construction of astrolabes. (Hauser 10. Without exact references.)

20. Description of a musical automaton, (E. Wiedemann. *Amari Festschrift* (1909) 169–80. Wiedemann made his translation from a text published by Professor Cheikho in Bayrut, but gives no details.)

The most important of these works apart from the *Book of Ingenious Devices* was known in the medieval west as *On the Measurement of Plane and Spherical Figures*, in a Latin translation by Gerard of Cremona, *Liber trium fratrem* etc. This was probably No. 14 above — see Suter 21. The Latin text was published by M. Curtze (*Nova Acta der Kaiserl. Leop. Carol Deutschen Akad. der Naturf.*, 49, Nr. 2, 1883)

The Book of Ingenious Devices (*Kitāb al-Ḥiyal*)

This work enjoyed a great reputation in medieval Islam. The earliest reference we have is the *Fihrist* 'Book of Ingenious Devices (*ḥiyal*) by the Banū Mūsà b. Shākir the astronomer, and it contains a number of movements'. (F.397.) Ibn al-Qiftī says that 'their book on *ḥiyal* is marvellous and famous' (Q.208). Bar Hebraeus is the first to mention Aḥmad as having been concerned largely with this subject: he says. 'Aḥmad was lower than him [i.e. Muḥammad] in all the sciences except in the construction of *ḥiyal*, in which things were revealed to him such as had been revealed to no-one [else]' (H.264). Ibn Khallikān, having mentioned at the beginning of his article that the Banū Mūsà were famous for their book on *ḥiyal*, says later 'their book on *ḥiyal* is wonderful and rare, and includes every strangeness. I have studied it and found it the best and most delightful of books. It is one volume.' (K.161–2.) Ibn Khaldūn (A.D. 1332–82) writes 'there exists a book on mechanics that mentions every astonishing remarkable and nice mechanical contrivance. It is often difficult to understand, because the geometrical proofs occurring in it are difficult. People have copies of it. They ascribe it to the Banū Shākir.'[16] The only comment we have by a technical man is that of al-Jazarī,[17] in the introduction to that section of his book (composed 1206) that deals with fountains. He says 'I did not follow the system of the Banū Mūsà, may God have mercy upon them, who in earlier times distinguished themselves in the matters covered by these subjects.' He goes on to make his criticism of the Banū Mūsà's fountains, which he considered to be unreliable; in particular he thought that the intervals between the changes of shape of the fountains were too brief. The only modern studies of major importance on the *Book of Ingenious Devices* are the papers published by E. Wiedemann and F. Hauser in the early years of the present century. Their joint paper deals with the large drink dispensers (Models 75–87),[18] and provides descriptions of the operation of the devices together with modified copies of the original illustrations with Roman lettering. Hauser's much longer work covers the remaining devices, and was based upon a translation and information on sources supplied by Wiedemann.[19] These papers are of considerable merit and allow anyone with technical training and a working knowledge of German to obtain a good understanding of the operation of all the devices. Hauser's work contains a great deal of information on the sources, on the life and works of the Banū Mūsà, a survey of similar works by Greek and Arab writers, a full description of the manuscripts, and transliterated Arabic equivalents of technical terms. Much of the credit for this presentation must be given to Wiedemann. Thereafter the treatment is much the same as in the joint paper — descriptions of the devices, modified copies of the illustrations, explanatory notes and drawings.

In general the work of Wiedemann and Hauser achieves the purpose for which it was presumably intended: that of informing an audience of historians of science, and engineers interested in the history of their profession. Even in this context the information they provide on source material is inadequate, since proper references, including the edition and date, are never given. For a wider public Hauser's explanations of the principles — hydrostatic, aerostatic or mechanical — that are embodied in the devices are seldom adequate for a layman. Finally it must be said that there is no real substitute for a proper translation, (which Hauser considered unnecessary — p. 28) with faithful photographic reproduction of the original illustrations. Only when this is available can readers make their own estimate of the value of the Banū Mūsà's contribution to the history of technology.

In view of the interest shown in the work in the Islamic world from the tenth to the seventeenth century it is somewhat surprising that we know at present of only three almost complete manuscripts and two fragments, compared for instance, with some fifteen copies of al-Jazarī's masterpiece. The fragments are in Leyden (catalogue Vol. 3, No. 1019 169 Gol) and New York (Spencer Collection, Risala'i Hakim Muhammad Yasin). One of the almost complete manuscripts is in the Vatican (No. 317) and another is in two parts, one in Gotha (Catalogue van Pertsch, Arab 1349) the other in Berlin (Catalogue Ahlwardt 5562). Fortunately the three manuscripts together permit full retrieval of text and illustrations of the one hundred devices called models (*shakl*) by the Banū Mūsà. The best manuscript is Topkapi Saray A.3474. This contains only seventy-seven models; in particular the fountains and lamps are omitted. The text is accurate and the drawings are well executed with reference letters correctly positioned and corresponding to the references in the text.* The Vatican text is complete up to and including Model 90, although some of the leaves are out of sequence. Model 91 is missing completely, as is the beginning of Model 92. The text ends with Model 93 (no illustration), with the words 'the work is complete'. The Gotha/Berlin manuscript has more lacunae than the Vatican — in particular Models 11–18, inclusive, are missing — but Models 91–100 are complete. Both manuscripts begin with the words 'this is the Book of IngeniousDevicesby the Banū Mūsà'. In the Vatican text, however, there occurs the sentence, before Model 24, 'this is the second book of the work by Abu'l-Ḥasan Aḥmad b. Mūsà', with similar remarks for the third and fourth books before Models 43 and 66 respectively. This therefore lends weight to the statements in our sources to the effect that Aḥmad was the principal author of this work. The first model is missing from the Topkapi MS. At the top of Folio 1 R are the words 'Book of

*I am grateful to Dr. David King (see Note 41) for calling my attention to this manuscript. Unfortunately a microfilm of the MS did not come into my hands until this paper was about to go to press and I was not therefore able to use it. The loss is not great because, of the five models I have described in detail, three are not included in A.3474. The other two — Models 42 and 77 — can be adequately understood from the Vatican MS. I shall, however, be using A.3474 as the principal basis of my forthcoming book.

Ingenious Devices and wonderful matters concerning water machines'.

It seems likely that the original work contained one hundred devices, the same number that can be retrieved from the manuscripts, but one of the Models, No. 94, is not by the Banū Mūsà, since its description includes remarks such as 'as the Banū Mūsà did'. Without further evidence, however, it must be assumed that the other ninety-nine models were from the hand of the brothers. All the manuscripts are written in legible *naskhī* (the most widely used type of Arabic script). In many cases the same errors occur in the Gotha/Berlin and in the Vatican MSS, and it is probable that both were copied from the same exemplar. There are far fewer errors in the Topkapi MS. Almost all the illustrations are sectionalized elevations, one to each model, no additional detailed drawings being provided. Once the conventions have been understood the drawings are quite easy to read. They are marked with letters of the Arabic alphabet, to which reference is made in the text, although in the Gotha/Berlin MS the letters are missing from the drawings. In the Topkapi MS the letters are accurately placed and correlate with those given in the text, which are themselves consistent throughout. In the other two MSS the lettering in both text and drawings is haphazard and inconsistent.

All the descriptions are fairly brief: first a description of the purpose and operation of the device is given; then, with the words 'this is its example' (*mithāl*), attention is drawn to the illustration and the construction is briefly described; finally a more detailed description of the operation is given — often it is only by reading this final section that the foregoing paragraphs can be fully understood. In many cases the chapter terminates with advice about the use of the device, for instance 'this device is suitable for use in baths, or places where taps are installed'. The most serious weakness in the work is the absence of any proper instructions for the manufacture of the devices. Dimensions are rarely given, materials of construction are mentioned in only one or two instances, assembly sequences, fittings and jointing are never mentioned, apart from general statements such as 'part A is soldered to part B'. The supports for tanks, installed inside vessels, are not shown — presumably the tanks rested on cross-pieces. This vagueness, which is in marked contrast to the meticulous instructions given by al-Jazarī, not only detracts from the value of the work as an historical document, but also at times makes understanding difficult. This is particularly so in cases where results depended upon the achievement of a delicate balance of pressures, when the dimensions and positioning of the various components were critical.

It is very probable, therefore, that the Banū Mūsà were not themselves skilled in metalwork, but that their devices were made by craftsmen under their guidance, no doubt with the feedback of advice from the craftsmen if they considered a design to be impracticable. It could even be argued that the devices are merely abstract illustrations of the principles of machanics, but this is going too far. Most of the devices would certainly work and, in any case, the Caliphs and their circle would not have been content with unrealized paper concepts. Some of the models, however, are merely variant versions of others, with small additions and alterations, and it is possible that these were not made, but were suggested as refinements. The Banū Mūsà's approach was empirical and demonstrates a sophisticated use of hydrostatics, aerostatics and mechanics, without a matching explanation of the correct physical and mathematical uses of the phenomena, many of which had yet to be explained. There is nothing new about this, since man has frequently made effective use of natural phenomena before their causes had been identified.

Many of the devices are intended to mystify the onlookers, and hence it was necessary to conceal from them the means by which the effects were produced. For instance, when two liquids were to be poured into a vessel, the first was poured in forcefully, the second gently; if it was desired to prevent a liquid discharging when a pitcher was tilted the operator had to cover a concealed air-hole with his finger or with a piece of wax. The operator is usually referred to as 'he who understands how the device works' in contrast to 'those who are ignorant of how it works'. In the ensuing descriptions the former is designated 'the Adept', for the sake of brevity.

Eighty-five of the one hundred devices can be classified as trick vessels, which produce a variety of effects, and embody a seemingly bewildering array of flow systems and assemblies. It would be impossible to give an adequate impression of the flavour of the book, and of the methods, simply by describing a number of selected examples, unless so many were chosen as to render this article overly long. It is felt, therefore, that the following method will provide a suitable substitute for fully representative sampling: first a full list of the models is given with brief descriptions of their functioning. Then descriptions of the ten most commonly used motifs are provided, with illustrations — close inspection of the work reveals that most of the devices use varied combinations of these motifs, with the addition of individual motifs and components, to produce the desired effects. An understanding of these concepts therefore permits rapid comprehension of most of the models. Finally, five representative models are described in detail, with illustrations, in one case with a full translation. In this way it is hoped to convey an appreciation of the Banū Mūsà's work — its value, merits and weaknesses.

List of Models

Those marked with an asterisk are described in detail later. Unless otherwise stated outlets are near the bottoms of the vessels.

1. A beaker (*ka's*): demonstrating the action of a concentric siphon.

2. A pitcher (*ibrīq*) with a spout: once interrupted, outpouring cannot be resumed.

3. A pitcher (*ibrīq*) with a spout: once interrupted, inpouring cannot be resumed.

4. A jar (*jarra*) with open outlet pipe: no water discharges while inpouring is in progress; discharge occurs whenever inpouring is stopped.

5. Two basins (*jām*): by one is the figure of a small wild animal, by the other the figure of a lion. When water is poured into the small animal's basin he drinks nothing; when the lion's basin is filled both he and the other animal drink.

6. A basin (*ijjāna*): the figure of a bull is placed by the basin with his muzzle in the water it contains and one then hears a sound, simulating the sound of a thirsty bull.

7. A trough (*ḥawd*) into which about 2 litres of water are poured. When twenty or more small animals drink from it the water-level is not lowered, but when a bull drinks all the water disappears.

8. A pitcher (*ibrīq*). Action as Model 3.

9. A pitcher (*ibrīq*). Action as Models 3 and 8, except that a single interruption is possible.

10. A pitcher (*ibrīq*). Action as Models 3 and 8.

11. A pitcher (*ibrīq*). Action as Model 9.

12. A pitcher (*ibrīq*) with a spout. When the steward tilts it over a person's hands he can allow water to discharge, or prevent it from discharging.

13. A pitcher (*ibrīq*) with a spout: hot and cold water are poured in, and do not mix inside the pitcher. According to the pourer's wish, hot, cold or mixed water can be discharged.

14. A pitcher (*ibrīq*): by pouring in one or two ounces of wine the Adept can deceive his audience that the pitcher is full; he can also fill it completely and make them believe it to be empty; he can dispense from it as much or as little as he desires.

15. A drinking vessel (*kūz ibrīq*). Action as Model 14.

16. A jar (*jarra*) with two open outlet pipes, one above the other. One liquid issues from an outlet during inpouring, the second liquid issues from the second outlet when inpouring is stopped.

17. A boiler (*milyār*) with an outlet tap: hot water issues from the tap only so long as cold water is poured into the top of the boiler.

18. A boiler (*milyār*) similar to Model 17, except that it can be filled while the tap is open.

19. A jar (*jarra*) having a tap with three borings in its plug. Different measured quantities of three different liquids are poured in to the jar, and each can be extracted separately from the tap.

20. A fountain (*fawwāra*) with variable discharge.

21. A jar (*jarra*). Action as Model 19, except that the liquid quantities are not measured.

22. A jar (*jarra*) with a tap. When the tap is opened it always discharges a known quantity.

23. A jar (*jarra*). Action as in Model 22.

24. A beaker (*qadaḥ*). The adept can deceive the audience into thinking it is full by pouring in one or two ounces of wine.

25. A jar (*jarra*) with a tap; two liquids are poured in but only the second can be drawn from the tap.

26. A beaker (*qadaḥ*). Action as in Model 14.

27. A jar (*kūz*) with an outlet pipe: during inpouring of a liquid, the pourer can, according to choice, allow the liquid to discharge, or prevent its discharging.

28. A jar (*kūz*, later *jarra*) with two outlet pipes, one above the other; during inpouring of a liquid, the pourer can allow discharge from either or neither of the outlets.

29. A whistle (*saḥḥāra*): the whistle is at the top of a spherical vessel in whose underside is a perforated plate. When the vessel is plunged into water the whistle sounds.

30. A pipette (also *saḥḥāra*): upon lifting from the water, the water in the pipette may or may not be released, depending upon how it was immersed in the first place.

31. A flask (*qinnīna*). Action as in Model 14.

32. A flask (*qinnīna*) with two inlets (say A and B): wine is poured into A, water into B; when the flask is inverted water discharges from A, wine from B.

33. A jar (*jarra*) with an outlet pipe, from which the contents flow in equal measured quantities, with a distinct pause between each discharge.

34. A flask (*qinnīna*) from which pure water, pure wine, or a mixture of both may be poured.

35. A flask (*qinnīna*) from which only a known quantity of wine can be poured each time it is tilted.

36. A flask (*qinnīna*) from which, at each outpouring, a known quantity, more than that quantity, or nothing, may be obtained.

37. A flask (*qinnīna*) with an open top, from which alternate known quantities of wine and water can be poured.

38. A jar (*jarra*) with two diametrically opposed taps: liquid will only discharge from the tap that is opened first.

39. A jar (*jarra*) with a tap: it is first filled with wine, but water

discharges, not wine, as long as the water is poured into the top of the jar.

40. A jar (*jarra*). Action as in Model 39.

41. A jar (*jarra*) with an outlet pipe, from which wine discharges when inpouring of wine is stopped; when water is then poured in continuously, it discharges in place of the wine.

42. A jar (*jarra*) with three outlet pipes, one above the other: while wine is poured in nothing discharges, when inpouring is stopped wine begins discharging from the middle outlet. When water is poured in, it discharges from the middle outlet, wine from the other two; when inpouring of water is stopped, discharge of water ceases and wine again flows from the middle outlet — and so on.

43. A jar (*jarra*), with a tap, into which three different liquids can be poured, without mixing. When the tap is opened the liquids discharge in the sequence in which they were poured in.

44. A jar (*jarra*), its action similar to Model 43, but with an open outlet, which begins the sequential discharge after inpouring of the third liquid is complete.

45. A jar (*jarra*). Action as in Model 43, except that there is a pause between the discharge of the liquids.

46. A vessel (*inā'*) or a jar (*jarra*). Action as in Model 43.

47. A jar (*jarra*), with an outlet pipe; as the liquid is poured in, it can be allowed to discharge, or prevented from discharging, as desired.

48. A jar (*jarra*) with two outlet pipes one above the other; according to the method of inpouring, the liquid can be made to discharge from one or other of the outlets.

49. A jar (*jarra*) with a tap: water and wine are poured into the same inlet. When the tap is opened a known quantity of water discharges, then the same quantity of wine, again the same quantity of water, and again the same quantity of wine.

50. A jar (*jarra*). Action similar to Model 49, except that it has an open outlet pipe.

51. A jar (*kūz*). Action as in Model 47.

52. A jar (*kūz*) with open outlet pipe: when water is poured in, wine previously poured in can be allowed to discharge or not, as desired.

53. A jar (*kūz*). Action as in Model 47.

54. A jar (*jarra*) with open outlet pipe: nothing discharges from it when a known quantity of wine is poured into it from a measuring vessel; when water is poured in a quantity of water, equal to the measured quantity of wine, discharges.

55. A jar (*jarra*), its construction similar to Model 54: wine, water, or wine mixed with water discharges when water is poured in.

56. A jar (*jarra*) with two outlet pipes one above the other:

when inpouring of wine and water is stopped the liquids issue in known quantities, switching outlets after each discharge, until the jar is empty.

57. A jar (*jarra*) with two outlet pipes, which can be made to discharge alternately.

58. A jar (*jarra*) for wine and water with two diametrically opposed outlet pipes: when inpouring is completed the liquid that was poured in last begins to discharge from one of the outlets, and when this discharge is stopped the other liquid discharges from the other outlet, and so on.

59. A jar (*jarra*). Action as in Model 58.

60. A jar (*jarra*) with an outlet pipe: after wine has been poured in, wine, water, or a mixture of both can be drawn off. The result depends upon the rapidity of the inpouring of the water.

61. A jar (*jarra*) with an outlet pipe from which water and wine discharge alternately, as long as water is poured in continuously after the wine has been poured in.

62. A jar (*jarra*) with an outlet pipe from which, after wine has been inpoured, wine discharges as long as water is poured in, then water discharges as long as wine is poured in.

63. A jar (*jarra*). Action similar to Model 62, except that four liquids are used.

64. A jar (*jarra*) with two outlet pipes. Water and wine are first poured in separately before the jar is set before the audience. Then a mixture of wine and water is poured in and wine discharges from one outlet, water from the other. The audience is led to believe that the two liquids have separated inside the jar.

65. A jar (*jarra*). Construction and action similar to Model 64, but in this case the jar can be shaken without anyone detecting that it already contains wine and water.

66. A jar (*jarra*) with two diametrically opposed taps (say A and B). When both taps are opened wine discharges from A and water from B. When A is closed water ceases discharging from B and wine discharges in its place; when A is again opened, the situation reverts to the original one.

67. A jar (*jarra*). Action as in Model 66.

68. A jar (*jarra*). Action similar to Model 66. But the wine and water must be poured in in measured quantities.

69. A jar (*jarra*). Action as in Model 68.

70. A jar (*jarra*). Action similar to Model 58 but having a single outlet instead of two. The action is controlled by the operator.

71. A jar (*jarra*) with two outlet pipes, one above the other, and with two holes in the handle. It is filled with wine and water. When both holes in the handle are open, a mixture of wine and water discharges from both outlets; when one of the holes is closed wine discharges from one outlet, water from the other; if this hole is opened and the other closed the wine and water switch outlets.

72. A jar (*jarra*) with two diametrically opposed outlet pipes. Wine and water are poured into the jar. When inpouring is stopped wine discharges from one outlet, water from the other. As long as the hole in the handle is kept closed, the wine and the water switch outlets continuously.

73. A jar (*jarra*) with a tap. Either wine or water will discharge, depending upon whether a concealed hole in the handle is open or closed.

74. A jar (*jarra*) with two diametrically opposed pipes. When water is poured in, it discharges from one of the outlets; when an oil is poured in it discharges from the other.

75. A trough (*ijjāna*) that always replenishes itself when men draw water from it or when animals drink from it.

76. A basin (*jām*) set beside a container. Wine is poured into the container and when inpouring is stopped, wine flows into the basin until it is full. When a quantity of wine is withdrawn, a like quantity flows into the basin. The same design can be used with water for a drinking trough for animals, or with oil for a self-feeding lamp.

77. A basin (*jām*) or a trough (*ijjāna*) beside a closed reservoir. When moderate quantities of liquid are taken from the basin, a like quantity runs into it from a pipe at the bottom of the reservoir; if, however, a large quantity is taken no replenishment occurs.

78. A basin (*jām*) or a trough (*ijjāna*). Action similar to Model 76, except that the replenishment is from a pipe at the top of the reservoir.

79. A basin (*jām*) or a trough (*ijjāna*). Action similar to Model 77 but replenishment is from a pipe at the top of the reservoir.

80. A basin (*jām*) or a trough (*ijjāna*), by a reservoir. Wine, water or a mixture of both may be made to discharge from the reservoir into the basin.

81. Two basins (*jām*). A double version of Model 77.

82. Two basins (*jām*). Action similar to Model 81.

83. A basin (*jām*) or a trough (*ijjāna*) set beside a reservoir. A pipe discharges from the reservoir into the basin. If liquid is taken from the basin until it empties, the downpipe automatically refills it.

84. A basin (*jām*) or trough (*ijjāna*). Action as in Model 83.

85. Two basins (*jām*) set on either side of a reservoir. From the upper part of the reservoir discharge pipes are directed into each of the basins. If wine (say) is poured into one basin the downpipe discharge wine into this basin, while water discharges into the other; and vice versa.

86. A trough (*ijjāna*) or a basin (*jām*) set beside a reservoir, from the upper part of which a downpipe is directed into the trough. When a liquid is poured into the top of the reservoir it discharges into the trough. When inpouring is stopped discharge ceases; discharge resumes when inpouring is resumed.

87. Another design to produce the action of Model 86.

88. A fountain (*fawwāra* — the same term is used for all the fountains). The water issues from the fountainhead in the shape of a shield, or like a lily-of-the-valley. That is to say, the shapes are discharged alternately — either a sheet of water concave downwards, or a spray.

89. A fountain that discharges a shield or a single jet.

90. Fountain operated by wind or water, it discharges a single jet or a lily-of-the-valley.

91. Fountain in which the action of Model 89 is doubled, i.e. there are two fountainheads one discharging a single jet the other a shield; they alternate repeatedly.

92. Fountain incorporating a worm-and-pinion gear. Action as in Model 90.

93. Fountain with one main fountainhead and two or more subsidiary ones. When the main fountainhead ejects a shield the subsidiaries eject single jets, and when the main fountainhead ejects a single jet the subsidiaries eject shields.

94. Fountain with two fountainheads. Action is double that of Model 90.

95. Lamp (*sirāj*) with automatic feed.

96. Lamp with automatically extruding wick.

97. Lamp combining actions of Models 95 and 96.

98. Lamp with self-positioning windshield.

99. Bellows for removing foul air from wells.

100. Grab for extracting objects from underwater.

Motifs

The motifs described below are those most commonly used to obtain the desired effects. Usually two or more of these components are used in combination.

1. This is a concentric siphon; pipe (a–cc) was probably joined to pipe (bd) by copper wires soldered to each pipe. Pipe (bd) passes through the floor of container (f), the joint being airtight. When liquid is poured into the container it rises to point (b), the discharge of the first drop of water reduces the air pressure in space (ab) sufficiently to cause a continuous flow through pipe (bd) until the liquid level reaches (cc). This siphon is very similar to those used in lavatory cisterns, except that in the latter case the partial vacuum is caused by the sudden raising of the bell covering the downpipe. The writer tested this device using plastic pipes of 9 mm. and 30 mm. internal diameter. It operated immediately, with rapid discharge; variations in distance (ab) had no noticeable effect on its operation. In Arabic this siphon is called *ka's al-ʿadl* which can be translated approximately as the *Cup of Equivalence*. The

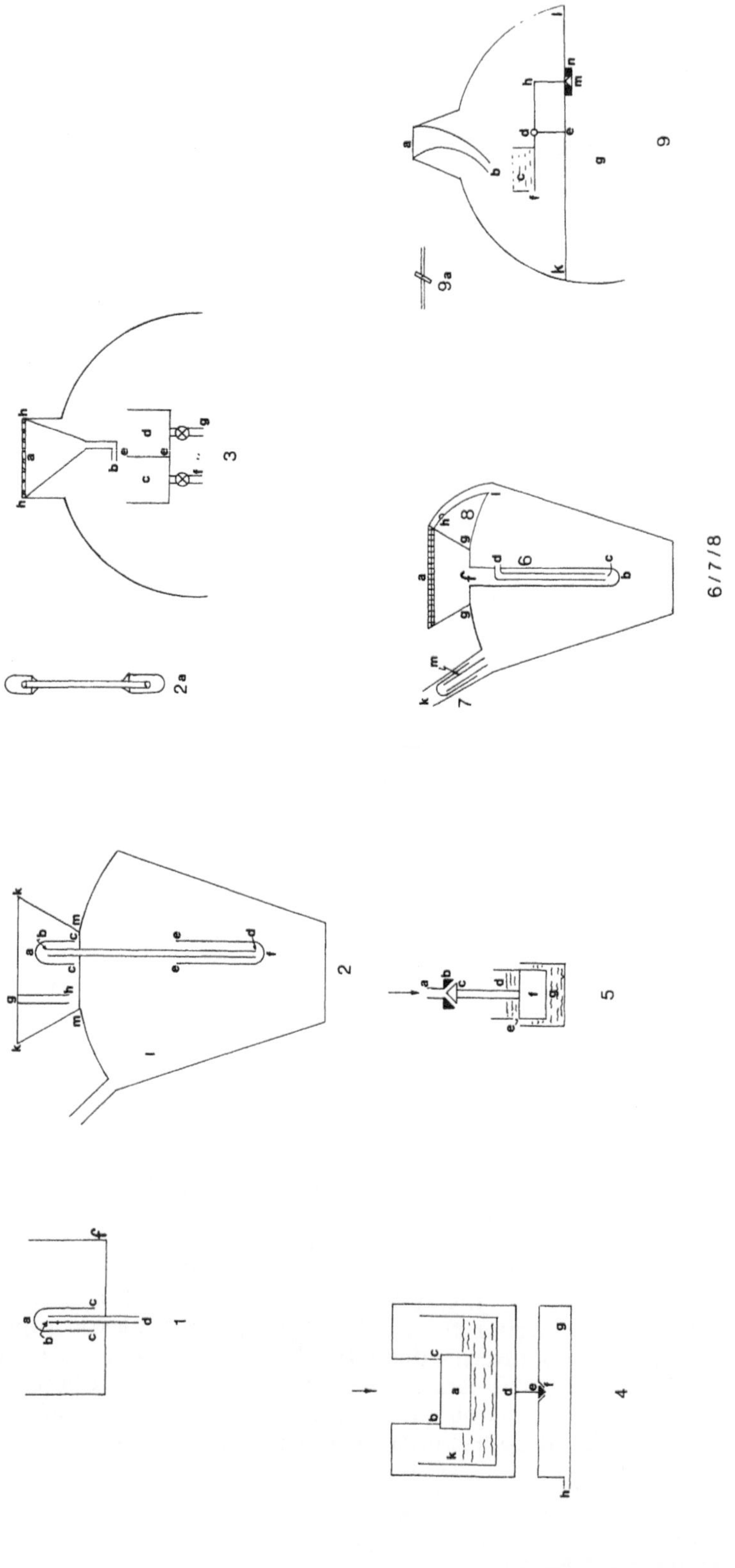

word ʿ*adl* contains the concepts of equality, balance and justice and the expression therefore implies that the purpose of the siphon is to produce a balance of pressures. This idea is more readily explicable when considering the double concentric siphon (see Motif No. 2). The ordinary bent tube siphon is used by the Banū Mūsà, but the concentric siphon is much commoner.

2. The double concentric siphon: used to prevent a liquid being poured into a container after the first pouring is stopped (Fig. 2a shows the conventional representation used in the MSS). The neck of vessel (l) is closed at its top and its root by plates (kk) and (mm). Joints between plates and neck, between plate (mm) and pipe (bd), and between plate (kk) and pipe (gh) are airtight. Vessel (l) is open to atmosphere. Liquid is poured in through pipe (gh), which must not be wide enough to allow air to enter. The liquid rises to point (b) (the air emerging through route (bde) into the vessel), pressure in space (ab) is lowered and the liquid flows through route (bde) since the liquid head (b–ee) is sufficient to maintain the flow. When pouring is stopped the liquid level in the neck falls and a liquid/air pressure balance is established; a second pouring is impossible because the end of pipe (bel) is closed, and the head in pipe (gh) is insufficient to break the air-lock in the neck of the vessel.

3. This is a device to permit two different liquids (usually wine and water) to be poured into the same vessel without mixing. The neck of the vessel is covered by a sieve-like plate (hh) beneath which is a soldered funnel (ab), whose end (b) is bent to the horizontal. Below (b) tanks (c) and (d) are installed; they have a common dividing wall (ee) and open tops. Pipes (f) and (g) are drawn simply to indicate that the tanks have outlets, probably valved, into the next part of the system. Wine (say) is poured into the top of the vessel 'copiously and forcefully', and runs into tank (c). Water is then poured in 'gently, drop by drop' and runs into tank (d). The device is drawn as it is invariably shown in the manuscripts, and Hauser frequently says that (b) is too far to the left. This is probably not the case, since it would have been possible to pour in the second liquid slowly enough for it to be held by surface tension to the underside of the funnel and deposited into tank (d); only the last few drops would fall into tank (c). If, however, point (b) were to the right of plate (ee), then after pouring of the first liquid ceased, a large fraction of the contents of the funnel would fall into tank (d).

4. The assembly shown here demonstrates a method of opening a valve which is directly below the receiving vessel. Float (a) is in tank (k), into which the liquid is flowing. Rods (bd) and (cd) are soldered to the top of the float, and bent round the outside of tank

(k) as shown, and joined at point (d). A short valve rod connects the plug (e) of the valve to point (d). As the float rises valve (e) lifts from its seat (f). In the case shown here this would permit air to enter the sealed tank (g) and liquid could discharge from the narrow outlet (h).

5. This assembly has the same purpose as device No. 2. Liquid is poured in at (a) and flows through the open valve seat (b) into the small tank (d) from which it overflows into the large tank (g). Tank (d) is soldered to the top of float (f). The valve rod carrying plug (c) is soldered to the top of the float and passes through tank (d). Tank (d) has a small hole (e) in the bottom of one of its sides. As long as tank (d) is full its weight prevents the float (f) from rising, but when pouring is stopped tank (d) empties through hole (e), the float then rises and closes the valve so that pouring cannot be resumed. This assembly is often used in conjunction with Motif No. 4, so that when the float rises it also actuates a valve below tank (g).

6. 7. 8. The illustration combines three motifs. 6. The vessel is provided with a perforated cover (a) and the root of its neck is closed by a plate (gg). and a wide pipe (fb) closed at end (b) is soldered to the centre of the plate. All the joints are airtight. A narrow pipe (cd) runs vertically up pipe (fb); the top of it is bent and protrudes from pipe (fb) near its top — the joint between the two pipes is airtight. When liquid is poured in at (a) it flows along route (afcd) into the vessel and because end (c) of pipe (cd) is then blocked no air can enter the vessel from its top. 7. The vessel has a spout (k) to the root of which a concentric siphon (m) is fitted, its narrow pipe penetrating the wall of the vessel. This is to prevent air entering the vessel when liquid is poured out. 8. It also has a hollow handle that opens into it at (l) and in the handle is a small concealed hole at (h). If the server secretly covers hole (h) with his finger when making the motion of pouring there will be no discharge because no air can enter the vessel. (In fact one drop will come out, sufficient to lower the air pressure in the vessel slightly below atmospheric.)

9. A vessel is divided by a plate (kl) to which is fitted the seat (n) of a conical valve, the plug (m) which is soldered to the end of a rod (mhdc). The horizontal section of the rod turns about on an axle (de) which is supported by a vertical stanchion (de) whose lower end is soldered to plate (kl). To the left-hand side of the rod is fixed a small tank (c). in the underside of which is a small hole (f): the lever arm is so designed that when tank (c) is empty the right-hand side is heavier and valve m/n is open. The tank is filled through the curved funnel (ab) and the liquid overflows from the tank and collects in the upper part of the vessel. When pouring is stopped tank (c) empties through hole (f), valve m/n opens, and the liquid

the number of components installed they cannot have been much smaller than this. It will be noticed from the brief descriptions given in the List of Models, that there are a number of recurrent effects, that several models may produce the same effects, or that one model is simply a more complex version of another. The ten motifs indicate how many of the results were obtained. Motif 2 was used so that inpouring could not be resumed once it had been interrupted, and if a second double siphon, its top higher than the top of the first, were introduced then one interruption is possible. Motif 5 could also be used to prevent a second filling and it had another purpose, also effected by Motifs 4, 5 and 9, of timing the opening of a valve at some point in the flow system. Motif 3 was the commonest method for permitting the inpouring of different liquids without their mixing inside the vessel, but Motif 5 was also used for this purpose. The drawing of Motifs 6, 7 and 8 is actually that of Model 12. When hole (h) is covered with the finger or with wax, no liquid can be poured from the spout. The action can be doubled by running two pipes into the handle, each terminating at one end in a hole in the handle and at the other in a closed tank. The outlets from the tanks lead either into a spout or into an open or trapped outlet. Let the two holes in the handle be A and B, corresponding to liquids A and B: then when hole A is open liquid A discharges, when hole B is open liquid B discharges, when both are open a mixture of A and B discharges, and when both are closed nothing discharges. By considering the List of Models together with the Motifs it is possible to infer the construction of many of the models. A fuller idea of the Banū Mūsà's style and presentation can, however, be obtained from the description of a typical Model, No. 42. The illustration is traced from one of the MSS, the only alteration being that the Arabic letters have been replaced by Roman ones, as far as possible in accordance with the instructions in the text. A reconstruction has also been supplied. (A similar system is followed for the other four models described below.)

Model 42. Fig. 42 is from the Vatican MS, Folio 27. A jar (*jarra*) with three open outlet pipes. As long as wine is inpoured nothing discharges; when inpouring is stopped wine begins to discharge from the middle outlet; when water is poured in it discharges from the middle outlet, wine from the other two, when inpouring of water is stopped no more water discharges and wine again discharges from the middle outlet. And so on. A tank (a) is installed in the jar below the inlet. (Too far to the right in Fig. 42, moved to the centre in Fig. 42a). In the tank is a float (ṭ) to which are connected bent rods joined together at (s), from which point a vertical valve rod (sz) leads to the downward opening valve (z) (Motif 4). The seat of this valve is in a small tank (w) from which the middle outlet (k) protrudes. (Tank (w) is too large in Fig. 42,

runs into the lower part (g) of the vessel. (Fig. 9a sho
conventional representation of an axle.)

10. Conical valves. These are shown in Figs. 4, 5 and
occur frequently throughout the book. The Banū Mūsà w
our knowledge, the first to use this component as a matter of
and in a wide variety of applications. It does not occur as a c
device in the works of the Greek writers on mechanics, wi
important exception of a treatise on a water-clock attribu
Archimedes,[20] extant only in Arabic manuscript versions
earlier part of this treatise, including the description of a fee
control system incorporating a conical value, may indeed
Greek origin, whereas the second part was probably addec
Muslim author. Al-Jazarī uses a similar system and acknowl
his debt to 'Archimedes' (p. 17). Writing towards the cl
the tenth century Abu ʿAbd Allah al-Khuwārizmī, in
encyclopaedia of the sciences entitled the *Keys of the Sc*
(*Mafātīḥ al-ʿUlūm*),[21] gives the following description o
component: 'The ground valve (*bāb maṭḥūn*) has a male
female—the male enters the female; it opens and closes. Whe
closed it is tight-fitting and has no outlet. Most of them are
shaped.' (p. 253 f). Al-Jazarī tells us that they were made o
bronze, and that the plug was ground into the seat with emer
adds 'every tap and ground valve is made in this way'. (p. 20 f
Khuwārizmī's statement that 'most of them are cone-sh
implies that other profiles were used, and indeed the one i
'Archimedes' clock is drawn with a curved profile. In fac
operation of the valve depends mainly upon the position o
plug relative to the plane of the seat at the discharge point –
remainder of the seat is unimportant, but no doubt they were r
as frustrums of cones to facilitate machining. In modern pra
the shape of the plug is chosen to suit the required pressure-
characteristics, and may be conical, parabolic or another typ
curve. Conical valves first appear in the West in the work
Leonardo da Vinci, and entered the general vocabulary
European engineering in the machine book of Agostino Ran
published in 1588. They are an essential component in macl
design and instrumentation, since they provide a sensitivity
control systems that is unobtainable with plate-valves or cla
valves.

Models

(A) SMALL TRICK VESSELS

With the exception of Models 5, 6, 7, 20 and 29, the first seven
four of the devices are small trick vessels. Since many of these h
to be lifted by an operator we can conjecture that the average s
was of the order of 40 cm high by 30 cm in diameter; consider

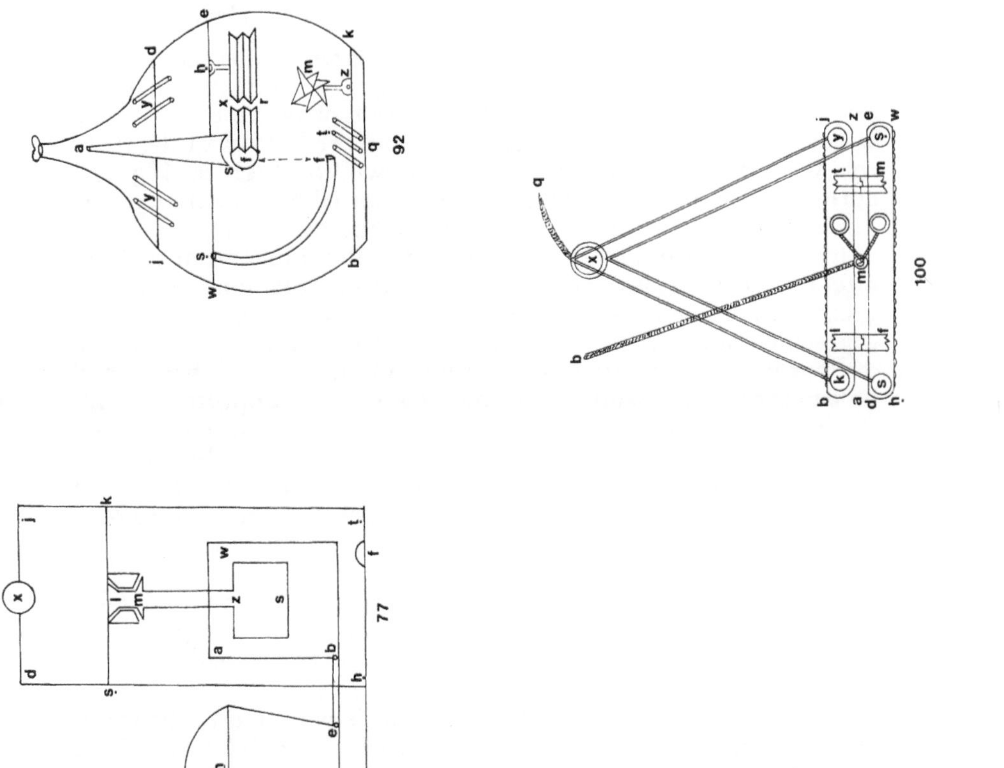

Figure 2. Models.

correctly drawn in Fig. 42a.) Tank (a) is connected to tank (w) by a pipe (edw) that comes out of tank (a) slightly below its top. (Too low in Fig. 42, correct in Fig. 42a). From tank (a) pipes (ab) and a wider pipe (jḥb) are led out, the first from the bottom of the tank, the second from near the centre (shown too high in Fig. 42, correctly in Fig. 42a). The purpose of pipe (jḥb) is to prevent the liquid from rising during inpouring to the outlet of pipe (edw). The purpose of pipe (ab) is to discharge the liquid from tank (a) after inpouring is stopped, but its rate of discharge should not be so great as to prevent float (ṭ) from rising during inpouring. The two pipes have a common outlet in the seat of valve (b). Valve (b), discharges into tank (m) (badly drawn in Fig. 42) which has a small hole in the bottom of one of its sides. Tank (m) is supported on float (l) which rests in tank (k). Two rods connected to float (l) are joined below tank (l) to a valve rod which terminates at the plug of valve (x). The two rods, the two valves (b) and (x), tank (m), float (l) and tank (k) thus constitute Motifs 4 and 5. Valve (x) opens into tank (th), from which outlets (f) and (t) are led out. (Pipe (th t) is incorrectly drawn in Fig. 42 — it should penetrate tank (th) as shown in Fig. 42a.) A wide pipe (ṣ sh) encloses tank th, and valve (z) is in its underside. (Incorrect in Fig. 42, correct in Fig. 42a.) A hole (sh), twice the diameter of outlets (t) and (k), is made near the bottom of pipe (s sh), into the interior of the jar.

The operation is as follows: first wine is poured into the jar. It runs into tank (a), float (ṭ) rises and closes valve (z). The wine flows out of tank (a) through pipes (ab) and (jḥb) and then through valve (b) into tank (m). It discharges from the small hole in tank (m) and from the rim of tank (m) into tank (k) whence it overflows into the interior of the jar and also into pipe (ṣ sh), where it collects. Because pipe (ṣ sh) and the interior of the jar are connected by hole (sh) the wine reaches the same level in both containers. No discharge can take place, because float (l) cannot rise while tank (m) is full of water and hence valve (x) remains closed. When inpouring is stopped tank (a) empties through pipes (ab) and (jḥb), float (ṭ) descends, valve (z) opens and the wine in pipe (ṣ sh) runs into tank (w) and discharges from outlet (k). Some wine runs from the interior of the jar into pipe (ṣ sh) through hole (sh), but insufficient to bring its surface to the level of valve (x), before tank (m) empties, lifting float (l) and opening valve (x). When water is poured into the jar it runs into tank (a) and raises float (l), so closing valve (z). The wine rises in pipe (ṣ sh) until it is about the level of valve (x). It runs through valve (x) and discharges from outlets (f) and (t). The water cannot flow from tank (a) through pipes (ab) and (jḥb) because valve (b) is closed; it therefore flows through pipe (edw) into tank (w) and discharges from outlet (k). When inpouring of water is stopped, the water-level in tank (a) falls due to the discharge through pipe (edw), and this fall is sufficient to allow float (ṭ) to descend far enough to open valve (z). Wine therefore again discharges from outlet (k), while nothing discharges from the other two outlets.

This model is fairly typical. There are an average number of errors on the drawing but the text is reasonably clear. The relative proportions and positioning of the various components is crucial to the proper operation of the device. Lacking adequate dimensions we cannot be sure that Fig. 42a shows the exact relationships. It is intended, as are the other reconstructions, to assist the reader to understand how the device worked.

(B) LARGE TRICK VESSELS

Models 5, 6, 7, 75–87 are larger types of trick vessel, although the first three incorporate some of the features of the smaller vessels. The general design of Models 75–87 consists of an open trough or basin placed beside a cylindrical vessel that contains the supply reservoir and the machinery. In the simpler type, when liquid is drawn from the trough or basin it is replenished up to its original level either through a concealed pipe or from a pipe discharging from the top of the main vessel. A refinement permits replenishment if moderate quantities are withdrawn, but cuts off all further supply if a large quantity is withdrawn. The most elaborate of these models has this double-action, together with two basins instead of one. Model 77 is an example of a double-acting device with single basin. A full translation is given first.

Model 77. (Fig. 77 is from Vatican MS, Folio 63 R.) Construction of an empty basin or a trough, mounted on a plinth, into which two or three *raṭls* [about 6–9 kg] of wine are poured and from which several times that [amount] is drawn without its diminishing. And if the Adept draws [wine] from it [in small amounts he can draw] many times [the amount poured in]. And if water is poured in instead of wine and a hundred, or more than that up to a thousand or two thousand, animals approach it, then provided the animals approach it one by one, whenever each of them quenches its thirst [the quantity it has drunk will return to the trough]. And if two or three animals approach the trough and drink from it simultaneously, all the contents of the trough will be exhausted.

We make for that the illustration of a vessel of handsome construction, circular like a cylinder if we wish, square [in cross-section] if we wish, or any other pattern whose shape we find pleasing. It is marked (jd ṭḫ). We divide it by a plate (k ṣ). The basin from which one drinks is marked (sh.e) and we install it in the position we have illustrated. Inside the vessel we install a tank (wba) opposite the basin, and the top of this tank is level with the top of the basin, or approximately so. We connect the bottom of the basin to the bottom of the tank by a pipe (eb) and its end (e) in the basin should be concealed, unnoticed and unseen. In tank (wba) we install a float (sz) that is immersed in the wine, as we have done elsewhere. Facing tank (wba) we make a hole (l) in plate (k ṣ) and we suspend to this hole two ground valves, as we have illustrated, with two plugs, one (l) opening upwards, and the other (m) opening downwards. We connect the two plugs by a rod (lm) and the rod is extended until its other end is soldered to the float at position (m). Position (w) is opposite position (sh) and they should be in the [same] horizontal plane, and slightly below the top of the basin. The layout should be such that when the float rises to position (w) the plug (m) blocks the seat, and if the float approaches the bottom of tank (wba) the plug (l) blocks the seat. At position (f) we make a hidden hole for the escape of the air.

It should be clear that if we pour any quantity of wine we wish into [upper] vessel (jkd Ṣ), which is the reservoir, then place the basin in the assembly, and pour into it about two *raṭls* of wine, the wine flows into hole (e) and through pipe (eb) into tank (wba). Float (sz) rises, valve (l) opens and the wine discharges into tank (wba) and flows through the pipe into the basin until the basin fills up to mark (sh). At that juncture the valve plug rises and closes the valve (m) as we have explained. However if anything is drawn from the basin the float sinks and the wine flows to the amount that was extracted from the basin. And if the person is aware of how it works, he extracts the wine with a large cup (*qadaḥ*), so that the amount that he drinks exceeds that which flows from valve (l), whereupon float (sz) sinks, valve (l) is closed by plug (l) and everything remains in the basin [i.e. nothing more flows into the basin]. For this reason if the wine is replaced by water, and the animals drink one by one, whenever [one of them] quenches his thirst [the basin will be replenished]. If two or three approach and drink simultaneously, they will drink more than the discharge from valve (l), so float (sz) will sink, valve (l) will be closed from above, and nothing will flow from it, whereupon everything in the basin or the trough will be exhausted. And that is what we wished to explain.

There are a number of lacunae in the Arabic text that corresponds to the first paragraph of the translation, as indicated by the additional phrases in square parentheses. Such omissions are not common in the Vatican MS. The only real fault with the illustration is that the valve is shown with insufficient travel; a more workable arrangement is shown in Fig. 77a. The operation is fairly clear from the text. When a small amount is withdrawn valve (m) opens and a like amount runs into tank (wba) and pipe (eb) balances the levels in tank (wba) and basin (*sh*.e). When a large amount is withdrawn plug (l) drops into its seat and the reservoir is sealed.

This model has been selected because it includes perhaps the earliest example of a double-acting conical control valve. In a similar model the double valve is replaced by a tap: a vertical rod soldered to the float has a ring on its upper end that passes over the handle of the tap. The tap is open when the handle is a certain angle above the horizontal and closed when the handle is at a certain angle below the horizontal. When only replenishment, and not cut-off, was required the valve or the tap was single-acting.

(C) FOUNTAINS

The fountains — Models 20, 88–94 — are in most cases not described or illustrated clearly, particularly the design of the fountain heads. The single vertical jet is obvious enough; 'the Lily-of-the-valley' is discussed below under Model 92. The 'shield' is nowhere clearly described and we must therefore refer to al-Jazarī (p. 164) where the shape is produced by the impinging of a jet on the lower, concave side of a dished plate. The water then descends into the pool in the shape of a dish, concave downwards. Model 20 is a simple fountain delivering a single jet whose height can be altered by varying the static head above the outlet. Model 88 describes the method of producing a shield (*turs*) or a lily-of-the-valley (*susān*). The remaining models embody the idea of alternation: the fountain emits a given shape for a period, then changes over to emit a second shape, and so on. One way of doing this is to lead out two pipes from the fountain head, one inside the other. At the supply installation these pipes are separated, the narrow one entering one tank, the inside one entering another. As a refinement a second fountain head is added. Let the fountain heads be A and B, the shapes (a) and (b): then A emits (a) while B emits (b) and, when they switch over, A emits (b) while B emits (a). This is done by using two double pipes instead of one, but is essentially the same principle as the simpler version. A second basic method was the use of worm-and-pinion and special taps — this is described in detail for Model 92. For the first type the problem is to direct the flow now into one tank, now into the other. In model 89* this is achieved by running the water from supply channel (a) into a small tank (b) fixed to the end of a pipe balanced on an axle. While the axle is horizontal the pipe (bg) discharges into tank (k). On the balanced pipe, towards end (g) two small tanks (e) and (f) are suspended, and a small pipe from the main supply pipe drips water into tank (e). Tank (f) is connected to tank (e) by a small hole and has a narrow discharge pipe (w) in its bottom. When tank (e) is full the balanced pipe tilts, trough (b) lifts away from channel (a) which then discharges directly into the second main tank (t). Meanwhile tank (e) empties into tank (f) which in turn empties through pipe (w). When both are empty the pipe tilts back to the horizontal and

*Models 89 and 90 are not illustrated here: however in the descriptions which follow the manuscript letter notations which cross-reference drawing and text have been left in.

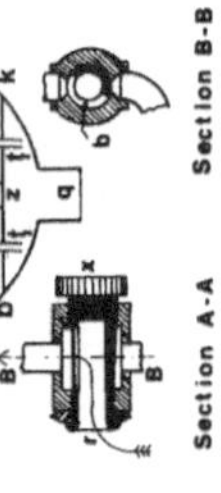

92a

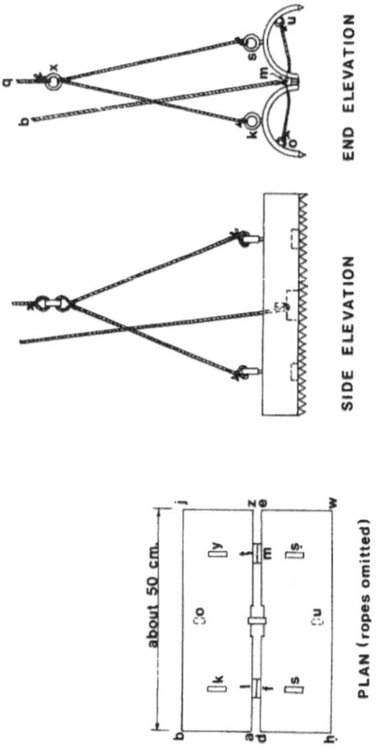

100 a

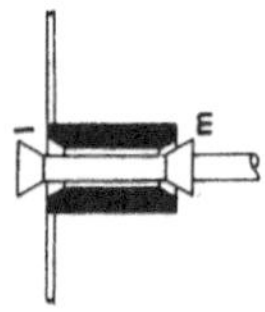
77a

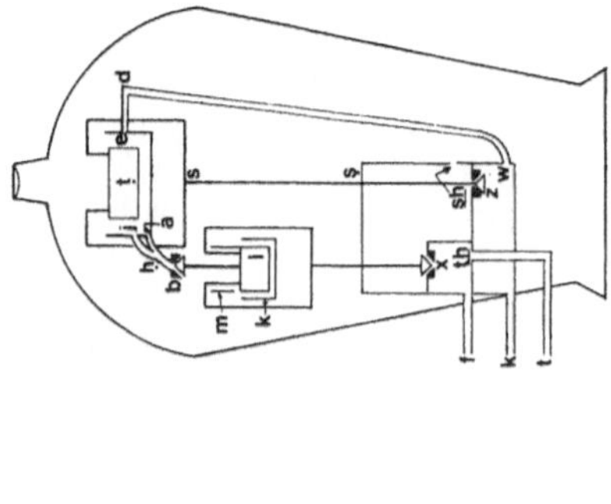
42a

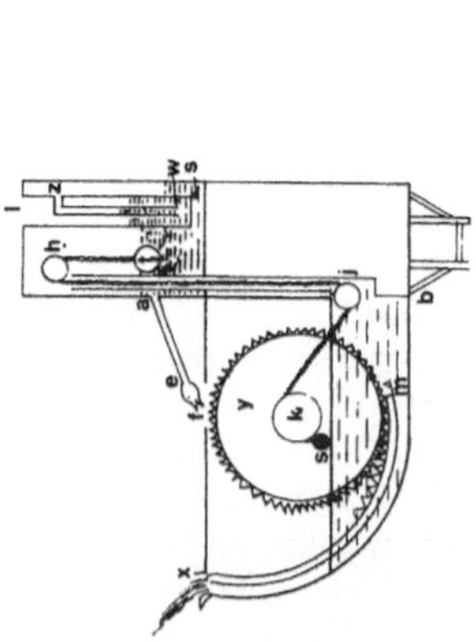
97a

Figure 3. Reconstructions.

the discharge is again into tank (k). In Model 90 a water-wheel or a wind-wheel is mounted at the top of a vertical axle upon which a tank (q) is centrally mounted; (q) is over the two main tanks (g) and (b) and has a discharge hole at one side. As the wind or water rotates the wheel, tank (q) rotates and discharges now into main tank (g), now into main tank (b). Model 91 is a doubled version of Model 88, except that the heads emit 'lilies' and single jets, whereas Model 88 emits a 'lily' and a 'shield'. Models 93 and 94 work on the same principle as Model 92, described below, except that Model 93 has one main fountainhead and several subsidiaries and Model 94 has two fountainheads, and the shapes emitted differ.

Model 92. Fig. 92 is from the Berlin Ms, Folio 131 Verso. Fig. 92a is from Hauser, Table XX, p. 159. A fountain that emits a straight jet or a lily-of-the-valley. Neither the text nor the illustration are clear, and reference should be made to Fig. 92a, which is a credible reconstruction. Even allowing for the use of conventions, the draughtsman seems to have found it difficult to show the combination of screw, pinion and tap. The action is as follows: the water enters through inlet (q) and pipes (ṭ), the jets from which rotate vaned wheel (m). (The water is presumably under a sufficiently high static head for it to fill chamber (webk) under pressure.) Wheel (m) turns screw (k) which rotates the pinion together with the plug of the tap. When slit (b) in the plug of the tap (see Section B–B) is opposite the opening in the body that coincides with pipe (sa) the fountain emits a vertical jet. When the plug of the tap rotates, bringing slit (b) in line with the opening in the body that coincides with pipe (fṣ), the fountain emits a lily-of-the-valley. In this and another fountain the purpose of pipes (y) and the shape of the lily-of-the-valley is not clear. As drawn, the most likely shape would be a spray. It is possible, however, that pipe (sa) passed through the top of the fountain and closed it completely; pipes (y) — say six of them — were then passed to the outside through the wall of the vessel to produce separate fine jets.

(D) LAMPS

There are four lamps, Models 95–98. Model 97, described below, combines the designs and functions of Models 95 and 96. Model 98 was a kind of 'Hurricane lamp' The lamp is installed inside a vertical half-cylinder that rotates in axles mounted on a stanchion. A vertical triangular wind-vane projects from the side of the half-cylinder so that the convex side of the half cylinder is always facing the direction from which the wind is blowing.

Model 97. Fig. 97 is from the Berlin MS. Folio 136 Verso. Fig. 97a is from Hauser, Table XXII, p. 171. The explanation in the text is quite clear, but the illustration is not. Toothed wheel (y) is too small, and both this and pulley (k) are erroneously shown on a vertical, instead of horizontal axle. The vertical pipe leading from (j) is mentioned but not shown, and there is no indication that the reservoir had to be at a higher level than the lamp. Discharge pipe (ae) should not be curved upwards, and should show a bird's head at end (e). The illustrations in the Topkapi MS show the figures of animals, birds, gargoyles etc., which are mentioned in the text, whereas the illustrations in the other two MSS nearly always omit these. No doubt, in the environments where the MSS were copied, religious considerations prevented the representation of such figures. Finally, the two circles at the left of the illustration seem to serve no purpose. Inside the lamp the large toothed wheel (y) and the pulley (k) are mounted in the same horizontal axle. The toothed wheel meshes with the curved rack (mx) which runs in a groove at the bottom of the lamp; the wick is tied to a hole in the rack at (x). At (j) there is a second pulley and a hole through into the space below the reservoir. At (j) a vertical pipe is erected that passes through to the oil reservoir. Above this pipe is a pulley (ḥ). A chain passes over pulley (k), under the pulley at (j) and over pulley (ḥ). To its end in the lamp a weight (s) is attached, and to its end in the reservoir a float (ṭ). The supply pipe to the lamp (aef) should not be wide enough to allow air to enter the reservoir while oil is flowing. On the right of the reservoir are pipes (ls) and (wz) constituting Motif 6. The reservoir is filled at point (l), float (ṭ) rises and the rack is drawn in direction (m). When the oil reaches (a) it begins discharging through pipe (aef) into the lamp and continues to do so until it covers hole (j), at which juncture discharge ceases because no air can enter the reservoir. The wick is then lit, the oil in the lamp falls until hole (j) is uncovered, whereupon there is another discharge from pipe (aef). Float (ṭ) descends and the rack moves in direction (x). And so on.

(E) MECHANICAL GRAB

Model 100. Fig. 100 is from Berlin MS, Folio 143. The description in the text is very clear but the illustration is not; reference should be made to Fig. 100a. Two copper half-cylinders are connected by hinges (lf) and (ṭm). Four rings (k, y, s, ṣ.) are soldered to the half-cylinders at the top of their convexities, ropes are tied to these rings and brought together at (x). Rope (xq) is the lowering rope and can be of any length, as required. Rope (bm) passes through a slot at point (m); short ropes (mu) and (mo) are tied or spliced to rope (bm) and tied to rings (u) and (o) respectively. (Rings (o) and (u) are not shown on Fig. 100, nor are the slight recesses in the half cylinders at (m).) Teeth, which mesh together when the half-cylinders close, are fixed along edges (jb) and (ḥw) of the half-cylinders. The device is lowered into the water by rope (qx), rope (bm) being slack. When the bottom is reached rope (qx) is slackened and rope (bm) is pulled, drawing the half-cylinders together. The device is then withdrawn from the water by rope (bm) and its contents examined on dry land. (The ends of the half-cylinders were probably covered by plates but these have been omitted from Fig. 100a for reasons of clarity.)

Conclusion

If the Book of Ingenious Devices had been written in modern times the authorship would probably have been assigned to Aḥmad, with due acknowledgements to Muḥammad and al-Ḥasan for their assistance and encouragement. The Arabic commentators are unanimous in ascribing the work to Aḥmad and, after the mention of all three brothers in the opening of the MSS, only Aḥmad is mentioned thereafter. It is reasonable to suppose that Aḥmad was the engineer of the family, al-Ḥasan the geometer, while Muḥammad took a general interest in all the sciences, was the main force behind the collection and translation of earlier works, and also involved himself in politics and public works. Aḥmad's book was not intended to define and demonstrate basic principles of pneumatics, hydrostatics and mechanics, and should not be judged by its failure to do so. We can be fairly sure, from the list of works written by the brothers or commissioned by them, that they were not unfamiliar with mathematical and scientific principles, and that Aḥmad's method of presentation was therefore deliberate and not due to ignorance. What he was concerned to do was to describe the machines that he had designed: whether they were constructed by himself or by a craftsman working under his guidance is not particularly relevant. Compared with the Greek works, they contain far less theory and a much greater degree of engineering inventiveness. There is no reason to suppose that any of the one hundred devices included in the two main manuscripts, with the exception of Model 94, was not the work of Aḥmad, although Hauser suggests (p. 27) that the last eight were later additions. Lamps similar to those described by Aḥmad occur in Hero's *Pneumatica*, and since we are told that Model 97 was used in Christian churches and Magian temples, its use had probably been widespread for centuries before the Abbasid period. The bellows and the grab were probably designed to meet needs that arose during Aḥmad's involvement in civil engineering projects. There is no doubt that he was justifiably proud of his ability as an engineer, and that, as Bar Hebraeus implies, many of the devices embodied techniques and mechanisms that were Aḥmad's own inventions. The book represents a marked advance in mechanical engineering, and this should be borne in mind when we consider its derivations.

The documentary sources that may have been available to

Aḥmad were written in Greek, the most important being the *Pneumatics* of Philo,[22] the *Mechanics*[23] and *Pneumatics*[24] of Hero, and the treatise on a water-clock attributed[25] to Archimedes. In addition, there was a centuries old tradition in Syria, unbroken by the Arab conquest, for the construction of water-clocks and associated automata.[26] The knowledge of mechanics therefore came to Baghdad in Hellenistic guise, but it would be an over-simplification to postulate a Greek written tradition as the sole inspiration of the Banū Mūsà and their successors, although it was perhaps the main inspiration. The pre-eminence of the Greeks in the pure sciences cannot be questioned, and we are only becoming aware of their contribution to mechanical technology,[27] but in identifying the Greek elements in early Arabic mechanics we are faced with a number of problems, most of which can be attributed to our lack of knowledge of the scientific and technical advances made between the time of Hero and the time of the Banū Mūsà, a span of about eight hundred years. Our information is particularly meagre for Byzantium and Sassānid Iran. We know, of course, that the sciences flourished in Alexandria under the Roman Empire; one has only to cite the names of Hero (first century), Ptolemy (second century), Diophantos and Pappos (both third century) as evidence that Greek science was still full of vigour during this period. We have less knowledge about technology in Roman times —Vitruvius, writing during the reign of Augustus, is still the main documentary source.[28] There were also a number of anonymous improvements in the construction of machines for practical purposes. For instance, the chain-of-pots drawn by animal power through a set of gears became a really effective machine in the fourth or fifth century with the introduction of the ratchet-and-pawl[29] and the siege engines operated by men hauling on ropes entered the Islamic area from the Far East towards the end of the seventh century.[30] The main flow of ideas, on present evidence, seems to have been from west to east. Thus the centre of learning at Gondēshapūr (Arabic: Jundaysābūr) in south-western Iran was probably founded by Greeks, although it did not rise to eminence until the arrival of Nestorian Christians from Edessa and Nisibis in the sixth century. The language of instruction was Syriac, but there is a tradition that the introduction of medical studies, the main subject taught at Gondēshapūr, was due to an Indian. It may be considered as the centre through which Byzantine learning passed to Baghdad.[31] Of importance also were the Sabeans of Ḥarrān in upper Mesopotamia, of whom Thābit b. Qurrà was one.[32] They had preserved a large corpus of Greek scientific writings, also in Syriac, and were to be very influential in the development of Arabic science in Baghdad. It cannot be doubted, however, that there were significant advances in Iran and further east, although attestation is scanty. Writing in A.D. 1203, Riḍwān

b. al-Saʿātī quotes a tradition to the effect that some of the techniques used in water-clocks were transmitted to Greece from Iran and were incorporated in clocks built in Byzantine Syria.[33] Certainly, many of the traditional crafts of Iran are of ancient origin[34] and it seems likely that metalworkers, for instance, would have moved from Iran into Baghdad when the city's cultural life began to flourish.

We cannot say with certainty which of the Greek works were known to the Banū Mūsà, with the exception of Hero's *Mechanics*, translated by Qustà b. Luqā during their lifetime.[35] It is probable that other works of Hero were also available to them, since his reputation among the Arabs was already high in the tenth century. ʿUmar Ibn Muḥammad al-Kindī (*ca.* 970) mentions Hero's writings on pneumatics, and that he constructed clocks;[36] Ibn al-Nadīm says that he wrote a book on automata (p. 397). The Arabic text of Philo's *Pneumatics*, as it exists in A.S. 3713, probably dates from the fourteenth century and contains late Hellenistic and Islamic additions, but the chapters by Philo himself may have been known in Arabic much earlier than this. Ibn al-Nadīm knew of the 'Archimedes' water-clock, which he calls 'a book on the construction of an instrument that casts balls, by Archimedes' (p. 397). Again the existing manuscripts contain Islamic additions, but is possible that the earlier chapters were known to the Banū Mūsà. It is symptomatic of the difficulties that arise when attempting to untangle the Greek and Arabic traditions that the penultimate section of the 'Archimedes' contains 'snakes-and-birds', *parerga*, very similar to the motifs described in Chapter 40 of A.S. 3713 (195–8 in Prager), but the language in the 'Archimedes' section indicates that it was composed in Arabic. Problems of this kind, together with the Banū Mūsà's omission of any acknowledgement of the works of predecessors, makes identification of their sources and isolation of their own contribution a matter of some difficulty. It may be best to clarify the problem, for a start, by listing those models of the Banū Mūsà that have close counterparts in Philo and Hero; chapter numbers are taken from Hauser, Prager and Woodcroft.

Banū Mūsà	*Description*	*Hero*	*Philo*
1	Concentric Siphon	2	10
13	Hot and cold water dispenser	7	—
14, 15, 24	Vessels that can be made to appear filled when unfilled, and vice versa	—	43–8
18	Boiler that discharges hot water when cold water is poured in	74	—
21	Jar from which several kinds of wine can be drawn from a single tap	32	21
29	Water-operated whistle	48	60
30	Water-lifting pipette	—	11
34	Vessel that discharges wine or water separately or mixed	59, 64, 65	23

Banū Mūsà	*Description*	*Hero*	*Philo*
75–87	Vessels that are constantly refilled	20, 61	17, 18, 19
95	Lamp with automatic feed	71	20
96	Lamp with automatic extrusion of wick	33	—

These are not all the models where there are resemblances, but only those where the construction and the effects produced are very similar. Many of the Banū Mūsà's devices are elaborations of basic ideas contained in the works of Philo or Hero, or both, and these have not been listed above. Moreover, each of the three writers developed themes that were ignored, or dealt with only briefly, by the other two namely: the early chapters of Philo, demonstrating pneumatic theory; Hero's pyrotechnic and sound-producing devices; the Banū Mūsà's fountains, and mechanical grab. Some components are common to all three. For instance, taps with multiple borings, bent-tube and concentric siphons, air-vents. The double concentric siphon (Motif 2) does not appear in the Greek books, neither does the funnel with bent end (Motif 3) for pouring in different liquids — instead Philo used a rotary valve (Chs. 43, 44).

It should be evident from the foregoing, and from the high reputation enjoyed by Hero among the Arabs that the Banū Mūsà had access to the works of Philo and Hero and possibly to the 'Archimedes' treatise and to later Hellenistic works. In the light of recent investigations, it is permissible to speak of a Greek school of mechanical engineering, stemming from Ctesibius (300–270 B.C.) and continuing through Philo and Hero, probably into Byzantium, whence it passed into Sassānid Persia. The school is represented, not only by writings that have survived, probably only a small part of the original corpus, but also by archaeological evidence such as the 'Tower of the Winds' in Athens and the Antikythera geared computer.[37] In Sassānid Persia the Greek system may have merged with Iranian, Indian and Chinese elements, and it was this composite tradition that came to the attention of the Banū Mūsà in Khurāsān and later in Baghdad. Without knowing the full content of this system, we cannot be certain how many of Aḥmad b. Mūsà's original ideas are embodied in his work, but we can at least say that it differs markedly from the known writings of the Greek school. Apart from a greater concern with engineering at the expense of scientific theory, the main difference is the greater degree of automatic control, achieved by the use of conical valves and delay systems. This preoccupation with automatic controls and timing came to be important in the west only in comparatively recent times. There are a number of parallel examples in the history of technology of ideas having lain dormant for centuries, to reappear when the practical need for them arose. This is not to say, however, that the work of the Banū Mūsà was without influence on the development of western machine technology in medieval times, but the influence was probably indirect.

Over three hundred years later than the Banū Mūsà, al-Jazarī composed his work, which represents the culmination of the Arab achievement in mechanical technology in the Middle Ages. Al-Jazarī's work incorporates all the significant devices and techniques of his predecessors, whose works he acknowledges in the appropriate places: 'Archimedes' for the water-machinery of his monumental water-clock (p. 17); Ibn Yūnus for the design of a candle-clock (p. 87); the Banū Mūsà for fountains (p. 157); Hibat Allah b. al-Ḥusayn[38] (d. 1139 or 1140) for musical automata (p. 170). Al-Jazarī then describes the improvements — and they were real ones — that he made to the work of his predecessors and, in addition, describes a number of devices, techniques and components that are not to be found in their works.

Almost all the ideas and techniques, usually thought of as typically Islamic, which later re-appeared in the West, can be found in al-Jazarī. A direct transmission of his work to Europe cannot, however, be postulated. There is no known medieval translation of it into a European language but, in any case, it is probable that the most important transmissions occurred before his book was composed. It is important to observe that al-Jazarī records the assistance of a continuing tradition of machine technology from the time of the Banū Mūsà to his own day. He prides himself on being a part of this tradition (p. 15), and he mentions the engineering writings of such men as Hibat Allah, who otherwise is known to us only as maker of astrolabes. He also describes techniques, such as metal casting in closed mould-boxes in green sand, as having been long established in the Islamic world, although many of these were not known in Europe until later times.[39] He is therefore a valuable witness for the unbroken activities of Arab engineers throughout the medieval period, and a necessary witness, because as yet very few Arabic manuscripts have come to light. This continuity makes possible the assumption that Arabic knowledge of machines *could* have passed to the west at any time between the ninth and the thirteenth century. In fact, the most important transmission seems to have occurred in Spain during the twelfth century.

It has been accepted for some time that the introduction of the water-clock into Christian Spain was due to Muslim influence, but the actual passage of knowledge was probably earlier than the Alfonsine tradition of the mid-thirteenth century,[40] and could have occurred by the inspection of existing Islamic clocks or by gaining access to Arabic writings — actual translation would have been unnecessary, because many scholars in this period were bilingual. The discovery of a hitherto unknown Andalusian MS on water-clocks and other machines is therefore of crucial significance. This Arabic MS is attributed to Abu ʿAbd Allah, known as Ibn Muʿadh, who worked in Cordova in the eleventh century — the copy itself is dated 1266.[41] It has not yet been fully studied, but it contains descriptions of thirty machines, including water-clocks, with illustrations that appear to be quite advanced; for instance, they show well-constructed gear-trains. There is no reason to doubt that this document, and others like it, was known in the Christian parts of Spain.

Since the work of the Banū Mūsà was widely known and appreciated in medieval Islam it is probable that elements of their work were embodied in the writings and constructions of later Islamic engineers. The water-clock, with its gears, pulleys, cams, jackwork and automata is essentially similar to the tower clocks that appeared in Europe in the thirteenth century, undoubtedly as derivations from the water-clock; one has simply to consider the weight and escapement as replacements for the heavy float where fall is controlled by a uniform fluid discharge. The mechanical clock exerted a potent influence upon the development of machine technology. Although the Banū Mūsà did not construct water-clocks we can therefore assume an indirect influence by them, rather by Aḥmad, upon European technology, as follows: (1) Aḥmad's ideas adopted by later Arabic engineers who constructed water-clocks; (2) transmission of Islamic water-clocks to Christian Spain followed by their dissemination throughout Europe; (3) invention of the weight-driven clock in Europe early in the thirteenth century; (4) influence of the techniques and mechanisms of the mechanical clock upon the development of machine technology. Other contributions, perhaps more direct, may be identified as research progress.

ACKNOWLEDGEMENT

The author wishes to express his gratitude to the Leverhulme Trust Fund for a Research Award to assist him in his research into medieval Islamic technology.

Notes

1. The best general view of the growth of Arabic science from the ninth century onwards is given in: George Sarton, *Introduction to the History of Science*, Vol. I, Baltimore, 1927.

2. Article, *bayt al-ḥikma* in *Encyclopaedia of Islam*.

3. Ibn al-Nadīm, *Kitāb al-Fihrist*, ed. G. Flugel, Leipzig, 1871–2, 2 vols. Reprinted Cairo, A.H. 1348.

4. Ibn al-Qiftī, *Kitāb ikhbār al-ʿulamā' bi akhbār al-hukamā'*, Cairo, A. H. 1326.

5. Ibn Khallikān, *Wafayāt al-aʿyan*, ed. Ihsān Abbās, Beirut, Vol. 5, undated.

6. Bar Hebraeus, *Tārīkh mukhtasar al-duwal*, ed. A. Ṣālihānī, Beirut, 1890.

7. Abu' l-Fida', *Al-Mukhtasar fī akhbār al-bashar*. Lithographic reproduction of Arabic text in School of Oriental and African Studies, London. From undated Cairo edition.

8. Ṭabari, *Annales quos scripsit* . . . etc, ed. M. J. de Goeje, Leyden, 1879–1901.

9. Ibn Abī Usaybiʿa. *ʿUyūn al-anbā'*, ed. A. Muller, Cairo, 1882, 2 vols.

10. Ibn Yūnus, *Le Livre de la Grande Table Hakémite*, translated C. Caussin, Paris, 1804.

11. Al-Bīrūnī, *The Chronology of Ancient Nations*, trans. and ed. Edward Sachau, London, 1879.

12. These are the first four letters of the Arabic alphabet.

13. H. Suter, *Die Mathematiker und Astronomen der Araber und ihre Werke*, Leipzig, 1900. From a scholarly point of view Suter is more reliable than Sarton, but he gives no general summary.

14. See also Suter, *op. cit.*, 20–1.

15. This should probably read Menelaos. See Hauser 13.

16. Ibn Khaldūn, *Muqaddimah*, trans. F. Rosenthal, Routledge and Kegan Paul, London, 2nd Edition, 1967.

17. Al-Jazarī, *The book of knowledge of Ingenious Mechanical Devices*, trans. and ed. Donald R. Hill, Reidel, Dordrecht, 1974.

18. E. Wiedemann and F. Hauser, 'Uber Trinkgefässe und Tafelaufsätze nach al-Gazarī und den Banū Mūsà', *Der Islam* 8, (1918), pp. 268–91.

19. F. Hauser, 'Uber das '*Kitāb al-Ḥiyal*' — das Werk uber die sinnreichen Anordungen der Banū Mūsà.' *Abhandlungen zur Gesch. der Naturwissenschaften und der Medizin*, Erlangen, 1922. (Annotated translation of complete work by Donald R. Hill forthcoming.)

20. Donald R. Hill, *Construction of a Water-Clock attributed to Archimedes*, Turner and Devereux, London, 1976.

21. Abu Abd Allah al-Khuwārizmī, *Liber Matatih al-Olūm*. Arabic text with Latin critical apparatus, ed G. Van Vloten, Leyden, 1895.

22. The *Pneumatica* of Philo exists in latin MSS., fragments of Greek Mss, an Italian Ms, an Arabic Ms, and there are fragments preserved in another Arabic Ms. The references for the last two are A.S. 3713, Aya Sophia, Istanbul and Eastern Manuscript 954, Bodleian Library, Oxford, respectively. The work has been dealt with by several modern writers, as follows:
(a) Carra de Vaux, 'Le Livre des Appareils Pneumatiques et des Machines Hydrauliques par Philon de Byzance,' *Paris Académie des Inscriptions et Belles Lettres*, 38 (1903), Pt. 1. (b) W. Schmidt, 'Physikalisches und Technisches bei Philon von Byzanz', in *Bibliotheca Mathematica* 3, Series II, (1901). (c) W. Schmidt, 'Aus der antiken Mechanik' in *Neue Jahrbucher fur das klassische Altertum*, 1904, pp. 329–51. (d) A. G. Drachmann, 'Ktesibios, Philon and Heron; a study in Ancient Pneumatics', *Acta Historica Scientiarum et Medicinalium*, (ed. Bibliotheca Universitatis Hauniensis, Copenhagen), 4, (1948), 1–197. (e) F. D. Prager, *Philo of Byzantium-Pneumatica*, Ludwig Reichert, Wiesbaden, 1974.

Carra de Vaux's work includes the full Arabic text of A.S. 3713 with a French translation. The translation is good although in places it strays from the literal meaning, with no indication that this has occurred. The illustrations are reconstructions by Carra de Vaux — the original drawings from the manuscript are not reproduced. This important omission aside, the work can be used with some confidence. Prager's work has serious defects. His historical analysis of the Philo traditions, his presentation of the Western side of that tradition, and his attempt to isolate Philo's own contributions, all show sound scholarship. His high regard for theoretical science, however, is accompanied by apparent distaste for engineering applications that leads him to undervalue and even deprecate the work of the Arabic writers, such as Aḥmad bin Mūsà, al-Jazarī and al-Kindī, who 'show an almost exclusive interest in the demonstration of assorted contraptions' (p. 54). This attitude leaves out of account the value of the technical innovations made by the Arab engineers and the fact that al-Jazarī in particular described his constructional techniques in scrupulous detail. Moreover, Prager makes heavy weather of interpreting certain devices. For instance, A.S. 3713 includes one of al-Jazarī's devices, a slave girl who emerges from a cupboard (Category II, Ch. 10 in al-Jazarī; Ch. 30, A.S. 3713). Prager comments (p. 177, note 520) that the description in both texts is so vague that a 'graphic reconstruction of the mechanism is impossible'. This is nonsense: the construction and operation of the device, one of al-Jazarī's simplest, is made perfectly clear by his text and illustrations. By far the most serious defect in Prager's work, however, lies in his treatment of the photographs reproduced from A.S. 3713. He says (pp. 125–6) that the reference letters and legends are explained beneath the reproductions of the first forty-three illustrations, but from then on the explanation is omitted, 'in general because we found the legends undecipherable, as in Chapter 58 (Fig. 58) or because we found them difficult and the reference letters obvious'. This is bad enough, since in fact the legends can almost everywhere be read with comparative ease, including those in Fig. 58. But the reader is led to expect that the legends on the earlier reproductions are translated fully and accurately, and this is not the case. They are often either inaccurate or completely wrong, and are in many cases omitted. To cite only two instances: Fig. 12, p. 134 is upside down and the English legends, all incorrect and incomplete, do not therefore tally with the Arabic ones; Fig. 43, p. 197 has eight legends, all legible, but only one of these is translated, incompletely. The evidential value of the photographs to a non-Arabist is therefore minimal or rather, if he is unaware of these defects, negative.

23. Hero, 'Les Mécaniques ou L'Elevateur de Heron d'Alexandrie sur la Version Arabe de Qostà ibn Lûqâ'. Translated and edited by Carra de Vaux in *Journal Asiatique*, 9e Série (1898), Tome I, pp. 386–472; Tome II, pp. 152–92, 193–269, 420–514.

24. *The Pneumatics of Hero of Alexandria.* A facsimile of the 1851 edition by B. Woodcroft. Introduced by Marie Boas Hall, London and New York, 1971.

25. See note 20.

26. H. Diels, 'Über die von Prokop beschriebene Kuntsuhr von Gaza', *Ablaudlungen d. preuss. Akad. Wiss., Berlin,* (Phil.-Hist. Klasse), No. 7, 1917.

27. Derek de Solla Price, *Gears from the Greeks,* Science History Publications, New York, pp. 51 f.

28. Vitruvius, *On Architecture,* ed. and translated F. Granger, Loeb Classical Library, London and New York, 1931 (reprinted 1970).

29. Thorkild Schiøler, *Roman and Islamic Water-Lifting Wheels,* Odense University Press, 1973, p. 169.

30. Donald R. Hill, 'Trebuchets', *Viator 4,* (1973), 101-2.

31. Article, 'Gondēshapūr', in *Encyclopaedia of Islam.*

32. Bar. Hebraeus, *op. cit., 266.*

33. Two MSS of Riḍwān's work are known: Gotha, Arab 1348; MS Istanbul, Koprulu I. 949. The reference is in 1348, Folio 4, recto. An abbreviated translation was supplied by E. Wiedemann and F. Hauser in 'Uber die Uhren in Bereich der Islamischen Kultur', *Nova Acta, Abh. der Kaiserl. Leop. Carol Deutsch. Akad. der Naturforscher* 100, (Halle, 1915), 176–272. The reference is on pp. 179–80.

34. Hans E. Wulff, *The Traditional Crafts of Persia,* M.I.T. Press Cambridge, Mass., 1966, *passim.*

35. Suter. *op. cit.* pp. 40–2.

36. E. Wiedmann, *Aufsätze zur Arabischen Wissenschaftgeschichte,* ed. W. Fischer, Vol. I, Olms, Hildesheim, 1970, pp. 69–70. (This al-Kindī should not of course be confused with the famous philosopher.) This work of Wiedemann's is a collection in two volumes of his Beiträge for the Erlangen Society. The index to Vol. II should be consulted for other references to Hero, and other writers, Greek and Islamic.

37. Derek de Solla Price, 'Automata in History', *Symposium on Automata and Simulated Life as a Central Theme in the History of Science,* University of California Los Angeles, 1963, pp. 13–16.

38. Suter, p. 117.

39. The first description of this casting technique in Europe is in Biringuccio's *Pirotechnica* written in A.D. 1540. It was probably adopted in the west at the end of the fifteenth century. See Cyril Stanley Smith. 'The Early History of Casting, Moulds, and the Science of Solidification' in *Metal Transformations,* ed. Mullins and Shaw, New York, 1968, p. 23.

40. Price, 'Automata', (see note 37), p. 17.

41. I am indebted to Dr David A. King, of the Smithsonian Institution Project in Medieval Islamic Astronomy, in Cairo, for passing on to me information about this important work. The machine section in fols. 1v–48v is part of a larger work entitled *Kitāb al-asrār fī natā'ij al-afkār,* MS Biblioteca Medicea laurenziana Or. 152, No. 282 in Assemani's catalogue. See the article 'al-Jayyanī' in the *Dictionary of Scientific Biography.* Working from a microfilm of this MS, which is unfortunately in very poor condition, I have been able to publish a preliminary evaluation of this work. See Donald R. Hill 'A Treatise on Machines by Ibn Muʿ*adh* Abu ʿAbd Allah al-Jayyanī'in *Journal for the History of Arabic Science,* Aleppo University, I, 1977. In general, Ibn Muʿ*adh*'s water-clocks and other machines are large and robust, with advanced gearing and mechanical transmissions. His applications of hydrostatics and aerostatics are, however, less sophisticated than those of Banū Mūsà and al-Jazarī.

XIII

Qusṭā b. Lūqā al-Baʿlabakkī*

Qusṭā B. **LŪQĀ** AL-BAʿLABAKKĪ, mediaeval scientist and translator. He was of Christian origin, from the town of Baʿlabakk p. 529 [*q.v.*]. In Baghdad, where he worked for some time as a doctor, scientist and translator, his reputation was as high as that of Ḥunayn b. Isḥāq [*q.v.*]. He was fluent in Greek, Syriac and Arabic, being particularly noted for his excellent style in Arabic. The last part of his life was spent in Armenia, where he was induced to take up residence by the prince Sanḥarīb. According to ʿUbayd Allāh b. Jibrāʾīl, he came into contact with a certain Abu 'l-Ghiṭrīf al-Baṭrīq, for whom he composed a number of scholarly works. Qusṭā died in Armenia *ca.* 300 H (912-13); a shrine was erected over his grave, which was accorded the same honours as the graves of kings and other eminent personages.

It was, of course, usual for Arabic scholars to be well versed in a wide range of subjects, and Qusṭā was no exception. He is said to have been skilled in medicine, philosophy, geometry, arithmetic, astronomy, and music — all these subjects are included in the lists of his works given by the biographers. Ibn al-Nadīm (*Fihrist*, ed. G. Flügel, Leipzig 1871-72, pp. 410-411), having first stated specifically that he has excluded translations, lists over thirty of Qusṭā's original works and Ibn Abī Uṣaybiʿa adds a further thirty works

*In this new reprint of the *Encyclopaedia of Islam* article a slightly simplified transliteration system for Arabic has been used and place names have been given in their anglicised form whenever possible. All references in square brackets are to other articles in the *Encyclopaedia of Islam*. References in the indexes of this book are to the original pagination, here given in the margins.

to this list (*ʿUyūn al-anbāʾ fī ṭabaqāt al-aṭibbāʾ*, ed. A. Müller, Cairo & Königsberg 1882-84, I, pp. 244-245). Medical works, which preponderate, include the following treatises: on gout; infectious diseases; insomnia; knowledge of fevers, types of crises in illnesses, the pulse; paralysis-types, causes and treatment; the four "humours"; and phlebotomy. Non-medical works include several treatises on philosophy and logic; on astronomy, on the celestial sphere; two commentaries on Euclid's *Elements*, a treatise on algebra, a commentary on the book of Diaphantos on algebra; on the steelyard (*qarasṭūn* [see ḲARASṬŪN]); on weights and measures; and on burning mirrors. Some of Qusṭā's translations are extant, *e.g.* those of Diophantos, Theodosios, Autolykos, Hypsikles, Aristarchos, and Hero. For a list of his works, both originals and translations see Suter, *Die Mathematiker und Astronomen der Araber und ihre Werke*, Leipzig, 1900, pp. 40-42 and Index, and Sezgin, *Geschichte des arabischen Schriftums*, Leiden 1967 onwards, III, pp. 270-274, V, pp. 285-286, VI, pp. 180-182 and Indexes). Both writers give locations of extant manuscripts.

No comprehensive study of Qusṭā's works has yet been undertaken, nor has there been made any detailed evaluation of his contribution, certainly a significant one, to the progress of science. His services as a translator must surely rank at least equally with his original works. The biographers are unanimous in praising his skill as a translator of Greek works into Arabic, and in the light of the surviving translations their esteem seems to be fully justified. For example, although the original of Hero's *Mechanics* is lost, an examination of Qusṭā's Arabic version (Carra de Vaux in *Journal Asiatique*, 9ᵉ série (1893), tome 1, pp. 386-472, tome 2, pp. 152-269, 420-514 = Arabic text and French translation) leaves us with little doubt that this is a faithful and sensitive rendering.

3

Bibliography (in addition to the works mentioned in the text): Ibn al-Qifṭī, *Ikhbār al-ʿulamāʾ*, Cairo 1326 H (1908), pp. 173-74; H. Suter, "Die Abhandlung Qosṭa ben Lūqās und zwei andere anonyme über die Rechnung mit zwei Fehlern und mit der angenommenen Zahl", in *Bibliotheca Mathematica*, 3. Folge, 9 (1908-9), pp. 111-112; G. Gabrieli, "Nota biobibliografica su Qusṭā", in *Rendiconti dell' Accademia dei Lincei*, classe d. sci. mor. 21 (Rome 1912), pp. 341-382; G. Sarton, *Introduction to the History of Science*, I, p. 602 and Index; W. H. Worrell, "Qusta ibn Luqa on the Use of the Celestial Globe", *Isis*, 35 (1944), pp. 285-293; Brockelmann, *Geschichte der arabischen Literatur*, Leiden 1947-52, I, pp. 204-205, 512; A. G. Drachmann, *The Mechanical Technology of Greek* p. 530
and Roman Antiquity, Copenhagen 1963.

XIV

Al-Bīrūnī's Mechanical Calendar

1. Introduction

Abū-r-Rayḥān Muḥammad b. Aḥmad al-Bīrūnī was one of the greatest scholars of medieval Islam, and certainly the most original and profound.* He was equally well versed in the mathematical, astronomical, physical and natural sciences and also distinguished himself as a geographer and historian, chronologist and linguist, and as an impartial observer of customs and creeds. He was born in 362/973† in Kath, capital of Khuwarizm (the region of the Amu-Darya delta on the southern shores of the Aral Sea). He spent the first 25 years of his life in his homeland, and then spent some years in the northern provinces of Iran before returning to Khuwarizm. In 408/1017 his country was conquered by Sultan Maḥmūd of Ghazna and al-Bīrūnī, together with other distinguished scholars, was compelled to accompany the Sultan to Ghazna (now in Afghanistan). For the rest of his life he was based at the court at Ghazna, but went with Maḥmūd on several of his expeditions to north-west India, where he learned Sanskrit and closely observed the local customs and religious practices, a sum of knowledge which he put into his justly famous *Description of India*, completed in 421/1030. He died

* I acknowledge with gratitude the generosity of The Royal Society, The World of Islam Festival Trust and The Arab-British Chamber Charitable Foundation in providing me with grants to assist me in my researches into Islamic technology. I wish to express my sincere thanks to Mr Francis Maddison (Museum of the History of Science, Oxford) for his advice and assistance to me in the preparation of this paper, and also to Dr Judith Field and Mr Michael Wright (both of the Science Museum, London) for their help and co-operation.

† The first year given is the Hegira year, the second the Christian year.

in 440/1048, probably in Ghazna. With the exception of the last date, the foregoing account is taken from Boilot (1); for the date of al-Bīrūnī's death, see Rozenfel'd, Rozhanskaya, and Sokolovskaya, p. 18, who cite an article by Karimov.[1]

Al-Bīrūnī is known to have written 180 works; these are listed, with descriptions and provenances of manuscripts, if known, in Boilot (2).[2] They vary from short treatises of a few pages to long books embracing wide fields of knowledge. Only a handful have been published. The treatise that concerns us here is part of a much longer work entitled *Kitāb fī istī'āb al-wujūh al-mumkina fī ṣan'at al-asṭurlāb—Book on the Full Comprehensiveness of the Possible Methods for Constructing the Astrolabe.* The book has never been published, either in the original or in translation, but a number of commentaries have been written about it by modern authors (listed on p. 191 in Boilot (2)). It consists of detailed instructions for constructing the astrolabe—both northern and southern—with a number of variations for the design of the rete, some of them with fanciful names (see W. Hartner in *Encyclopaedia of Islam*, second edition, 1960, I, 722–28). There are also chapters on the spherical astrolabe, on conic sections and on the construction of a sophisticated type of compass. One of the most interesting features of the work is al-Bīrūnī's references to the work of earlier scholars of the third/ninth and fourth/tenth centuries. Among his immediate predecessors he mentions his teacher Abū Naṣr Manṣūr b. 'Ali b. 'Irāq[3] and his contemporary Abū Sa'īd as-Sijzī (d. *c.* 415/1024).[4] He also derived a number of constructions from Abū Ḥāmid as-Saghānī who was renowned as an instrument maker (d. 379/990).[5] Al-Bīrūnī mentioned these scholars, and others, mainly in order to acknowledge that constructions that he was describing were based upon their work. It is therefore significant that he makes no such acknowledgment for the geared calendar, which he calls 'The Box of the Moon',[6] and we can therefore infer that the instrument was well known from its Hellenistic or Byzantine origins and that the name of its inventor was not known. As we shall see, the calendar described in the main body of the treatise closely resembles the instrument dealt with in 'Gears from the Byzantines' by Field and Wright, but he then goes on to describe several variants devised by other craftsmen, presumably Muslims. It seems likely that the tradition for making this instrument was diffused to Islam from Byzantium in the first part of the third/ninth century, at a time when so much Greek learning was transmitted to Baghdad,[7] and that the Muslims then experimented with different designs in an attempt to improve its accuracy.

[1] D. J. Boilot (1), 'Al-Bīrūnī, *Encyclopaedia of Islam*, second edition, I (1960), 1236–38; B. A. Rozenfel'd, M. M. Rozhanskaya and Z. K. Sokolovskaya, *Abu-r-Raikhan al-Biruni 973–1048* (Moscow, 1973).

[2] D. J. Boilot (2), 'L'Oeuvre d'Al-Berūnī. Essai bibliographique', *Mélanges de l'Institut d'Études orientales du Caire*, 2(1955), 161–256, and 3(1956), 391–96.

[3] H. Suter, 'Die Mathematiker und Astronomen der Araber und ihre Werke', *Abhandlungen zur Geschichte der mathematischen Wissenschaften*, 10 (1900) (published as a separate volume); new impression (Amsterdam, 1981), p. 81.

[4] Ibid., 80.

[5] Ibid., 65.

[6] The calendar was the subject of a paper by E. Wiedemann: 'Ein Instrument das die Bewegung von Sonne und Mond dargestellt, nach al-Bīrūnī, *Der Islam*, 4 (1913), 5–13. This is a partial translation, and Wiedemann does not always distinguish between the original text and his own comments and additions. There are also some errors in his reconstruction of the instrument. These are discussed in the technical commentary.

[7] In the reign of the Caliph al-Ma'mūn, Muḥammad b. Mūsà is said to have travelled to Byzantine territory to collect scientific works. In the same period (early third/ninth century) eminent scholars and translators from Ḥarrān in Upper Mesopotamia moved to Baghdad. See the *Book of Ingenious Devices* by the Banū Mūsà, annotated and translated by Donald R. Hill (Dordrecht, 1979), p. 4.

Boilot (2), p. 191, lists twelve manuscripts of *Kitāb al-istī'āb*..., of which we were able to obtain three for the preparation of this paper. It is considered that the use of these manuscripts permitted a virtually complete reconstruction of the original text. The manuscripts are as follows:

Leiden Or 591

This is certainly the best of the three available MSS and consists of 88 folios. The Arabic numbering of the folios is out of sequence and the Roman page-by-page numbering has therefore been followed. The MS contains the following works:

(1) Astrolabic treatise of Sharaf ad-dīn aṭ-Ṭūsī.
(2) Treatise of Abū Naṣr b. Zarīr on the astrolabe.
(3) Treatise of Qustā b. Lūqā on the sphere.
(4) Astronomical treatise of Theon of Alexandria.
(5) Al-Bīrūnī's *Kitāb al-istī'āb*... This begins on p. 47. The Box of the Moon is from p. 162 to p. 167.

The colophon reads:

> Its compilation was completed by Ismā'īl b. Ibrāhīm b. Maḥmūd al-Khilāṭī in the city of Khilāṭ [on the shores of Lake Van in Armenia] on the morning of Jumādā II year 614 [11 September AD 1217].

The treatise is written in a clear *naskhī*, with most of the consonantal points inserted. In the Box of the Moon section, and by inference in the whole manuscript, the grammar is sound and errors and omissions are infrequent. The illustrations are good.

Staatsbibliotek Berlin; Catalogue W. Ahlwardt 5794– 5796 (Sprenger 1869).

This consists of 147 folios and contains the *Kitāb al-istī'āb* only. Boilot (1), p. 191 gives the number as 5795, but the complete text is assembled under the three numbers given above. The Box of the Moon is from f. 130 recto to f. 136 recto. The MS is written in *farsī* script and appears to follow the Leiden text fairly closely, but with many more mistakes. The illustrations are poor. There is no colophon.

British Library OR 5593

This contains only the *Kitāb al-istī'āb* and consists of 89 folios, the Box of the Moon being from 81 recto to 84 verso. It is written in a poor *naskhī* with many consonantal points omitted. It is full of grammatical solecisms and other errors. The illustrations are poor and there is no colophon.

In an earlier passage in the book al-Bīrūnī promises to describe the Box of the Moon and two other instruments[8] (this passage is on p. 94 in the Leiden MS, for example). He mentions the inventors of the other two instruments, but gives no attribution for the Box of the Moon, probably for the reasons discussed above. Nor

[8] The first of these is an eclipse predictor. Al-Bīrūnī says that it was developed by a certain Nasṭūlus and by al-Ḥasan b. Muḥammad al-Adamī (Suter, p. 27), before being completed by 'Utārid b. Muḥammad al-Ḥāsib (Suter, p. 67). The second instrument is for predicting when to look for the new Moon in the sky, by enabling its user to check the relative positions of the Sun and the Moon. Like the astrolabe it has a plate and a rete. According to al-Bīrūnī, several articles on its construction were written by Abū Dawūd Sulaymān b. Ghaṣama as-Samarqandī. It was clearly a useful device for those concerned with determining the commencements of the Islamic months.

does he give an indication as to the purpose of the instrument. It is worthy of note that, unlike the Byzantine calendar, it does not include a ratchet and this omission may, in fact, have made it more useful. If an event, the payment of a debt for example, was due at a particular new Moon, then the alidade could have been turned until that new Moon was displayed, and the day of the month read off from the sun's pointer.

2. The Text

In the Notes that follow the Arabic text the manuscripts are referred to by the following letters of the alphabet:

Leiden: *alif.*
British Library: *bā.*
Berlin: *jīm.*

It will be noticed that there are quite a large number of variants, but many of these are trivial, being concerned with such minor points as incorrect genders, particularly for numerals. Nevertheless, more readings would have remained uncertain had not all three manuscripts been available.

Typesetting of the Arabic was done on the Monotype Lasercomp typesetter at the Oxford University Computing Service. John Cooper, of St John's College, Oxford did the input.

Plate 1.

(*a*), on p. 143: Moon's volvelle (Leiden).
(*b*), on p. 145: Moon's volvelle (British Library).

The letters in the small circles, starting in the circle at 'nine o'clock', read as follows: left *ṭ*, centre *s*; top *ḥ*, centre *m*; right *l*, centre *f*; bottom *k*, centre *x*. The main centre-point is marked *e*.

Reading the numbers (in *abjad* numerals) in the divisions of the annulus, starting at the division at the top at the left of the diameter, we have 1 to 30, (anticlockwise) then 1 to 29, in the top right-hand division.

Note: the illustration from the Berlin manuscript has been omitted: it shows only the circles and diameters, without lettering or teeth.

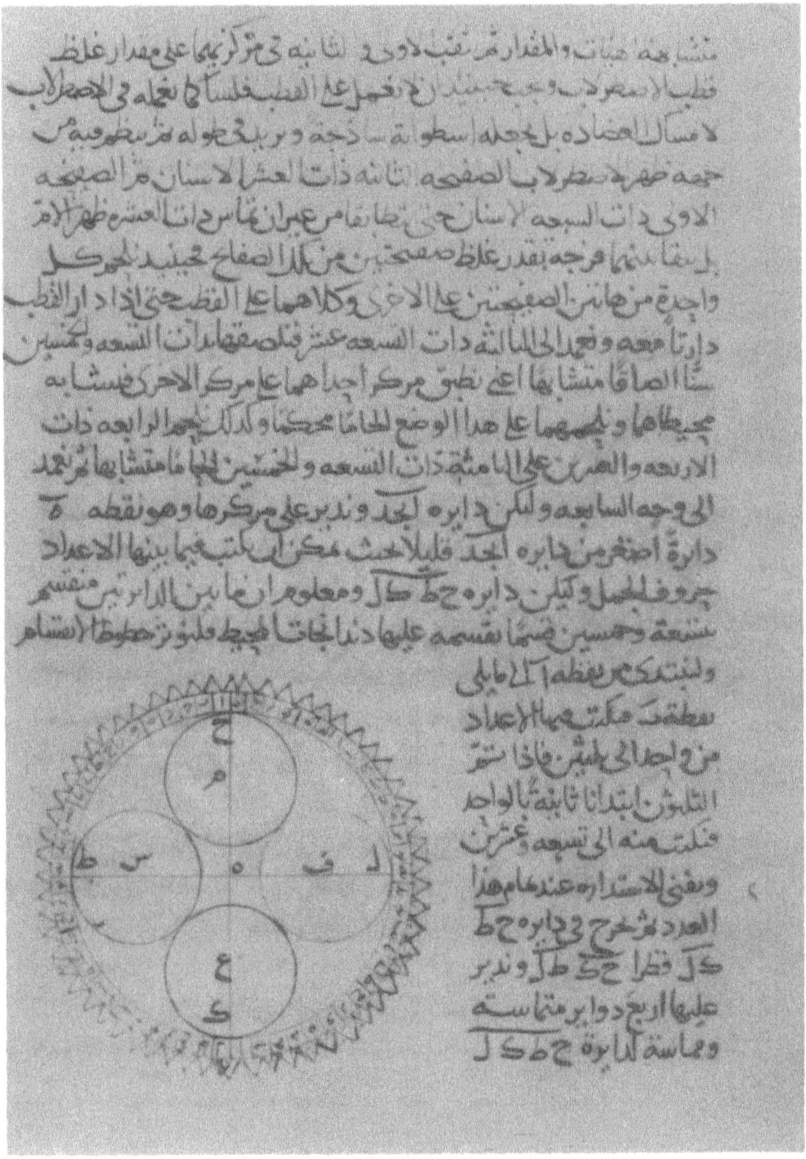

Plate 1 (*a*).

صنعة حق القمر

ونقول أن هذة الآلة ليس[١] الغرض فيها كألذى في الأصطرلاب[٢] من التحقيق المشفوع[٣] بالبرهان الواضح البين[٤] ولكن مغزى مستخرجه[٥] أن يرى زيادة القمر ونقصانه وما مضى من الشهر وموضع النيرين التقريب وإذا قصدنا لتركيبه على ظهر الأصطرلاب فلنجعل على حرف ظهره طوقاً كالحجرة بل أقل عرضاً منها وأكثر[٦] سمكاً ثم نقسم نصف قطر ما يقع من ظهر الأم داخل هذا الطوق[٧] بتسعين قسماً متساوية ونعمد إلى قطع شبه كالصفائح فى إستوا السطح لكنها أغلظ وامتن[٨] ونقطع منها ثمانى صفائح مستديرة قطر الاولى سبعة[٩] أجزاءً من تلك الأجزاء التسعين التي قسمناها[١٠] وقطر الثانية عشرة أجزاءً والثالثة تسعة عشر جزءاً[١١] والرابعة أربعة وعشرون[١٢] جزءاً والخامسة أربعون[١٣] جزءاً والسادسة ثمٰنية وأربعون[١٤] جزءاً وكل واحدة من السابعة وا الثامنة تسعة وخمسون[١٥] جزءاً ونقسم كل واحدة منها بالعدد المفروض لأقطارها ونبرد عليها فى مواضع الأقسام دندانجات متساوية مثلثة الشكل حادة الأطراف متشابهة الهيئات والمقدار ثم نثقب الأولى والثانية فى[١٦] مركزيهما على مقدار غلظ قطب الأصطرلاب ونجب حينئذ أن لا نعمل على القطب فلساً كما نعمله فى الأصطرلاب لإمساك العضادة بل نعمله إسطوانةً سادجةً ونزيد فى طوله ثم تنظم فيه من جهة ظهر الأصطرلاب الصفيحة الثانية ذات العشر[١٧] الأسنان ثم الصفيحة الأولى ذات السبعة[١٨] الأسنان حتى تتطابقا من غير أن تماس ذات العشرة ظهر الأم بل تبقى[١٩] بينهما فرجة بقدر غلظ صفيحتين من تلك السفائح فحينئذ نلحم كل واحدة من هاتين السفيحتين على الأخرى وكلاهما على القطب حتى[٢٠] إذا دار القطب دارتا[٢١] معه ونعمد إلى الثالثة ذات التسعة عشر فنلصقها بذات التسعة والخمسين سناً الصاقاً متشابهاً أعنى نطبق [مركز أحداهما على مركز الأخرى][٢٢] فنشابه محيطاهما ونلحمها على هذا الوضع[٢٣] إلحاماً محكماً وكذلك نلحم الرابعة ذات الأربعة والعشرين على الثامنة ذات التسعة والخمسين إلحاماً متشابهاً ثم نعمد الى وجه السابعة وليكن دايرة **ابجد** وندير على مركزها وهو نقطة **ه** دايرة أصغر من دايرة **ابجد** قليلاً بحيث نمكن أن نكتب فيما بينها الأعداد بحروف الجمل وليكن دايرة **ح ط**

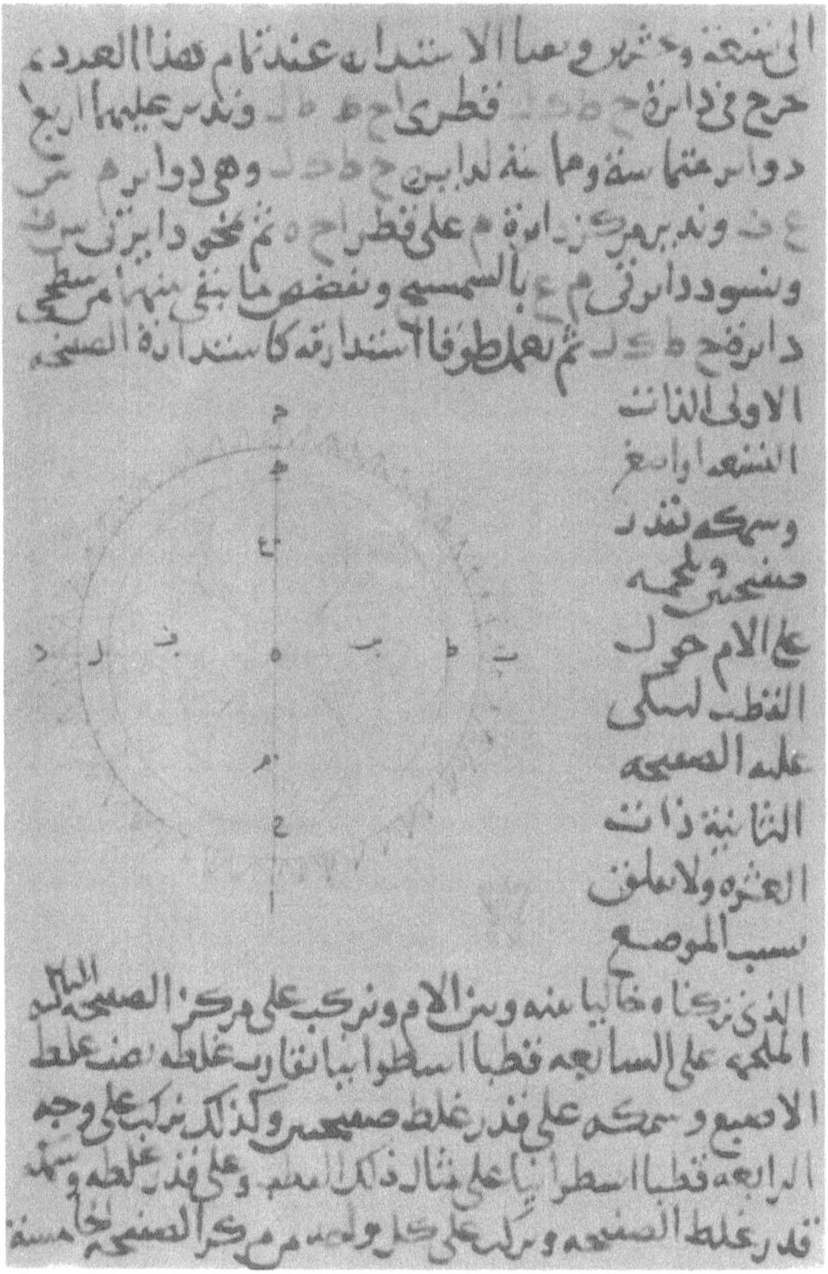

Plate 1 (*b*).

146

ك ل[٢٤] ومعلوم أن ما بين الدايرتين منقسم بتسعة وخمسين قسماً تقسمه عليها دندانجان المحيط فلنوثر خطوط الإنقسام ولنبتدىء من نقطة **ا** إلى ما يلى نقطة **ب** فنكتب فيها الأعداد من واحد إلى ثلثين فإذا تم الثلثون إبتدأنا ثانيةً بالواحد فنكتب منه إلى تسعة وعشرين وتفنى الإستدارة عند تمام هذا العدد ثم[٢٥] نخرج فى دايرة **ح ط ك ل** قطرى[٢٦] **ح ك ط ل** ونُدير عليهما[٢٧] أربع دواير متماسة ومماستة لدايرة **ح ط ك ل** وهى دواير **م س ع ف** وندير مركز دايرة **م** على قطر **ا ح ه** ثم نمحو دايرتى **س ف** وتسود دايرتى **م ع** بالسيمسُخْتج ونفضض ما يبقى منها من صطحى دايرة **ع ط ك ل** ثم نعمل طوقاً إستدارته كاستدارة الصفيحة الأولى ذات السبعة[٢٨] أو أصغر وسمكه يقدر صفيحتين ونلحمه على الأم حول القطب ليتكى عليه الصفيحة الثانية ذات العشرة فلا تقلق[٢٩] بسبب الموضع ألذى تركناه خالياً بينه وبين الأم ونركب على مركز الصفيحة الثالثة الملحمة على السابة قطباً إسطوانياً ثابتاً[٣٠] يقارب غلظه نصف غلظ[٣١] الإصبع وسمكه على قدر غلظ صفيحتين وكذلك نركب على وجه الرابعة قطباً استوانياً على مثال ذلك القطب وعلى قدر غلظه وسمكه قدر غلظ صفيحة[٣٢] ونركب على كل واحد من مركز الصفيحة الخامسة والسادسة قطباً على الهيئة المذكورة إلا أنا نجعل سمك قطب الخامسة على قدر غلظ صفيحتين وسمك قطب السادسة صفيحة واحدة مثل قطب الرابعة ونمدّ هذين القطبين أعنى الذى للخامسة والذى للسادسة إلى الجهة الأخرى من الصفيحتين مداً له قدر فأنا نريد أن نركب فيهما قوسين ونجعل لكل واحد من هذة الأقطاب طوقاً يسعه ولا يفضل منه شىء ولكن إذا وضع [على موضع][٣٣] من الأم والحم ذلك الطوق عليه دار القطب فيه مع سهولة حركة واستقرار على الأم ثم نلحم الطوق المحيط بقطب الثالثة على موازاة ستين جزءاً من أجزاء الإرتفاع فى موضع ليشتبك فيه أسنان السابعة بأسنان الأولى ولا يمانعها الطوق الأقص[٣٤] في الدوران ثم نلحم الطوق المحيط بقطب الرابعة أسفل من ذلك إلى نحو أوايل[٣٥] أجزاء لإرتفاع بحيث تشتبك فيه إسنان الثالثة بأسنان الثامنة ثم لا تتعلق شىء من أسنان الثامنة بأسنان[٣٦] الثانية ولا يمانعها الطوق الأقصى في الدوران ثم نلحم الطوق المحيط بقطب السادسة في موضع تشتبك فيه أسنانها مع أسنان الرابعة تحت[٣٧] الثامنة ونلحم الطوق المحيط بقطب الخامسة [من الجانب][٣٨] الأعلى من

السابعة بحيث تشتبك أسنانها بأسنان الثانية ونلحم أيضاً على الأم قسياً من الأطواق تتكى عليها ما تحتاج من الصفائح إلى متكأ ليلا تضطرب عند الحركة فتصير الاطواق ملحمة على الأم والأقطاب متحرك فيها فمتى أدير قطب الأصترلاب دار بدورانه ذات السبعة وذات العشرة اما ذات العشرة فتدير[39] ذات[40] ا الأربعين واما ذات السبعة فتدير ذات التسعة[41] والخمسين وتدور[42] معها ذات التسعة عشر لالتحامها بها وتدير التسعة عشر ذات التسعة والخمسين الأخرى وتدور[43] ذات الأربعة والعشرين وتدير ذات الاربعة والعشرين ذات الثمانية والأربعين فإذا فرغنا من ذلك هيأنا طبقاً ينطبق على الحجرة بهندام يستقر فيه قريباً من مماسة الصفيحة السابعة ولأجل هذا يجب اَن يكون سمك الطوق الأقصى الشبيه بالحجرة أرجح قليلاً من ثلثة اضعاف غلظ صفيحة ونثقبه فى ثلثة مواضع الأول هو الوسط على مسامتة قطب الأصطرلاب والثانى على مسامتة قطب الصفيحة الخامسة ذات الأربعين وهو للقمر والثالث على مسامتة قطب الصفيحة السادسة ذات الثمانية والأربعين وهو للشمس ثم نخرق هذا الطبق[44] أيضاً بالقرب من الطوق الأقصى خرقاً مساوياً بالدايرة م التى على وجه الصفيحة السابعة [بحيث إذا دارت الصفيحة السابعة][45] على قطبها[46] وافى[47] أحدى دايرتى م ع المسودتين ذلك الخرق وطابق حرفه حرفها[48] ونخرق فيه[49] ايضاً خرقاً مربعاً بين هذا الخرق[50] المستدير وبين الطوق المحيط على مسامتة حرف[51] تسعة وعشرين[52] المكتوب بالجمل وذلك حين تكون دايرة م[53] مطابقة للخرق المستدير فإذا فرغنا من ذلك أدرنا حول كل واحد من قطب الأصطرلاب وقطبى النيرين على هذا الطبق دايرة باى بعد شيئنا ونقسم إلتى فى الوسط[54] بسبعة اقسام متساوية نكتب فى كل واحد منها إسم يوم من أيام الأسبوع متوالية من اليمين إلى فوق إلى الشمال ونقسم كل واحدة من دايرتى الشمس والقمر باثنى عشر برجاً متساوية [كل واحد][55] منها ثلثين درجة إما مبسوطة أو مطوية[56] أنصافاً وأثلاثاً وأسداساً كما نعمله فى الأصطرلاب على حسب إحتمال الدايرة للقسمة ونكتب فى كل واحدة منها أسماء البروج متواليه إلى فوق من الشمال إلى اليمين ثم نركب فى قطبى النيرين قوستين تمرا ذنابها أو طرف آخر دقيق من اَطرافهما على درج البروج ويماسها فى الدوران وننظم عضادة الأصطرلاب فى قطبه فوق الطبق ونمسكها عليه بفرس

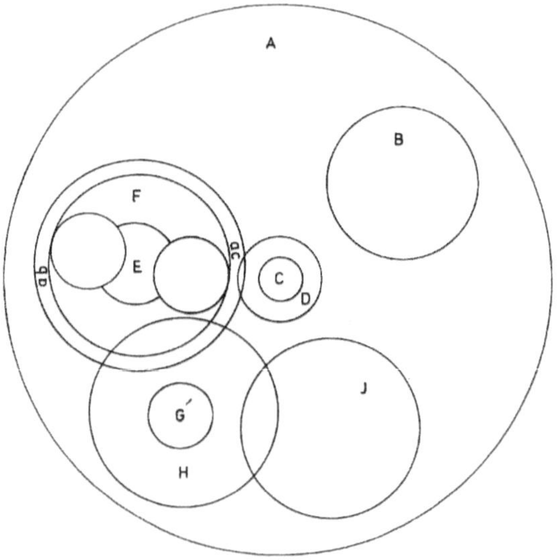

Key to Plate 2(*a*).

Plate 2(*a*), on p. 149.

(Refer to key for location of captions.)

A Any arc or tooth drawn in black is visible; anything drawn in red is below the disc as if it were seen from below the illustration.
B Having forty [teeth understood here and in all cases]; for the moon.
C Having seven.
D Having ten.
E Having nineteen; from inside.
F Having fifty-nine.
G Having twenty-four; from inside.
H Having fifty-nine.
J Having forty-eight; for the sun.
a 1
b 30
c 29

Plate 2(*b*), on p. 150.

The illustration is turned through 90 degrees compared with Plate 2(*a*). Caption A is at the bottom, but is otherwise identical to that of Plate 2(*a*). The other captions are also the same except that the figures are in numbers instead of words, and Caption C has 9 instead of seven.

Plate 2(*c*), on p. 151.

The differences from Plate 2(*a*) are as follows:
C–omitted; E–has only the word 'ten'; G–'from inside' omitted.

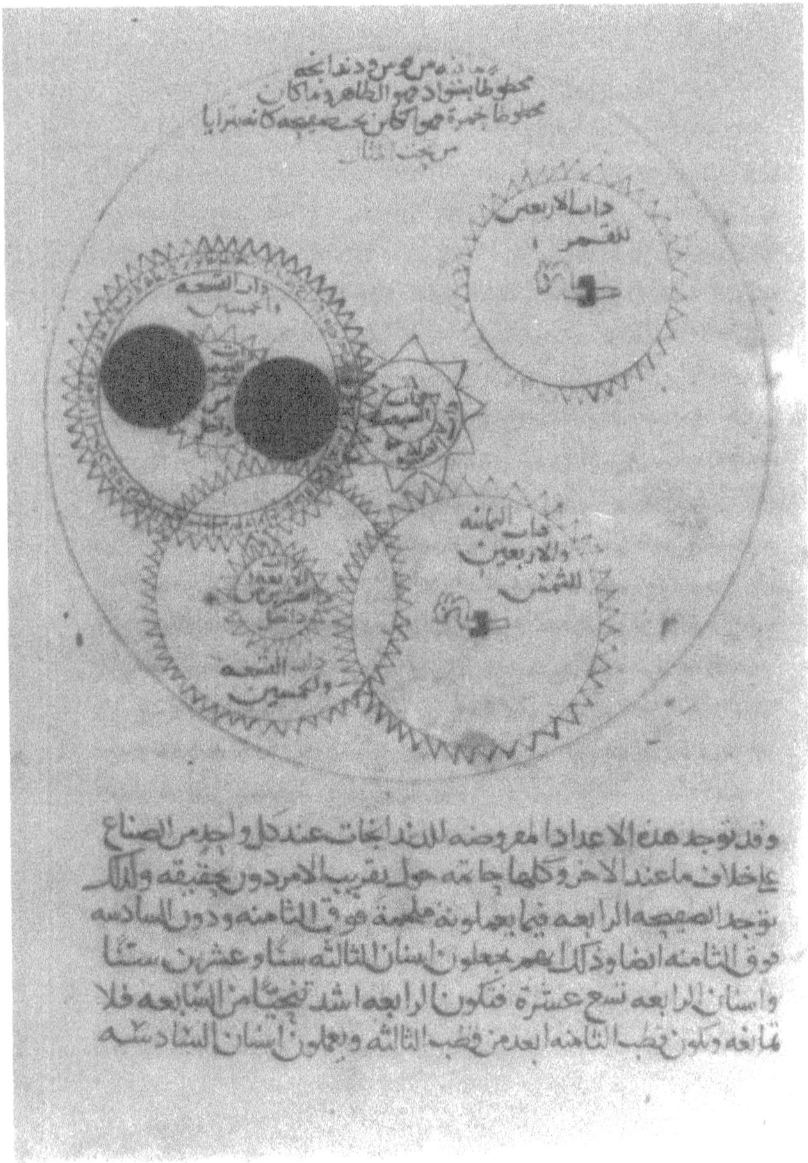

Plate 2(*a*).

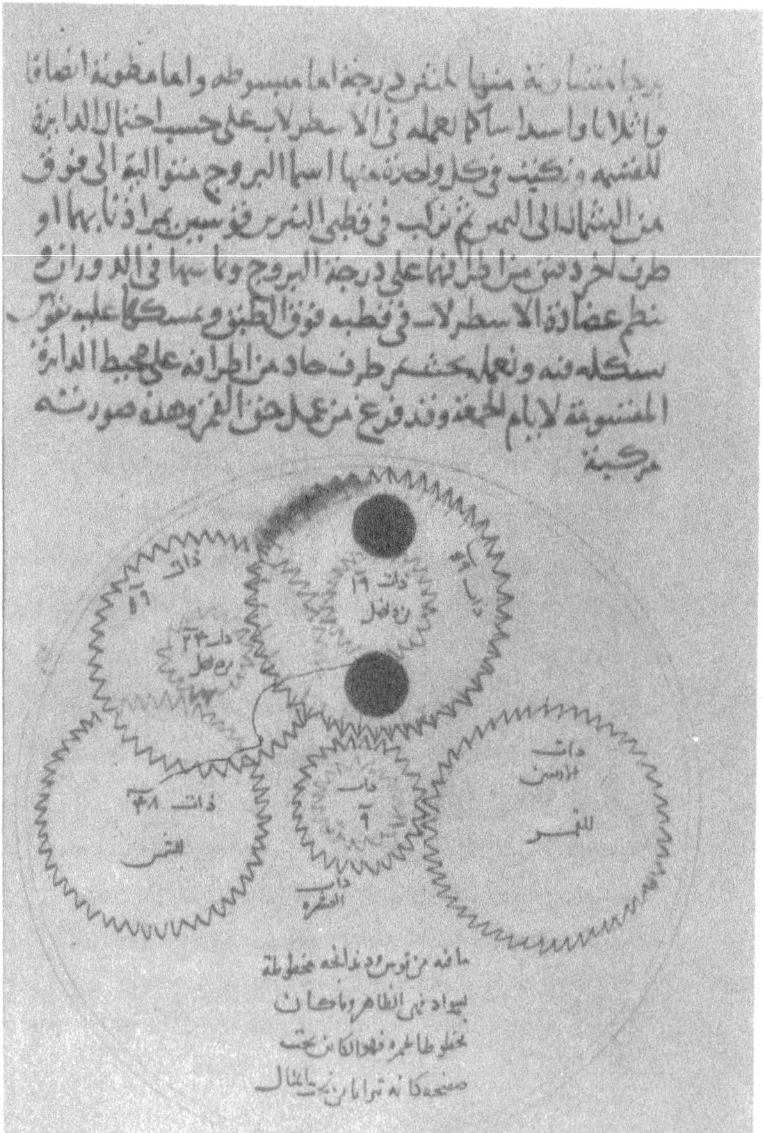

Plate 2(*b*).

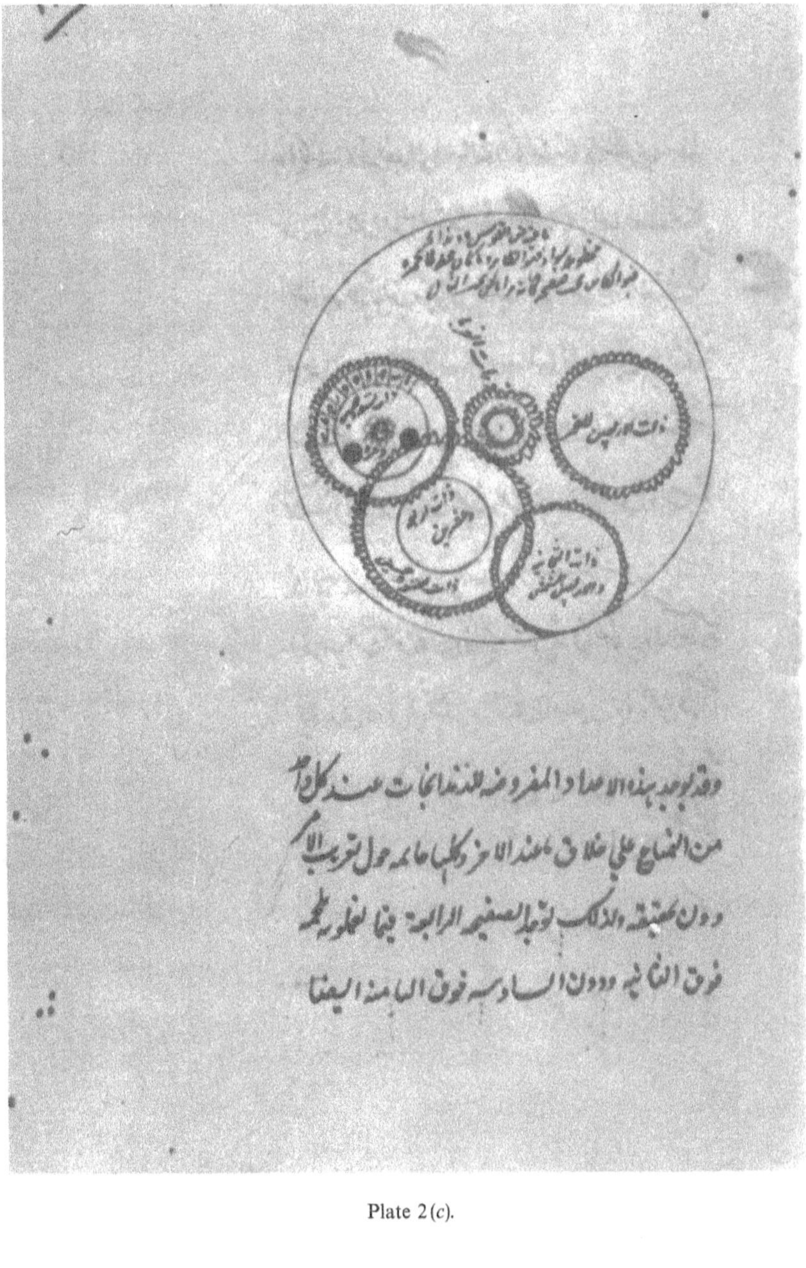

Plate 2(*c*).

نسلكه[٥٧] فيه ونعمله بحيث يمر طرف حاد من اطرافه على محيط الدايرة المقسومة لأيام الجمعة وقد فرغ من عمل حق القمر [والله علم][٥٨] وهذة صورته مركبة

وقد توجد هذة الأعداد المفروضة للدندانجات عند كل واحد من الصناع[٥٩] على خلاف ما عند الآخر وكلها حائمة حول تقريب الأمر دون تحقيقه ولذلك توجد الصفيحة الرابعة فيما يعملونه ملحمة فوق الثامنة ودون السادسة فوق الثامنة أيصاً وذلك أنهم يجعلون أَسنان الثالثة ستاً[٦٠] وعشرين سناً وأسنان الرابعة تسع عشرة[٦١] فتكون الرابعة أشد تنحياً من السابعة فلا تمانعه ويكون قطب الثامنة ابعد من قطب الثالثة ويعملون أسنان السادسة تسعاً[٦٢] واربعين والثامنة ستاً[٦٣] وخمسين ومنهم من جعل أسنان الثالثة تسع عشرة[٦٤] وأسنان الرابعة ثلثاً[٦٥] وعشرين وأسنان السادسة ستاً[٦٦] وأربعين وأسنان الثامنة سبعاً[٦٧] وخمسين وهذا أقرب من الحق لأن دورة الشمس تتم [فى هذا][٦٨] فى ثلثماية وأربعة وخمسين يوماً وفى الأول فى ثلثماية وتسعة وتسعين[٦٩] يوماً والحق فيما بينهما ويمكن أن نختصر أمر هذة الصفائح بأن نعمل صفيحتين قطر أحدهما أربعة أجزاء من الأجزاء [التسعين][٧٠] وقطر الأخرى أثنان وخمسون جزءاً منها ثم نركب ذات الأربعة على ذات الأربعين تركيباً متشابهاً ونلحمها عليها ونركب على مركز ذات الأثنين والخمسين قطباً إسطوانياً ممتداً فى الوجه الآخر غلظه يقدر غلظ ثلث صفائح ونجعل له طوقاً نركبه باللحام على الأم بحيث تشتبك أسنان ذات الإثنين والخمسين بذات الأربعة ونجعل لها متكاً ونثقب الطبق على سمت قطبها ونركب فيه الفرس فيكون لحركة الشمس وتنوب[٧١] عن قطب السادسة فيما يقدم ويكون أقرب الى الحقيقة مما[٧٢] إستعمل هناك لإستتمام الشمس دورتها فى ثلثماية وستين يوماً وهذا ما أردنا الأبانة عنة [والله لعلم][٧٣]

ملاحظات

١. جـ تنقص : [ليس]
٢. ب، جـ : الاسطرلاب (وكل ما تظهر هذة الكلمة)
٣. ب : المشنوع
٤. أ، جـ : النير
٥. جـ : يستخرجه
٦. أ : أكبر
٧. ب : الطرف
٨. جـ : متن
٩. ب، جـ : تسعة
١٠. ب : قسمناه
١١. ب : جز ؛ جـ : أجزاءً
١٢. ب : عشرين
١٣. ب : أربعين
١٤. ب : أربعين
١٥. ب : خمسين
١٦. ب : على
١٧. ب، جـ : عشرة
١٨. ب، جـ : تسعة
١٩. أ : تبقا
٢٠. ب تنقص : [حتى]
٢١. ب : دارا
٢٢. ب تنقص : [مركز أحداهما على مركز الأخرى]
٢٣. ب : الموضع

٢٤. ب : **ح ه ك ل**

٢٥. ب تنقص : [ثم]

٢٦. أ : قطرا

٢٧. أ : عليها

٢٨. ب : تسعة

٢٩. ب، جـ : تعلق

٣٠. ب تنقص : [ثابتاً]

٣١. جـ : صلع

٣٢. ب : الصفيحة

٣٣. ب تنقص : [على موضع]

٣٤. جـ : أخرى

٣٥. جـ : اوا

٣٦. جـ : الأسنان

٣٧. جـ : بحيث

٣٨. أ تنقص : [من الجانب]

٣٩. جـ : تدير

٤٠. ب تنقص : [ذات]

٤١. جـ : سبعة

٤٢. ب : تدير

٤٣. ب جـ : تدير

٤٤. أ، جـ : الطوق

٤٥. جـ تنقص : [بحيث إذا دارت الصفيحة السابعة]

٤٦. أ : قطبيها

٤٧. ب، جـ : وافا

٤٨. جـ : فرقه حرقها

٤٩. ب : فيها ؛ جـ تنقص : [فيه]

٥٠. ب : الخط

٥١. أ، جـ : خرق

٥٢. جـ : فرمابن

٥٣. جـ تنقص : [م]

٥٤. جـ : الأوسط

٥٥. ب تنقص : [كل واحد]

٥٦. جـ : معمومة

٥٧. ب : بقوس سلكه ؛ جـ : بقوس سالكه

٥٨. أ، ب تنقص : [والله علم]

٥٩. جـ : التمناع

٦٠. ب : ستة

٦١. ب، جـ : تسعة عشر

٦٢. ب : تسعة

٦٣. ب : ستة

٦٤. ب، جـ : تسعة عشر

٦٥. ب : ثلثة

٦٦. ب : ستة

٦٧. ب : سبعة

٦٨. جـ تنقص : [في هذا]

٦٩. جـ : خمسين

٧٠. جـ تنقص : [التسعين]

٧١. جـ : وجوب

٧٢. جـ : فيا

٧٣. أ، جـ تنقص : [والله لعلم]

3. Translation*

The translation is fairly literal, but two departures have been made. Al-Bīrūnī calls the gear-wheels 'plates' because they were cut from a sheet of brass; to avoid confusion the word 'disc' has been used here. He refers his vertical dimensions, e.g. the thicknesses of the gearwheels, to the thickness of the plates from which the wheels were manufactured. For example, he says 'the depth is equal to the thickness of two plates'. This has been abbreviated in all cases, so that the phrase would read 'the depth is 2P'.

Construction of the Box of the Moon

We say that the purpose of this instrument is not the same as that of the astrolabe, in verification accompanied by clear plain proof, but the purpose of its fabrication is to show the waxing and waning of the Moon and how much of the month has elapsed and the position of the two lights [i.e. the Sun and the Moon] approximately. If we intend to assemble it on the back of the astrolabe, we install on its perimeter a rim like the *ḥujra* but narrower and higher than that. Then we divide half the diameter of the back of the *umm* inside this rim into 90 equal parts. We then take a piece of brass like the plates [of the astrolabe] in smoothness of its surface, but thicker and more solid. We cut from it 8 round discs [lit. plates], the first having 7 of those 90 divisions into which we divided it, the diameter of the second is 10 divisions, the third 19 divisions, the fourth 24 divisions, the fifth 40 divisions, the sixth 48 divisions, and both the seventh and eighth 59 divisions. We divide each of them into the numbers determined for their diameters and we file on them in the places of the divisions equally spaced triangular teeth with pointed ends and of equal size and shape. Then we bore the first and second on their centres to the size of the thickness of the arbor of the astrolabe, but we should not put a washer (*fils*) on the arbor as we do with the astrolabe to retain the alidade, but we make it a plain cylinder and increase its length. Then we fit to it from the direction of the back of the astrolabe the second disc with 10 teeth then the first disc with 7 teeth so that they cover each other without the 10-tooth touching the back of the *umm*, but with a gap between them equal to 2P. Whereupon we solder each of these 2 discs to the other and both of them to the arbor— when the arbor turns they turn with it.

Now we take the third with 19 [teeth] and place it on the seventh with 59 teeth in a similar way, i.e. so that their centre-points coincide. And we make their rims parallel and solder them securely together in this position. Similarly we solder the fourth with 24 [teeth] to the eighth with 59 teeth.

Now we turn to the face of the seventh; let this be circle *abjd*. We draw upon its centre, point *e*, a circle slightly smaller than circle *abjd*, so that we can write the letters of the alphabet in the space between them; let this be circle *ḥṭkl*. It is known that the space between the two circles is divided into 59 divisions, upon which the teeth of the perimeter are distributed. We engrave the lines of the divisions. Let us start from point *a* towards point *b* and write in them the numbers from 1 to 30, and when the 30 are complete we start again from 1 and write from it up to 29, and the circumference is filled up at the completion of these numbers. Then we draw on circle *ḥṭkl* two diameters *ḥk*, *ṭl* and draw on them four circles touching each other and touching circle *ḥṭkl*. These are circles *m*, *s*, *x*, and *f*. We draw the centre of circle *m* on diameter *aḥe* and we erase the two circles *s*, *f* and blacken circles *m*, *x* with *sīm-i-sukhtaj*, and we silver the remainder of the surface of circle *ḥṭkl*.

*In the translation the transliteration system of the *Encyclopaedia of Islam* has been followed, with certain exceptions: 'q' not 'k' is used for *qāf* and 'j' not 'dj' for *jīm*; consonants that are single in Arabic and double in Roman, such as kh and sh, are not underlined.

Now we make a round bush the same diameter as the first disc with 7 teeth, or smaller, its length 2P, and solder it to the *umm* around the arbor so that the second disc with 10 teeth rests on it. It does not wobble because of the space we left between it and the *umm*. On the centre of the third disc, soldered to the seventh, we erect a cylindrical arbor about half as thick as a finger and 2P long. Similarly, we erect on the face of the fourth a cylindrical arbor like that arbor, of the same thickness, 1P long. On the centre of both the fifth and sixth discs we erect an arbor in the above-mentioned manner except that we make the length of the arbor of the fifth 2P and the length of the arbor of the sixth 1P, as for the arbor of the fourth. We extend these two arbors, i.e. those for the fifth and the sixth, in the other direction from the discs by a certain distance, since we wish to fit bows to them. We fit to every one of these arbors a bush wide enough for it but with no clearance, so that if it is placed on the *umm* and the bush is soldered on it the arbor turns in it with an easy movement while resting on the *umm*.

Then we solder the bush surrounding the arbor of the third level with 60 divisions of the elevation in a position so that the teeth of the seventh mesh with the teeth of the first without the outer rim impeding the rotation. Then we solder the bush surrounding the arbor of the fourth lower than this towards the first division of the elevation so that the teeth of the third mesh with the teeth of the eighth and so that the teeth of the eighth do not overhang at all the teeth of the second and the outer rim does not impede the ratation. Then we solder the bush surrounding the arbor of the sixth in a position where its teeth mesh with the teeth of the fourth below the eighth. And we solder the bush surrounding the arbor of the fifth above the upper side of the seventh so that its teeth mesh with the teeth of the second. We also solder to the *umm* from the bushes bows to give the necessary support to the discs so that they do not wobble when they move. The bushes are soldered to the *umm* and the arbors move in them.

When the arbor of the astrolabe is turned the 7-[tooth] and the 10-[tooth] turn with it, and the 10-tooth turns the 40-tooth and the 7-tooth turns the 59-tooth, and the 19-tooth turns with it because it is soldered to it. And the 19-tooth turns the other 59-tooth and the 24-tooth rotates, and the 24-tooth turns the 48-tooth.

When we have completed that we prepare a cover to fit tightly over the rim and rest on it almost touching the seventh disc. For this reason the outer rim like the *ḥujra* must be somewhat greater in depth than 3P. Then we make holes in it in three positions: the first in the centre opposite the arbor of the astrolabe; the second opposite the arbor of the fifth disc with 40 [teeth] which is for the Moon; and the third opposite the axle of the sixth disc with 48 [teeth] which is for the Sun. Then we also make in the cover close to the outer rim a hole the same size as circle *m* which is on the surface of the seventh disc so that when the seventh disc turns on its arbor, and that hole reaches either of the black circles *m*, *x*, its perimeter coincides with theirs. We also make a square hole in it between this circular hole and the outer perimeter, which coincides with the 29 letters of the alphabet when circle *m* coincides with the round hole.

When we have completed that we describe a circle of any size we wish on the cover around the arbor of the astrolabe and the arbors of the two lights. We divide the central circle into seven equal divisions and write in each of them the name of a day of the week in succession from the right up towards the left. And we divide each of the circles for the Sun and the Moon into 12 equal signs of the Zodiac each of 30 degrees, [the latter] either shown separately or grouped into halves, thirds, or sixths [of the signs], as we do with the astrolabe according to the capacity of the circle. We write in each of them the names of the signs in succession from the left upwards towards the right. Then we fit to the arbors of the two lights bows whose tails or other sharp ends pass over and touch the degrees of the signs as they rotate. We fit the alidade to the astrolabe on its arbor above the cover and retain it by a cotter (*faras*) which we insert in it. We make it so that its sharp end passes over the circumference of the circle divided into the days of the week. So we have completed the construction of the Box of the Moon and this is the picture of it assembled. God is all-knowing.

For these numbers prescribed for the teeth there are differences from one craftsman to another, and each of them approach an approximate solution without precisely meeting it. For instance, they solder the fourth disc above the eighth, and also the sixth rotates above the eighth. And they make the third with 26 teeth, and the fourth with 19 teeth; and the fourth is further away from the seventh and so does not interfere with it. The arbor of the eighth is further away from the arbor of the third. They make the sixth with 49 teeth and the eighth with 56. Some make the teeth of the third 19; of the fourth 23; of the sixth 46 teeth and of the eighth 57 teeth. This is nearer the truth because the rotation of the sum is completed in this [case] in 354 days and in the first [case] in 399 days and the truth is between these.

We can simplify the matter of these discs by making two discs, one of them with a diameter of 4 divisions of the 90 divisions and the other with a diameter of 52 divisions of them. Then we mount the 4 [-tooth] on the 40 [-tooth] as [we mounted the discs] previously and solder it to it. And we mount on the centre of the 52 [-tooth] a cylindrical arbor extending to the other face, its length 3P, and provide for it a bush fixed by soldering to the *umm* so that the teeth of the 52 [-tooth] engage the disc with 4 [teeth]. We make for it a support and we make a hole in the cover opposite its arbor and fix the cotter to it. It is for the movement of the sun and replaces the previous sixth disc. This is nearer to the truth than what we have used before, for the sun to complete its rotation in 360 days. This is what we wished to explain. And God is all-knowing.

4. Technical commentary

(a) Assembly

The moving parts are to be assembled 'on the back' of the astrolabe, i.e. on a circular piece of brass. Having used this expression, al-Bīrūnī then always refers to the working surface as the *umm*, which in the astrolabe is the recess into which the plates fit; it is surrounded by the rim, called the *ḥujra*, a fairly wide annulus graduated into hours and degrees. The same term is used here, but in this case the *ḥujra* is a narrow, relatively high, circular band that forms the lower side of the box. The rim is referred to later in the text as the 'outer rim' (*aṭ-ṭawq al-aqsà*), which is a little confusing since there is also a similar rim on the cover.

The layout of the wheels shown in Figure 1 is derived from al-Bīrūnī's instructions. The third disc is to lie on a radius at about 60 degrees from the vertical diameter and the fourth is to be close to the vertical, so that discs 8 and 3 mesh together. The first (and second) are on the centre-point. The other two discs virtually position themselves. The drawing on Plate 2 is not quite accurate, the main error being that wheel 5 does not engage wheel 2 on any of the illustrations. The drawing in the British Library MS is at 90 degrees to those in the other two MSS.

(b) The gearwheels

The wheels are all cut from a smooth brass plate 'thicker and more solid' than the plates of an astrolabe. We shall assume, in the absence of more precise information, that the thickness was about 2 mm, as in the case of the Byzantine instrument. The wheels were sized by using the semi-diameter of the *umm* inside the *ḥujra* as a standard. This

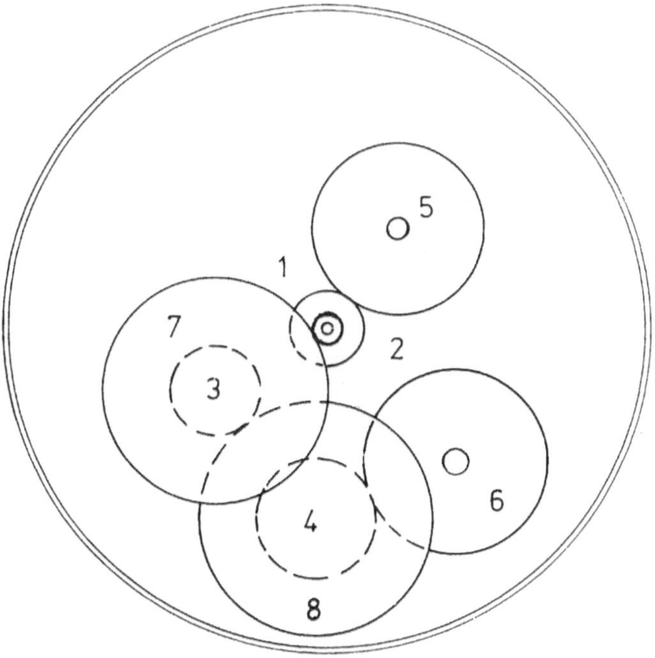

Figure 1.

Layout of base-plate; cover, pointers and teeth omitted.

semi-diameter was divided into 90 equal divisions and each wheel was sized by taking a predetermined number of these divisions—7 for the first, 10 for the second and so on. Since we are told that round plates of these diameters were cut from the brass plate, we can only assume that these were the outer diameters of the plain discs before the teeth were cut, not the diameters inside the teeth or the pitch circle diameters. We must also assume that the diameter of the *umm* was fairly large. Taking a diameter of 150 mm, the smallest wheel, before cutting, would have had a diameter of $150/2 \times 7/90 = 5{\cdot}83$ mm. This is probably too small to allow seven teeth in the shape of equilateral triangles to be cut into it. Also the phases of the moon would have been quite small and made an unimpressive display. The size may have been about 300 mm diameter, but it is clear from the text that the instrument was made in various sizes. For 300 mm, the outer diameter of the smallest wheel would have been 11·66 mm.

(*c*) *Gear ratios*

The diameters of the wheels in terms of the 90 divisions also determined the number of teeth. These teeth, in the shape of 'equally spaced triangular teeth with pointed ends and of equal size and shape' (a very exact specification!), were now cut into the perimeters of the wheels with a file. Following al-Bīrūnī's numeration of the wheels, the numbers of teeth were as follows:

Wheel	Teeth
1	7
2	10
3	19
4	24
5	40
6	48
7	59
8	59

The wheels for output are numbers 5 (for the sidereal month), 6 (for the sun's ecliptic) and 7 (for the moon's phases). The gear ratios produce the following results;

(i) Wheel 5 gives a sidereal month of 28 days (actual mean value 27·3217 mean solar days).

(ii) A single rotation of wheel 6, representing a tropical year, is derived from the following ratios: $59 \times \frac{59}{19} \times \frac{48}{24} = 366{\cdot}42$, which compares with the actual value of 365·2422 mean solar days.

(iii) Wheel 7 displays two successive synodical months, totally 59 days, against an actual average value of 59·06 mean solar days.

(d) The arbors, their mountings and attachments

The wheels are designed to rotate easily in their bearings, without oscillating: there are no upper bearings (see Figure 2). This is achieved by making the arbors of relatively large diameters compared with their lengths, and by making the journals as long as

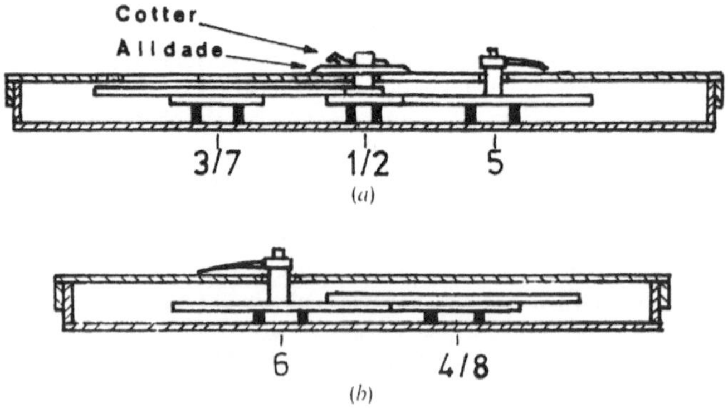

(*a*)

(*b*)

Figure 2.

(*a*) Section A–A on figure 3.

(*b*) Section B–B on figure 3.

possible. All the arbors except the central one are half a finger in diameter, i.e. about 10 mm, and the longer ones are about 4 mm in length. The central one—'the arbor of the astrolabe'— is 'the same diameter as the first disc with 7 teeth, or smaller'. It must have been no wider than the diameter inside the teeth of the wheel which in the case of an instrument 300 mm in diameter is about 7 mm. The worst case is for wheels numbers 3 and 7, which are supported on an arbor 4 mm long with a diameter of 10 mm, while wheel number 7 has a diameter of about 100 mm. Even so, they would have been quite stable.

The arbors all rotate in a sliding fit in bushes that are soldered to the *umm* and also apparently provided with side braces soldered to them and to the *umm* (not shown on Figure 2). (The drawing made by Hauser for Wiedemann, footnote 6, p. 9, figure 3, is incorrect in this respect: the diameters of the arbors are too small, and the central arbor has a much smaller diameter than that of its journal. It also passes through the *umm*, which is contradictory to the text.)

The arbor of the astrolabe had an alidade fitted to its upper end, retained by a cotter (*faras*) in the usual manner. The arbors of wheels 5 and 6 had 'bows' on their upper ends. These were presumably curved pointers whose sharp ends barely touched the graduated circles. It is assumed that these arbors were reduced in diameter above the wheels, and that their cross-sections were again reduced where they entered sockets at the ends of the pointers in a push fit. (It is possible that cotters and not bows were intended. These are shown in the illustration in the Leiden MS, and the words for 'cotter' and 'bow' in Arabic are very similar; one could easily have been mistaken for the other. Also, in the section on variants at the end of the treatise a *faras* is attached to the sun's arbor. The word 'bow' (*qaws*) is, however, used in all three MSS and we must accept this reading.)

(*e*) *Wheel 7—the Moon's volvelle*

For an instrument of 300 mm diameter this wheel would have had a diameter of about 100 mm. Around its perimeter were 59 teeth, inside which an annulus was engraved and divided into 59 divisions, one to each tooth. The numbers were engraved in the divisions from the top anticlockwise from 1 to 30 continuing from the bottom from 1 to 29, in *abjad* letters. (*Abjad* is a system in which the letters are assigned numerical values.) Four circles were drawn inside the annulus, touching each other and the inner circle of the annulus. The two circles on the horizontal diameter were erased and the other two were filled in with niello.[9] The remainder of the surface inside the annulus was silvered.

(*f*) *The cover* (see Figure 3)

Despite the text saying that the *ḥujra* was about 3P high it must in fact have been slightly over 4— two wheels soldered together, an arbor 2P long and an allowance for clearance; say about 10 mm in total. The cover fitted tightly over it. The cover was pierced in five places, three of these to allow the extensions to the arbors to pass through it. (Wiedemann, figure 3, p. 9, shows these as being much wider than the arbors but this

[9] *Sīm-i-sukhtaj* (*sic*) in all the manuscripts consulted. Wiedemann (footnote 6), p. 9, n. 1, says that it means 'something black', but *sīm-i sukhta*, i.e. 'burnt silver', is the Persian term used by Abū-l-Qāsim Kashanī (Tabrīz, *c.* AD 1300) for niello. See James W. Allan, *Persian Metal Technology 700–1300 AD* (London and Oxford, 1979), pp. 19–20, who quotes the methods of preparing niello given by Kāshānī and by al-Hamdānī (Yemen, with Persian contacts, *c.* AD 942); he points out that al-Hamdānī's niello was silver sulphide, whereas that of Kāshānī was a compound of silver, copper, and lead sulphides.

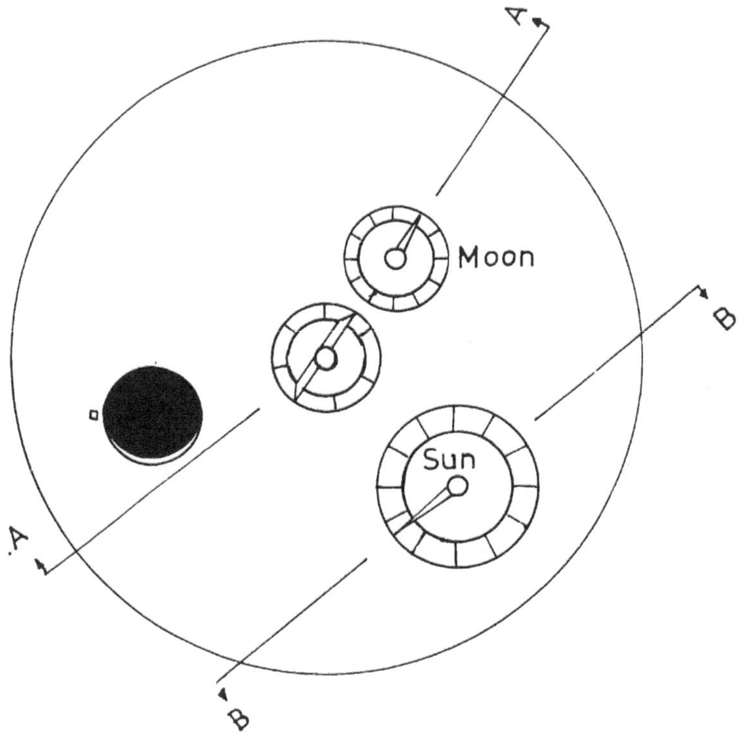

Figure 3.

Plan view of cover.

is incorrect—holes just a little wider than the arbors would have sufficed.) The fourth hole was of the same diameter as one of the two black circles on the face of wheel 7; it coincided with one or other of these at new moon. Close to this a small square hole was cut, the same size as one of the numbered divisions on the rim of wheel 7 and coinciding with the annulus, so that the day of the month could be read from the number under this 'window'.

Annuli were inscribed on the cover around the three arbors so that the tips of the pointers passed over them. The annulus around the arbor of the astrolabe was divided into seven compartments, upon each of which a day of the week was engraved. The annuli around the other two arbors were each divided into 12 compartments, in which the names of the Zodiacal signs were engraved, each subdivided into degrees. The degrees were shown singly if the annulus was large enough, otherwise they were grouped in halves, thirds or sixths of the signs.

(*g*) *Operation*

All the displays were set to their correct positions before the device was put to use. The alidade was then moved forward one division every day, so gradually altering the positions of the Sun and Moon and the shape of the Moon's phases. There was no ratchet, as there was in the Byzantine instrument, but this should not have caused any real problems. It would also have been possible to have used the device in reverse, as it were: turning the pointer until the Sun was in a particular degree of the Zodiac would have enabled a prediction for the Moon's phase for that night to be made.

It would have been necessary to have reset the instrument from time to time because of the inaccuracies incorporated in its construction. It would have been desirable to have had some means of moving the pointers upon their own arbors. We know this could be done with the alidade, but we are not told how the 'bows' were fitted – we have suggested a push-fit and spigot-and-socket joint. The position of the Moon indicated by wheel 5 would have become out of true more quickly than the other two displays, and with such a connection its pointer could have been moved on its arbor without dismantling the instrument. For a complete readjustment, however, it would have been necessary to have removed the pointers and the alidade, lifted the cover and reset all the wheels.

(*h*) *The variants*

Two variants used by other craftsmen are mentioned by al-Bīrūnī. In each of these the positions of wheels 4 and 8 on their arbor were reversed and wheel 6 was raised by 1 P to mesh with wheel 4 above wheel 8. This makes no difference to the gear ratios, but these are affected by the other changes. In neither case were the sizes of wheels 1, 2, 5 or 7 altered. For the other wheels the new numbers of teeth were as follows:

Wheel	Case 1 teeth	Case 2 teeth
3	26	19
4	19	23
6	49	46
8	56	57

These changes do not affect the periods for the sidereal or synodical months, but they alter the time for the Sun's rotation. In the first case the rotation of wheel 6 takes 327·72 days; the period is given as 399 days in the Leiden and London MSS, 359 in the Berlin one—both are of course incorrect. Al-Bīrūnī was usually very accurate in his calculations and such a wide margin of error is therefore surprising. It is possible that there is an error for the number of teeth in one or more gearwheels, in all the manuscripts. In the second case the rotation takes 354 days, as stated in the text.

Al-Bīrūnī then proposes a simplified design, omitting one wheel. Wheels 1, 2, and 7 remain as before. A wheel with 4 teeth is soldered to a wheel with 40 and the arbor of this double wheel shows the Moon's position in the Zodiac. The 40-tooth wheel meshes with wheel 2. A wheel with 52 teeth is made for the Sun's movement and mounted on an arbor so that it engages the 4-tooth wheel. (Some of this has to be inferred; although Wiedemann on p. 12 gives the impression that it is all explained in the text, this is not the case. We are told only that the 4-tooth wheel engages the 52-tooth wheel and that the latter is for the Sun's movement.) The output is 28 days for the sidereal month and 364 days for the solar year.

AL-JAZARĪ

AL-JAZARĪ, BADĪ^c AL-ZAMĀN ABŪ'L-^cIZZ ISMĀ^cĪL IBN AL-RAZZĀZ (*fl.* Diyār Bakr, 1206), *machinery, techniques of construction.*

All that we know of al-Jazarī's life is contained in the introduction to his work, *Kitāb fi Ma'rifat al-ḥiyal al-handasiyya* ("Book of Knowledge of Mechanical Devices"). He tells us that at the time of writing his book he was in the service of Nāṣir al-Dīn, the Artuqid king of Diyār Bakr, and that he had spent twenty-five years with the family, having served the father and the brother of Nāṣir al-Dīn. The Artuqids were a Turcoman dynasty who maintained a precarious autonomy during the twelfth century in Mesopotamia. By 1181, however, they had become vassals of Saladin.[1]

The book, which al-Jazarī wrote at the command of Nāṣir al-Dīn, is divided into fifty chapters, grouped into six categories: I, water clocks and candle clocks (ten chapters); II, vessels and figures suitable for drinking sessions (ten chapters); III, pitchers and basins for phlebotomy and ritual washing (ten chapters); IV, fountains that change their shape and machines for the perpetual flute (ten chapters); V, machines for raising water (five chapters); and VI, miscellaneous (five chapters): a large ornamental door cast in brass and copper, a protractor, combination locks, a lock with bolts, and a small water clock.

The book is clearly written in straightforward Arabic; and the text is accompanied by 173 drawings, ranging from rudimentary sketches to full-page paintings. On these drawings the individual parts are in many cases marked with the letters of the Arabic alphabet, to which al-Jazarī refers in his descriptions. The drawings are usually in partial perspective; but despite considerable artistic merit, they seem rather crude to modern eyes. They are, however, effective aids to understanding the text.

It is apparent from statements made by al-Jazarī, in the introduction and elsewhere, that he took pride in belonging to an international fraternity of craftsmen and in continuing the work of his predecessors. Thus he acknowledges that his monumental water clock (chapter 1 of category I) is based upon the clock of pseudo-Archimedes.[2] Al-Jazarī also mentions the Banū Mūsā in connection with their work on fountains and cites the treatise of Apollonius of Byzantium on an automatic musical instrument.[3] Others, about whom we know little or nothing, are also named; yet elsewhere he refers to devices made by earlier workers whom he does not name. Clearly, then, many of al-Jazarī's machines were derived from earlier models, mainly Islamic, some of which were improved versions of devices described by such classical writers as Hero of Alexandria and Philo of Byzantium. Almost certainly there was also transmission from India and the Far East.[4] Al-Jazarī was therefore not primarily an inventor but an engineer who saw his task as the perfecting of earlier work. And, although many of his devices were designed for entertainment rather than for utility, he obviously took his work very seriously.

Of all al-Jazarī's complete machines, perhaps only the double-cylinder slot-rod pump driven by a paddle wheel (chapter 5 of category V) is a significant contribution to the history of machines. Water clocks, candle clocks, and trick vessels, the descriptions of which occupy about three-quarters of the book, have little importance in the subsequent development of mechanical technology. It is the individual components and the constructional techniques, described by al-Jazarī in such scrupulous detail, that are of far greater importance, since centuries later many of them entered the general vocabulary of European engineering. Among the most important of these components and techniques are conical valves, casting of brass and copper in closed mold boxes with greensand, static balancing of large pulley wheels, use of wooden templates, use of a paper model in design, calibration of orifices, lamination of timber to minimize warping, use of true suction pipes rather than drowned suction (in the pump mentioned above), tipping buckets that discharge their contents automatically after a set time, and segmental gears.

The driving mechanism for the monumental water clock (chapter 1 of category I) provides a typical example of al-Jazarī's methods and gives a good idea of the character of the book as a whole. The design is not original—al-Jazarī acknowledges its derivation from the water machinery in the "Ar-

AL-JAZARĪ

chimedes" clock, and a similar construction was used by Muḥammad al-Khurāsānī al-Saʿātī for the large water clock at the Jayrūn Gate in Damascus (described by his son Riḍwān).[5] Al-Jazarī, however, undoubtedly made improvements upon the other designs. For instance, the conical valve, as described by "Archimedes," is a rather crude device; and neither he nor al-Saʿātī made an accurate flow regulator.

The water in reservoir *A* (see Figure 1) sank at a constant rate, and the large float *B* operated most of the time-recording automata by the pull of the string attached to it. The static head was kept constant in the float chamber *D* by means of the conical valve plug on top of float *C*, which entered the valve seat *H* at the end of tap *F*. When water issued from orifice *G*, with tap *F* open, it flowed momentarily from the reservoir into the float chamber, whereupon valve *H* closed momentarily.

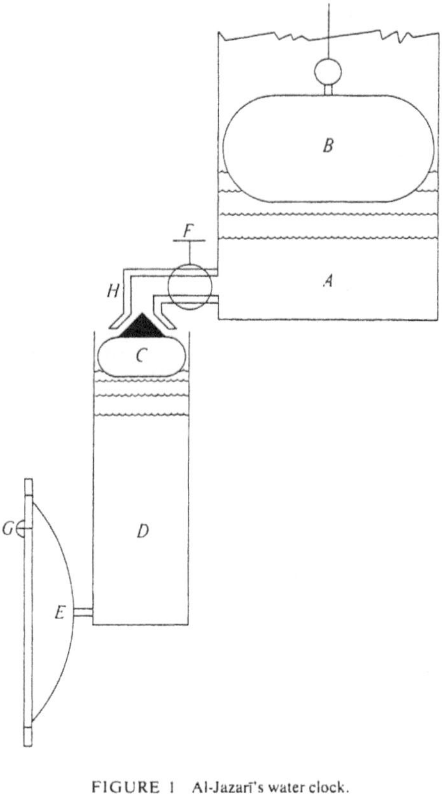

FIGURE 1 Al-Jazarī's water clock.

Thus there was only a very slight fluctuation in the static head in the float chamber. The clock recorded the passage of solar (or temporal) hours: the hours of daylight and the hours of darkness were divided by twelve, giving "hours" of varying length throughout the year, and from daytime to nighttime. The purpose of the flow regulator was to obtain these variations by altering the static head above the orifice. The system of pipes and channels inside the regulator was constructed in such a manner that the disk *E* carrying the orifice could be rotated through 360 degrees.

The reservoir (about 5.5 feet high by 1.0 foot in diameter), the float chamber (about 1.25 feet high by 3.5 inches in diameter), and the two floats were all made from sheet copper, soldered along the seams. Great care was taken to ensure that the vessels were of uniform cross section. The tap, valve seat, and valve plug were made from cast bronze. Plug and seat were ground together with emery on the lathe until the valve was watertight when closed, but the plug was free to slide easily out of the seat.

For the flow regulator al-Jazarī first describes how he tested various earlier designs—for example, an equally divided semicircle and an equally divided full circle[6]—but rejected them when he found that they were inaccurate. His own solution was to calibrate the orifice, a piece of drilled onyx, in the topmost position, for the required rate of flow for the day of the summer solstice. The other positions—zodiacal "houses," five-degree, and single-degree divisions—were determined by trial and error. These were engraved on an annulus surrounding the rotatable disk. A pointer attached to the disk (on a line passing through the orifice) extended over the annulus and was set to the appropriate degree on a given day, and to the diametrically opposed degree on that night. After manufacture, the pieces of equipment were carefully assembled on firm foundations and were brought to the vertical by using a plumb line.

The foregoing exemplifies al-Jazarī's main virtues: careful manufacture and assembly of components, and the ability to devise real improvements on the work of his predecessors. His main faults were a tendency to be inconsistent in his dimensions and some vagueness about the positioning of equipment. Nevertheless, taking drawings and text together, it can be said that he fulfilled his declared intention of describing the devices so that they could be reconstructed by a successor. Indeed, the clock described above was reconstructed in the Science Museum, London, for the 1976 World of

Islam Festival. It works perfectly, exactly in accordance with al-Jazarī's intentions.

Many of the components and techniques used by al-Jazarī and his Islamic congeners reappeared in Europe, apparently as reinventions, centuries later. Casting in closed mold boxes with greensand began in Europe about 1500.[7] The first mention of conical valves is by Leonardo da Vinci; and a float-controlled regulator for steam boilers, similar to the system described above for the water clock, was patented in England in 1784.[8] (It is important to distinguish between the use of automatically operating conical valves for feedback control and the simple lift-and-release "bath plug" type, which occurred as early as Hero of Alexandria[9] and was also used extensively by Islamic craftsmen, including al-Jazarī.)

It is hoped that further research will tell us whether European engineers had access to Islamic ideas; and if so, how they were transmitted. A book of the Banū Mūsā (*ca.* 850) was translated into Latin at the end of the twelfth century by Gerard of Cremona.[10] Versions of the book by al-Jazarī, and works by other Islamic engineers, may therefore also exist. It is possible, however, that by the thirteenth century, when the impetus of translation was slackening in Europe, ideas from Islam and other cultural areas were transmitted by travelers' reports and by personal contacts among craftsmen, rather than by written descriptions.

NOTES

1. See Claude Cahen, "Le Diyār Bakr au temps des premiers Urtukides," in *Journal asiatique*, **227** (1935), 219–276.
2. D. R. Hill, *On the Construction of Water-Clocks* (London, 1976), an annotated translation of Archimedes made from three Arabic MSS. E. Wiedemann and F. Hauser, "Uhr des Archimedes und zwei andere Vorrichtungen," in *Nova acta Academiae Caesarae Leopoldino-Carolinae germanicae naturae curiosorum*, **103**, no. 2 (1918), 164–202. This includes a German trans. made from two of the MSS.
3. F. Hauser, "Über das Kitāb al-Ḥiyal," in *Abhandlungen zur Geschichte der Naturwissenschaften und der Medizin* (Erlangen, 1922), nos. 89–94; E. Wiedemann, *Aufsätze zur arabischen Wissenschaftsgeschichte*, II (Hildesheim, 1970), 50–56, which includes a German trans. of the Apollonius MS.
4. In ch. 4 of category V, al-Jazarī refers to the chain of pots as a "Sindī wheel." Several times he mentions a metal that he calls *isfādruh*. According to al-Dimashqī (quoted by Wiedemann in *Aufsätze*, II, 120) this metal originated in China. At the present state of knowledge, other cultural borrowings cannot be identified with certainty.
5. E. Wiedemann and F. Hauser, "Über die Uhren in Bereich der islamischen Kultur," in *Nova acta Academiae Caesarae Leopoldino-Carolinae* . . . , **100** (1915), 167–272.
6. The pseudo-Archimedes used the semicircular regulator and mentions the full circle. Al-Sa'ātī used the full circle. These designs incorporate two errors: that the variation of daylight throughout the year follows a sine curve, and that the rate of flow is proportional to the static head.
7. Cyril Stanley Smith, "The Early History of Casting, Moulds and the Science of Solidification," in W. W. Mullins and M. C. Shaw, eds., *Metal Transformations* (New York, 1968), 23. Al-Jazarī makes it clear that the technique was well-established in Mesopotamia during his lifetime.
8. F. M. Feldhaus, *Die Technik* (Wiesbaden, 1970), 499; Otto Mayr, "The Origins of Feedback Control," in *Scientific American*, **223** (Oct. 1970), 113.
9. *The Pneumatics of Heron of Alexandria*, facs. of 1851 Woodcroft ed., with intro. by Marie Boas Hall (London, 1971), 37.
10. The MS is Madrid, Biblioteca Nacional, 10010. This is a mathematical treatise, not the machine book mentioned in Note 3. An English version of the latter is being prepared by the present author.

BIBLIOGRAPHY

Al-Jazarī's book has not been published in Arabic, although a number of MSS are extant. There is, however, a full English ed.: Donald R. Hill, *The Book of Knowledge of Ingenious Mechanical Devices* (Dordrecht, Netherlands, 1974). This is a complete trans. of al-Jazarī's work, made mainly from MS Graves 27 at the Bodleian Library, Oxford. All the original illustrations are included as unaltered photographs with the text, and in addition there are thirty-two plates of the miniature paintings from two Istanbul MSS. Notes are provided for each chapter, in many cases with explanatory drawings; and the "General Notes" discuss al-Jazarī's machines, components, and techniques in a historical context. The MSS are described, and their locations and references given, on 3–6 (to which should be added Topkapi Seray, nos. A3350 and H414).

The only earlier work of any significance is a series of seven articles in German periodicals, written by E. Wiedemann and F. Hauser between 1908 and 1921. Partly translations and partly paraphrases, they deal with the complete book and are listed by Hill (see above) in the bibliography and by Coomaraswamy (see below) on 20–21.

Two of the five MSS in Istanbul were dated A.D. 1315 and A.D. 1354. Miniature paintings from both of these are in public and private collections, and some articles have been written about them from an artistic viewpoint. The most important of these are A. K. Coomaraswamy, *The Treatise of al-Jazarī on Automata* (Boston, 1924); and Freer Gallery (Washington, D.C.), folder sheets 30.71 and 32.19.

XVI

Notice of an Important al-Jazarī Manuscript

This notice refers to a manuscript of al-Jazarī's book on machines entitled *The Book of Knowledge of Ingenious Mechanical Devices*, or *A Compendium on the Theory and Practice of the Mechanical Arts.* It had previously been thought that the manuscript of this work dated 715/1315 had been completely dispersed; see, for example, Hill *op. cit.* p. 5, and Ahmad Y. al-Hassan in Vol. 1, No. 1 of this journal. Happily, this assumption proved to be incorrect, since about two thirds of the original manuscript, the property of the Hagop Kevorkian Fund, was included in Sotheby's Spring Islamic Sales on 3rd April 1978 in London. As reported in "The Times" of 4th April 1978, the manuscript was purchased by Messrs. Spink and Son of St. James's, London for a little over £160,000. I would at the outset like to express my sincere gratitude to these two highly respected and responsible companies for the courtesy and co-operation that they have extended to me in the furtherance of my researches. Sotheby's gave me access to the manuscript before it was sold, and sent me a number of colour transparencies of the illustrations. I had fruitful discussions with members of the staff of Spink and Son, who provided me with a complete black-and-white photocopy of the manuscript. I am also very grateful to both companies for having given me permission to publish this paper, and to include in it, as I saw fit, any of the material that they so generously provided.

Some time ago a number of the illustrations were removed from the manuscript and found their way into various public and private collections. In my book I published twenty-one of these illustrations, these being all that I was able to trace. There are, however, 66½ illustrations missing from the manuscript sold by Sotheby's, so clearly a number remain undiscovered. Most of the lacunae in the manuscript can be accounted for by the removal of these illustrations, since either the accompanying demifolios of text or the surrounding text were removed with the illustrations.

The manuscript is written on thick polished paper, 314 mm. by 219 mm. to the page, in very fine *naskhī* script with 21 lines to the page. The colophon on page 207a (Persian pagination – see below) gives the name of the scribe as Farkh ibn ᶜAbd al-Laṭīf and the date of his copy as the end of Ramaḍān

715, or December 1315. This is therefore the third oldest known copy of al-Jazarī's work, being pre-dated by Topkapi Ahmet III 3472 and Topkapi H. 414. The list of Mss given by Hassan (*op. cit.* 60-62) brings up-to-date the the list in Hill 3-6. Of these fourteen Mss I have now examined the originals of eight, namely: Bodleian Library, Oxford, Graves 27 and Fraser 186; University of Leiden Or. 117 and Or. 656; Bibliothèque Nationale, Paris, Fonds Arabe 2477 and 5101, Suppl. Pers. 1145 and 1145a. I have also seen pages from Hagia Sophia 3606. I have photocopies of Topkapi Ahmet III 3472 and Chester Beatty Library, Dublin, No. 4187. I have yet to examine, therefore, Topkapi Hazine H 414 dated 672/1274, Topkapi Ahmet III 3350 dated 863/1459, and Topkapi Ahmet III 3461, date unknown. Hassan assesses the first of these three as a very good copy, the second as inferior with regard to the illustrations, and the third as uneven, with some sections poorly written and illustrated but most of it of good quality. Leaving aside the question of completeness, it seems that the 715/1315 manuscript can be ranked among the best. The calligraphy is excellent, the illustrations are very fine, and the text, though not without errors, is free from major blemishes.

The manuscript has Persian pagination, added at some time after the copy was made. Unfortunately, this pagination is unreliable: some pages are unnumbered, some are out of order, and in certain cases the lengths of the lacunae indicated by the gaps in the numeration do not match the lengths of the missing text. The expedient was therefore adopted of numbering the pages (i. e. demifolios), starting with the cover, and then sorting the pages so numbered into order. This produced the following sequence: 1 to 10, 69, 70, 11 to 36, 59, 60, 37 to 68, 71 to 95, 98, 96, 97, 99 to 113, 130, 131, 114 to 129, 132 to 189, 265 to 268, 190 to 264, 269 to 296.

In the following analysis this new pagination is used, and the illustrations are given the Figure Nos. from Hill. In the right-hand column the abbreviation 'S' is used for Section. Al-Jazarī provided one main illustration for each chapter and numbered them from 1 through to 50; these are marked with an asterisk.

Category	Chapter	Pages in new numbering	Original contents	Omissions: Illustrations	Omissions: Text
Cover		1	–	–	–
Intro-duction		2 to 4	Same	–	–
I	1	4 to 10, 69, 70, 11 to 30	10 Sections, Figs. 1 to 33	4*, 6, 7, 8, 10, 11, 13, 14, 15, 16, 18, 19, 20, 22 to 25, 28 to 33	End of S.1, centre of S.2, start and end of S.3, first few words of S.4, most of S.6 - only the start remains, all S.7, start and end of S.8, start and end of S.9
	2	30 to 36	5 Sections Figs. 34 to 40	34*, 39, 40	Most of S.1 - only the start remains, first few words of S.2, last few lines of S.3, all S.4, most of S.5
	3	59, 60 37 to 45	6 Sections, Figs. 41 to 47	41*, 42	All S.1, all S.2
	4	45 to 68, 71 to 76	15 Sections Figs. 48 to 59	48*, 51, 52, 59	End of S.1, last few lines of S.5, all S.6, start of S.7, end of S.12, start of S.13, end of S.15
	5	77 to 82	3 Sections Figs. 60 to 62	60*, 62,	Start of S.1, middle of S.3
	6	82 to 95 (86 is a duplicate of 85)	6 Sections Figs. 63 to 70	63*, 66, 67	Most of S.1 - only start remains, most of S.3 - only start remains, and some lines at the end, all S.4
	7	98, 96, 97, 99 to 103	3 Sections Figs. 71 to 74	74*	Last part of S.3
	8	104 to 107	3 Sections Fig. 75	75*	Start of S.1, most of S.3 - only the start remains

Category	Chapter	Pages in new numbering	Original contents	Omissions	
				Illustrations	Text
I	9	108 to 110	2 Section Fig. 76*	None	All S.1 except last two lines
	10	110 to 113, 130, 131	2 Sections Figs. 77, 78*	None	End of S.2
II	1	114 to 117	2 Sections Figs. 79, 80	80*	All S.1 except last three lines
	2	117, 118	1 Section Fig. 81	81*	Middle of Section
	3	119 to 129, 132 to 139	5 Sections Figs.82 to 86	82*	End of S.1
	4	139 to 145	3 Sections Fig. 87	87*	None
	5	146 to 155	3 Sections Figs. 88 to 93	88*	All S.1, start of S.2
	6	156 to 159	2 Sections Figs. 94 to 97	94*, 97	Start and end of S.2
	7	160 to 162	3 Sections Fig. 98	98*	All S.1, centre of S.3
	8	162, 163	2 Sections Fig. 99	99*	End of S.2
	9	164 to 167	2 Sections Fig. 100	100*	Title, end of S.2
	10	168 to 170	2 Sections Figs. 101, 102	102*	Start of S.1, middle of S.2

Category	Chapter	Pages in new numbering	Original contents	Omissions	
				Illustrations	Text
III	1	170 to 175	2 Sections Figs. 103, 104*	None (Chapter	None is complete)
	2	175 to 183	2 Sections Figs. 105 to 108	None (Chapter	None is complete)
	3	183 to 188	2 Sections Figs. 109 to 111	111*	Middle of S.2
	4	188, 189, 265, 266	1 Section Fig. 112*	None (Chapter	None is complete)
	5	266, 267	2 Sections Fig. 113	113*	Most of the chapter
	6	267, 268, 190 to 193	2 Sections Figs. 114, 115	114, 115*	Last few lines of S.2
	7	194 to 196	2 Sections Fig. 116	116*	Start of S.1, middle of S.2
	8	196 to 201	2 Sections Fig. 117	117*	End of S.2
	9	202 to 207	4 Sections Fig. 118	118*	Start of S.1, end of S.4
	10	208 to 211	2 Sections Fig. 119	119*	Start of S.1, end of S.2

Category	Chapter	Pages in new numbering	Original contents	Omissions	
				Illustrations	Text
IV	Introduction	212	Text only	None	All except last two lines
	1 to 10	212 to 245	10 Chapters Figs. 120 to 133, Pl. 32	This Category is complete, with all the illustrations, except for the following part of the text: end of chapter 7, start of chapter 8	
V	1	245 to 248	1 Section Fig. 134	134*	None
	2	248 to 250	1 Section Fig. 135	135*	None
	3	250, 251	2 Sections Fig. 136	136*	End of S.1, start of S.2
	4	251 to 253	1 Section Fig. 137	137*	None
	5	253 to 260	3 Sections Figs. 138 to 141*	None	None
				(This chapter is complete)	
VI	1	261 to 264, 269 to 273	3 Sections Figs. 142 to 148	Half of 142*	Introduction, start of S.1 (section 2 is complete but is misnumbered as section 1)
	2	273 to 277	3 Sections Figs. 149 to 152	None	None
				(This chapter is complete)	
	3	277 to 286	2 Sections Figs. 153 to 166	None	None
				(This chapter is complete)	
	4	287 to 290	2 Sections Figs. 167 to 172	172*	End of S.2
	5	291 to 293	1 Section Fig. 173	None	Last paragraph

The colophon is on page 293, after the end of VI. 5. On page 294 the letters of the 'secret' alphabet are given with their equivalents from the normal Arabic alphabet; the few lines of explanatory text are in a different handwriting from the rest of the manuscript - it is a badly written *naskhī*. There are three illustrations on page 295. The largest of these is in the centre of the page and is the main illustration of II.1, i.e. Fig. 80. It is clearly by a different draughtsman than the one who drew the illustrations for the main part of the manuscript; it is badly drawn and would be almost useless for explanatory purposes. Lower down, to the right and left of the page are two small, crude sketches of devices similar to those described by the Banū Mūsā, indeed the one on the right can be identified as their Model 79. Above either sketch there are passages describing these two devices (there is no text relating to the al-Jazarī device). These descriptions are not in the words of the Banū Mūsā. The handwriting is different again, a somewhat better *naskhī* from that on page 294, but not as good as as that of the main manuscript. At the bottom of the page there is the start of a passage in Fārsī, which continues on page 296. This passage, in yet another *naskhī* hand, is part of a treatise on weighing; at the bottom of the page there is a drawing of a balance with five pans. The contents of page 294 appear in most of the al-Jazarī manuscripts, and this page was probably added to supply an obvious omission. Pages 295 and 296, however, are quite extraneous.

Summarising from the foregoing tabulation, the only absolutely complete chapters are: III,1. - pitcher for dispensing hot and cold water; III,2. - pitcher for dispensing water; III,4. - peacock for dispensing water; IV, 1 to 6. - fountains; IV, 9 and 10. - musical automata; V,1. - pump; VI,2. - protractor; VI,3. - combination locks. All the illustrations are included in these fourteen chapters with, of course, all the main illustrations. There are also another five main illustrations which appear in incomplete chapters, namely: I,9., Fig. 76 - monkey candle-clock; I, 10., Fig. 78 - candle-clock of the doors; IV, 7 and 8., Figs. 130 and 131 - musical automata; VI,5. Fig.173 - water-clock of the sailor. Half of the illustration of the door, VI, 1., Fig. 141, remains. This fine miniature was originally on one full folio, and it seems likely that the other demifolio is lost for good. Fourteen of the dispersed main illustrations were published as plates in Hill, *op. cit.*, so we now have a record in two documents of 33½ of the original 50 main illustrations.

There are three chapters on water-raising machines, namely V, 1, 2 and 4, for which the text is complete although there are no illustrations.

In Hill, p. 238, Plate XXXII shows a device for which there was no description in the manuscripts that were available to me when I made the translation. There is, however, an addendum, to the text in the manuscript under review and in Topkapi 3472; in both cases this addendum occurs at the end of Chapter 10 of Category IV. The first part describes a woodwind instrument having 'fingers' that are raised and lowered in succession over its holes by cams fixed to the axle of a water-wheel. There is no illustration for this instrument. The second part of the addendum describes the device shown in the illustration mentioned above. This is a rocker-arm consisting of two flume-beam swapes joined at right angles, and oscillating about an axle located at the joint. The oscillation is produced by the 'bleeding' of water through a narrow pipe from the main supply channel into whichever of the flumes is temporarily horizontal; when the scoop at the end of the flume fills, this side tilts and the other flume comes to the horizontal and its scoop begins to fill, and so on. Al-Jazarī says that this device was not only used for fountains and musical automata, but that it was incorporated in many different machines.

References

1. A. K. Coomaraswamy, *The Treatise of al-Jazarī on Automata*, Museum of Fine Arts, Boston, 1924.

2. A. Y., Hassan, *A Compendium of the Theory and Practice of the Mechanical Arts.* Arabic text of al-Jazarī's work, collated from three of the best manuscripts. (Aleppo, Institute for the History of Arabic Science, 1978).

3. *Idem*, 'The Arabic Text of al-Jazarī's "A Compendium of the Theory and Practice of the Mechanical Arts"'. *Journal for the History of Arabic Science*, Aleppo, Vol. 1, No. 1, 1977. 47-64 in English, 129-165 in Arabic.

4. Donald R. Hill, *The Book of Knowledge of Ingenious Mechanical Devices.* An annotated translation of al-Jazarī's work. (Dordrecht, Reidel, 1974).

5. Sotheby, *Spring Islamic Sales*; catalogue for sale on 3rd April, 1978, 122-127.

6. E. Wiedemann, and F. Hauser, 'Über die Uhren in Bereich der Islamischen Kultur', *Nova Acta Abh. der Kaiserl. Leop. Deutschen Akademie der Naturforscher* 100, Halle 1915, 1-168.

Illustrations

(The numbers in parentheses give the Category, Chapter, and Figure Nos. from Hill)

Fig. 1.

(IV. 1, Main illustration, Fig. 121)

A fountain made to alternate by bleeding water from the main supply pipe, which is free to oscillate about an axle, into one or other of the tipping-buckets. As shown here, the discharge is into the right-hand side. When the tipping-bucket fills, it tilts and a vertical rod soldered to its rear face pushes the supply pipe, causing it to tip towards the left . The vertical rods are not shown in this illustration but appear in the parallel illustration in Topkapi 3472.

The fountainhead emits a straight jet when supplied from the left-hand tank, and several curved jets when supplied from the right.

Fig. 2.
(IV.2, Main illustration, Fig. 122)

This is a doubled version of IV.1. One fountain-head emits a single jet when the other is discharging several. When they change over, the situation is reversed. Again, the rods attached to the tipping-buckets are omitted.

Fig. 3.
(Addendum to IV.4, Fig. 125)

A fountainhead having three possible shapes. It is supplied from two tanks, with a switching system similar to that described for Fig.4 below. Three concentric pipes enter the fountainhead, the two inner ones from one tank, the outer one from the other. Water from the innermost pipe emerges as a straight jet; it emerges from the second pipe as a 'tent' - the water impinges on a convex plate and descends as a curved sheet. After change-over it emerges from the outermost pipe as a number of curved jets.

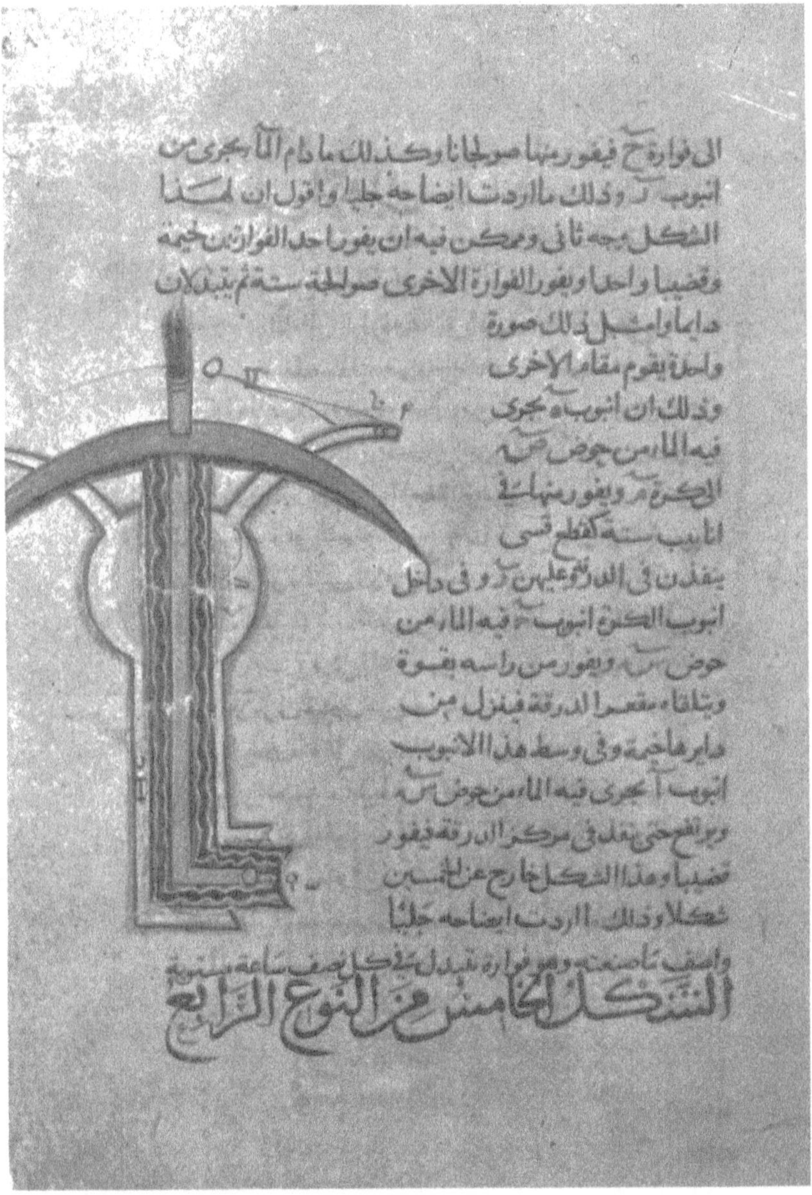
الى فوارة ح فيفور منها صولجانا وكذلك ما دام الما يجرى من
انبوب د وذلك ما اردت ايضاحه جليا واقول ان لهذا
الشكل وجه ثانى ويمكن فيه ان يفور احد الفوارتين خيمة
وقضيبا واحدا ويفور الفوارة الاخرى صولجانة ستة ثم يتبدلان
دايما وامثل ذلك صورة
واحدة يقوم مقام الاخرى
وذلك ان انبوب ه يجرى
فيه الما من حوض ص ه
الى الكرة م ويفور منها في
انابيب ستة كقطع قسى
ينفذن في الدرقة عليهن ز و في داخل
انبوب الكرة انبوب ح فيه الما من
حوض س ه ويفور من راسه بقوة
ويلقاه مقعر الدرقة فينزل من
دايرها خيمة وفي وسط هذا الانبوب
انبوب ا يجرى فيه الما من حوض س ه
ويرتفع حتى ينفذ في مركز الدرقة فيفور
قضيبا وهذا الشكل خارج عن الخمسين
شكلا وذلك ما اردت ايضاحه جليا
واصف ما صنعته وهو فوارة تتبدل في كل نصف ساعة ستية

الشكل الخامس من النوع الرابع

Fig. 4.

(IV. 10, Main illustration, Fig. 133).

Musical automaton. The whistles are at either side, at the top; air-pipes lead into them from the two tanks. The water runs into one side from the balanced supply pipe, causing the whistle to sound, until the float in the chamber at the top of the tank rises. The vertical rod on top of the float causes the supply pipe to tilt and discharge into the other tank. A siphon evacuates the first tank. The cycle repeats itself as long as water flows into the system.

Figs. 5, 6, and 7
(V.5, Figs. 139, 140, 141)

These illustrations are of the second of the alternative designs of the well-known slot-rod pump.

Fig. 5.

The two meshing cogwheels, the upper one driven from a paddle-wheel, the lower one having the slot-rod on its face. The orientation of this sketch is the same as in Topkapi 3472, whereas in Oxford Graves 27 it is turned through 90 degrees. The orientation shown here is almost certainly correct. The cogwheels are shown disengaged, otherwise the illustration is good.

١٧٨

من المحور في ارض الصندوق يدور على سكرجة وتحت القرص
حلقة يدور فيها المحور وعلى دائرة القرص دندانات بارزات
عن الصندوق وعلى القرص يب داخل الصندوق آ وعلى
الدندانات وهي خارجة من جانب الصندوق س وعلى وجه
القرص وتد منتصب عند محيطه ثم نتخذ سهماً واحداً في احد طرفيه
ثقب فيه مسامرات عند زاوية ح من الصندوق والطرف الاخر
مخروق طولا خرقا طوله
قطر دائرة يوترها راس
وتد القرص وهو يدخل في
الخرق ليصير الوتد في
غاية بعد عن زاوية ح
من الصندوق وطرف
الخرق وعلى وسط الخرق
نهاية ل وعلى وسط
الخرق يشابه طاق سهم
ت لا يميل الى جهة ح
ولا الى جهة س وهو
بالحقيقة متوسط بينهما

ومتى دار قرص آ من جهة ح الى جهة س ربع دورة فان وتد
القرص الى ناحية س ويميل معه سهم ت وهو غاية ميله هناك
وبحركة قرص آ دائماً حتى يدور ربع دورة و يصير الوتد الى جهة

Fig. 6.

A drawing of one of the two cylinders — the ring at the end of the connecting rod is attached to the side of the slot-rod.

XVI

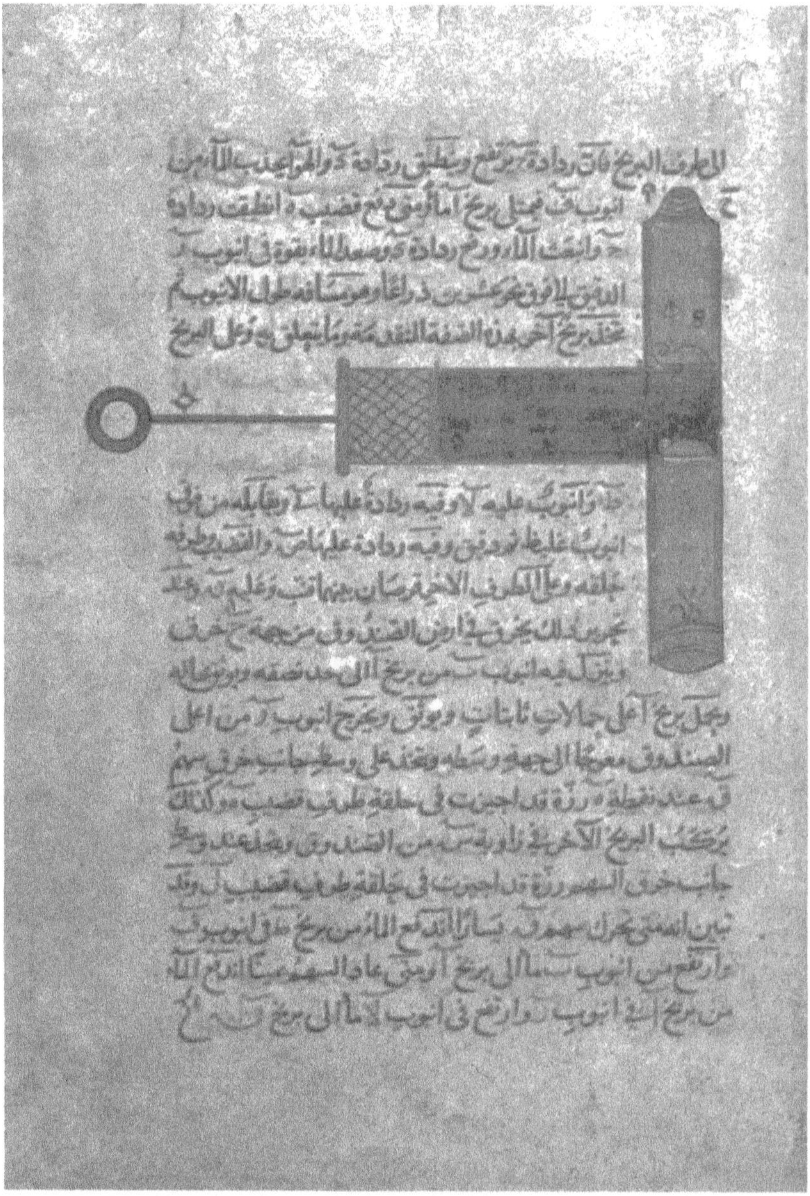

Fig. 7.

The main illustration of the complete machine; it is probable that this is a plan view, rather than an elevation (see Hassan op. cit. 56-59).

A Treatise on Machines by Ibn Mu^cādh Abū ^cAbd Allāh al-Jayyānī

The work that concerns us here is part of a manuscript preserved in the Biblioteca Medicea Laurenziana in Florence[1]. Numbered Or. 152 (formerly 282), it is catalogued as *anonymi tractatus de mechanicis*. The treatise is entitled *kitāb al-asrār fī natā'ij al-afkār*, 'The book of secrets about the results of thoughts', and it occurs among a number of mathematical treatises attributed to Abū^cAbd Allāh known as Ibn Mu^cādh[2]. The *nisba* al-Jayyānī refers to Jaén, the capital of the Andalusian province of the same name. Little is known of his life. He was born in Cordoba in 379/989-90 and died later than 1 July 1079 A.D., since he wrote a treatise on the solar eclipse which occurred in Jaén on that date. He was primarily an astronomer and a mathematician. Two of his astronomical works, not known to be extant in Arabic, were translated into Latin by Gerard of Cremona. In the *Libros del saber* (II 59, 309) he is quoted as considering the twelve astrological houses to be of equal length. Several of his mathematical works are extant in Arabic, and he was regarded as an eminent and progressive mathematician by Ibn Rushd (Averroes). He was a fervent admirer of Euclid, and was the most successful of those scholars who attempted to reconcile Euclid's doctrine of proportions with Arabic concepts and methods.

The attribution of the treatise on machines to Ibn Mu^cādh cannot be fully verified but in the absence of evidence to the contrary he may be considered as its author. It might be objected that the treatise differs essentially from the mathematical and astronomical works which form the bulk of Ibn Mu^cādh's writings, although there is no reason why a mathematician should not turn his attention to mechanics. It is perhaps more to the point, however, that the writer was obviously a scientist and not an engineer. It is quite evident that he is describing machines that he had seen by a method that he understood well, i.e. that of geometry, and that he had virtually no knowledge of, or interest in, the messy business of contruction. This point is

1. I wish to thank the Director of the Biblioteca Laurenziana for kindly providing me with the necessary photographic material, and for giving permission for the use of this material in the preparation of this paper.

2. David A. King, "Medieval Mechanical Devices", in *History of Science* 13:22 (1975) 284-289. For additional biographical material on Ibn Mu^cādh, see the article "al-Jayyānī" in the *Dictionary of Scientific Biography*, Vol. VII (New York, 1973), pp. 82-83.

discussed more fully below. Also, the length of daylight is given in the treatise as fifteen hours, for the day of the summer solstice. This corresponds to the latitude of 40 degrees, whereas the latitude of Cordoba is 37 degrees 53 minutes. If, however, the apparent times of sunset and sunrise, due to refraction, were used then the time would be very close to fifteen hours. It is almost certain, therefore, that the treatise was written in Cordoba by someone well versed in mathematics and astronomy, but with little engineering knowledge. This, together with the location of the treatise among other works attributed to Ibn Mu^{c}ādh, puts the burden of the argument upon those who would ascribe its authorship to another hand. For the remainder of this paper it is assumed that the author was indeed Ibn Mu^{c}ādh.

My purpose here is to draw the attention of scholars to the existence of this primary source for the study of medieval Arabic technology, to list briefly its contents, and to emphasise some of its most significant contents. Because of the poor state of the manuscript a long and careful study will be required before a more thorough assessment of the work's unquestionable value can be made. It is important however, that this preliminary notice be written, not least because some of the concepts embodied in the treatise have been thought not to have occurred in Arabic writings, and their introduction has been assigned to other cultural areas.

The manuscript is dated 644/1266, and although the text is probably a faithful copy of the original, the illustrations are almost certainly inferior to those that were drawn by, or under the guidance of, Ibn Mu^{c}ādh. It is particularly unfortunate that all representations of human and animal automata have been omitted (see below), an indication that the austere influence of the Almohades *(al-muwaḥḥidūn)* was still powerful in western Islam in the middle of the 7th/13th century. One or two of the leaves in the first part of the manuscript seem to be out of order, but it has not yet been possible to re-arrange these with confidence. This minor fault in the sequence is probably of recent occurrence but there is also some slight disorder from an earlier time. On folio 5v, for instance, there is a note in Arabic by the side of the illustration in different handwriting from that of the main text, saying that the text and illustration refer to Model 2, whereas they are paginated under Model 3. This is indeed the case.

The treatise, on folios 1v-46r, consists of 31 Models of which Nos. 1-5 are essentially very large toys, similar to clocks in that automata are caused to move at intervals, but without precise timing. Nos. 6-20 and 27-30 are water-clocks, all of which record the passage of the temporal hours by the movements of automata. Nos. 21-24 are war machines in the form of towers which can be raised and lowered by a scissors action like that of 'lazy tongs'. Nos. 25 and 26 are machines for raising water from wells, and No. 31 is a universal sundial. In this paper discussion will be confined to the first five machines and the water-clocks, pending a more thorough examination of the entire treatise.

Every page is partly obliterated, either being cut or affected by dampness. On average about 40% of each page is affected: the verso pages are worse than the recto, since usually about half the page to the right of the main diagonal is missing; the recto pages have been affected by damp along their right-hand margins, to the extent that about 30% of the text is missing or illegible. This means that no page can be read in its entirety. By taking a selection of passages from different Models, all describing a similar mechanism or technique it is sometimes possible to make a plausible reconstruction of what is being described, particularly when the concept is known from the writings of other Arabic authors. A careful and protracted study on these lines may one day permit us to make a fair assessment of the treatise, but failing the discovery of a better manuscript, a full explanation of each of the Models may never be possible. In the first place, some passages describing similar mechanisms are partly obliterated wherever they occur. Thus none of the descriptions of the very significant gear-trains in Models 2-5 is complete, and some vital concepts embodied in this gearing must therefore be conjectured. Secondly, some systems are described fully only once, and subsequent brief mentions of the same system refer back to the key passage. A notable example of this is the basic water machinery of the clocks, which is described in detail only in a partly obliterated section of Model 8. Thirdly, the descriptions are far too short in view of the size and complexity of the machines. For instance, the length of the descriptions for each of Ibn Muᶜādh's clock is only a small fraction of the lengths of the descriptions in Archimedes[3], Riḍwān,[4] and al-Jazarī,[5] for clocks of equivalent complexity. Finally, the illustrations are badly drawn and quite inadequate, even when they are intact. Identifying letters are often omitted, the drawing of the transmission systems is haphazard and almost impossible to unravel, and no detailed drawings are provided. The omission of all drawings of human and animal figures adds a further element of uncertainty. All these defects, except the omission of detailed drawings, are probably the fault of the copyist rather than of the author. We know, from the extant manuscripts of the works of the Banū Mūsā[6] and al-Jazarī, for example, that

3. Donald R. Hill, *On the Construction of Water-clocks* (Turner and Devereux, London, 1976).

4. Eilhard Wiedemann and Fritz Hauser, "Über die Uhren im Bereich der islamischen Kultur", *Nova Acta Abh. der Kaiserl. Leop. Carol Deutschen Akademie der Naturforscher* 100 (Halle 1915) 176-266.

5. Donald R. Hill, *The Book of Knowledge of Ingenious Mechanical Devices* (Reidel, Dordrecht 1974) (All the references to al-Jazarī in this paper are taken from this book).

6. Fritz Hauser, "Über das Kitāb al-Ḥiyal - das Werk über die sinnreichen Anordnungen der Banū Mūsā", *Abh. zur Gesch. der Naturwissenschaften und der Medizin*, Erlangen, 1922.

Wiedemann and Hauser, "Über Trinkgefässe und Tafelaufsätze nach al-Gazarī und den Banū Mūsā", *Der Islam* 8 (1918), 268-291.

(An annotated English translation of the Banū Mūsā's *Kitāb al-Ḥiyal* has been prepared by the present writer and will be published by Reidel of Dordrecht).

the quality of the illustrations can vary enormously among the various manuscripts, to the extent that one copy may contain all the faults listed above, and another none.

The manuscript is written in a clear *maghribī* hand and is easily legible; consonantal points are usually provided. The grammar is accurate. The vocabulary presents little difficulty, although some of the expressions used are different from those used for the same objects in the works of other Arabic writers and translators, but the differences are relatively few when compared with the parallels. In particular, nearly all the words used for the mechanical parts are the same as those occurring in the works of the other writers.

Ibn Muᶜādh's method of presentation demonstrates clearly his inclination towards geometry rather than engineering. First, the operation of the device is described in three or four lines: typically, a man holding an astrolabe turns his head towards a slave-girl, whereupon she ejects a pellet from her mouth (a less elegant variation of the birds used by 'Archimedes' and al-Jazarī). Then the Model is described in geometrical terms. Usually the main structure is a box *(ṣandūq)*, rectangular in cross-section, with the shorter dimension two thirds of the longer. Lines are drawn on the illustration, these lines being identified by letters, many of which, as mentioned above, are omitted from the illustrations. Nominally, the drawings are elevations, but parts are shown from any view which suits the draughtsman's purpose. Thus all wheels are drawn as full circles, including cogwheels meshing at right angles. When these lines have been drawn the axles, wheels, and other components are placed on the drawings with reference to the lines, somewhat in the manner of a mapmaker plotting features on a basic grid. Throughout the treatise, details of construction are almost totally lacking: all we are given are phrases such as 'we place wheel (kl) on axle (ty)' or 'the ends of the axle are on points (ew)', and so on. Occasionally materials such as wood and copper are mentioned, but with no instructions about how they are to be cut, formed and jointed. On the other hand, dimensions and weights are often supplied, either directly or in relation to other components. Care is taken, for example, to specify the relative sizes of meshing cogwheels and the number of teeth on each of them.

The following sequence for the descriptions of the various parts of the machines is usual: the box-like structure with its lines is described first, then the placing of the main wheels and axles, followed by the means of imparting motion to them, and finally the description of the systems of ropes, pulleys and cams that activate the automata. Unfortunately, Ibn Muᶜādh does not follow the example of the Banū Mūsā and al-Jazarī who, having given a description of the construction of a machine, conclude with a clear summary of how it works and how the various mechanisms interact with one another. This lack of a summary is particularly frustrating, because such a summary would have helped to make up for the obliterated passages.

The important ideas incorporated in the prime movers can be identified with some confidence, but it has proved impossible so far to discover precisely how the motion was transferred to the automata. We can be fairly sure, however, that these mechanisms were basically similar to those described by Riḍwān and al-Jazarī. Boards were mounted in the box to carry the pulleys and other mechanisms for transferring power. On the axle of the last wheel in the driven chain there was often a cross *(ṣalīb)* whose purpose is not yet clear to me. The automata themselves were also similar to those that are found in other Arabic works, e.g. doors that open each hour to reveal standing figures, figures that move their heads and limbs, etc. A novel feature in this treatise, however, is Ibn Muᶜādh's use of mercury in balances (e.g. Model 1,2 r). He says, apparently assuming that his readers were familiar with such devices, that the tube *(qaṣaba)* of the balance has mercury *(zībaq)* inside it. Perhaps its purpose was similar to that of the lead ball placed by al-Jazarī in the tube of a balance-arm. (e.g. Category IV, Ch. 5). When the balance tilts, the ball or the mercury runs to the lower end and holds the balance in that position until it is made to tilt in the opposite direction. This use of mercury suggests a link between this treatise and later Christian works on similar subjects, since the *Libros del saber* contain a clock operated by mercury[7].

Ibn Muᶜādh's machines are notable for their sheer size and power. This feature, which is brought out in the ensuing discussion, distinguishes the work sharply from that of the Banū Mūsā, with its emphasis upon delicate mechanisms and controls, and from those machines of al-Jazarī which embody similar concepts. In Ibn Muᶜādh's work there are no conical valves, delay systems, feedback controls, or use of small variations in atmospheric pressure — all the ideas, in fact, that have until now been regarded as typical of Arabic mechanical technology. The element of intermittent operation is not of course absent from the present treatise – indeed it is of the utmost significance – but it is achieved by different means. Delicacy is replaced by ruggedness: we have ropes instead of strings; large wheels up to 3 spans in diameter (about 72cm/28 inches); spanning weights up to 1½ *raṭls* (say at least 3 kg); with other weights and dimensions in proportion. Gearing is important, and in addition to the special gears discussed below we find all the usual types: parallel meshing, meshing at right angles, worm-and-pinion. For parallel meshing the lantern pinion *(dawlāb dhū ᶜawāriḍ)* is used for one of the two wheels, as it was by al-Jazarī. On page 28 we have the important statement that the teeth of cogwheels were cut to the shape of equilateral triangles, proving that they were true teeth and not wooden pegs. As usual, the word *dandānja*, of Persian origin, is used for a tooth.

We now come to the kernel of this paper, namely the prime movers of the automata machines and the water-clocks. Of the first five models, No. 1

7. See Wiedemann and Hauser, "Über die Uhren", 18-20.

is different from the others and I have so far not been able to establish how it works. The remaining four, from page 3v to 11v, are all similar and contain features of the first importance for the history of mechanical technology.

Figure 1 is the illustration from Model 5, which is the most complex of these machines. The others contain some, but not all, of the mechanisms shown in Figure 1. (It has, however, been necessary, because of obliterations, to study all the four Models in order to deduce their essential features). The purpose of Model 5 was to cause a set of doors, set in a row, to open at successive intervals, revealing jackwork figures. These doors were on one side of the usual boxlike structure that contained the working parts. The prime mover was a waterwheel, mounted in a stream outside the box – in Figure 1 this is represented by the circle with a double perimeter to the left of the illustration. This is called a *falak* pl. *aflāk*, a term which requires some explanation. *Falak* usually means the celestial sphere and is used by al-Jazarī, for instance, for the large disc carrying representations of the sun, moon, and zodiacal signs which rotates at constant speed at the top of his monumental water-clock (Category 1, Ch. 1). Ibn Muᶜādh does not use the word in this sense only, and it is important to emphasise that most of his *aflāk* are not for display, and are not analogous to the geared planetaria displayed on the outside of de Dondi's clock.[8] The *aflāk* in this treatise are anything that simulates celestial motion by mechanical means, namely the prime movers and their associated gears and wheels. A *falak* may be a gear-wheel, a water-wheel, or even a float! (See below). The word is often qualified, e.g. 'a *falak* like a wheel *(ᶜajala)*' and in the present case it is like a noria: "we make a *falak* on the pattern of the noria *(nāᶜūra)* and if the flow of water is scanty it is like the noria (naᶜūra) which turns the mills" (5v) In the text it is.

فلنعمل فلك شبيه النعورة (وهو فلك زح) فان كان الماء قليلا كان على شبيه الناعورة التي تدير الارحى

(Note that the spelling of noria varies; this happens with other words — e.g. *dawlab* and *dawlāb* are both used for a wheel). This remark seems to indicate that Ibn Muᶜādh was recommending the use of an overshot wheel if the flow was small, since this type of wheel is more efficient than the undershot type in such conditions. This implies that the engineers of the 5th/11th century were aware of the necessity for varying the design of the water-wheel to suit the flow conditions. Of greater significance, however, is the use of a full-size water-wheel to power a large machine. This brings us directly to the monumental water-clock of Su Sung, who flourished in China some decades later than the time of Ibn Muᶜādh. Now, admittedly, Su Sung's water-wheel was furnished with an escapement and it provided the power for a clock, but it is nevertheless evident that Ibn Muᶜādh's devices are basically similar to Su Sung's

8. Silvio A. Bedini and Francis R. Maddison, "Mechanical Universe, the astrarium of Giovanni de Dondi", *Transactions of the American Philosophical Society*, N.S., lvi, 5 (1966).

clock. We must therefore now qualify Needham's statement that there is no evidence for any Arabic influence on Chinese developments, and that the Arabic material indicates the passage westwards of certain Chinese elements.[9] As far as water-wheels and their use are concerned, it would be premature to draw any conclusions at all about transmissions between cultural areas. The question of the diffusion of water-wheels, horizontal and vertical, from Roman times onwards, is still hedged with confusion, and there is no space here to enter into a discussion of this topic. It is worthwhile, however, to take note of three points :

1. The use of both types of water-wheel was widespread in Islam from the 3rd/9th century onwards.[10]
2. The machines described by Ibn Muᶜādh show that water-wheels were used by the Arabs for other purposes besides the driving of corn-mills.
3. Al-Jazarī, writing in 602/1206 mentions an old machine which he inspected, in which a musical automaton was powered by a vertical water-wheel. In his comments on this machine he clearly implies that he knew how to control the speed of such a wheel by means of an escapement. (p.170)

On present evidence, therefore, it seems just as likely that the idea of using an escapement-controlled water-wheel to drive a clock moved from Islam to China as that the transmission was in the reverse direction. In any case, the idea was never developed in Europe, where the importance of water-wheels lies in their application to milling and industrial uses. The spread of the water-wheel and its application has yet to be intensively studied with reference to all cultural areas. At present it seems a reasonable hypothesis that the spread of both types of wheel was continuous from Roman times to the Middle Ages in Europe, Western Asia and North Africa, and that the uses to which the wheels were put varied according to social and economic needs. There is also reason to suppose that the selection of the type of wheel to be used was not haphazard, but rather that the choice was consciously made to suit the local hydraulic conditions.

Returning to Figure 1, the noria was mounted on an axle that passed into the box and rested in bearings fixed to its walls. The main central gear-wheel was on this axle. This wheel had 64 teeth on half its perimeter, and meshed with the two outer cogwheels, each of which had 32 teeth around its complete perimeter, and had a diameter equal to one quarter of the diameter of the large wheel. Each of the smaller wheels therefore made two rotations for one rotation of the large wheel. The description of the wheels inside the main wheel is badly defaced. The description of the main wheel for Model 4, however, which is very similar to Model 5 except that there is no central cog-

9. Joseph Needham, *Science and Civilisation in China*, Vol.4, Pt. 2 (Cambridge University Press, 1965) p. 536.

10. Norman Smith, *Man and Water* (Peter Davies, London, 1976), p. 142.

wheel, is almost intact. The setting of three sets of teeth is described (page 7): one set is on the outer perimeter, and two are on the inside, 'facing the axle'. Without any question, therefore, these machines contained segmental gears. It is not quite so certain that they contained epicyclic gearing, but taking the illustrations and the surviving parts of the text together there seems little room for doubt. Surely, no-one interested in the history of machines and clockwork can examine Figure 1 without a sense of excitement, since it shows a system of gears for transmitting torque that is much more complex than any other power-driven gears known to have existed so early. Its most obvious ancestor is the Antikythera geared calendar[11], but this was a delicate manually operated device, not a water-powered machine in which the main cogwheel was three spans in diameter.

The clocks, Models 6-20 and 27-30 are all water-clocks except No. 29, which is a clock having lamps, one of which lights up every hour. The water-clocks all have the same basic driving system, together with a set of mechanisms for transferring the movements to the automata. The automata are of the usual types, already familiar from the works of other Arabic writers: doors that open to reveal figures, figures that discharge pellets from their mouths, figures that move their heads and arms, etc. Model 8 has a set of mirrors of 'fine, white glass', in the original:

(ويكون في كل جزء) مرآة من زجاج أبيض رقيق

one of which is illuminated every hour. The figures are on the outside of large box-like stuctures that house the mechanisms and the water reservoir. All the clocks record the passage of temporal or solar hours, i.e. the hours of daylight and darkness are divided by twelve to give hours that vary in length from one day to the next. The water-machinery is described most fully, although with some obliterations, for the mirror clock (Model 8, 15r-v). It is a cruder device than that used by al-Jazarī, being simply an outflow clepsydra of constant cross-section in which the outlet is a small concentric siphon *(kaʾs al-ʿadl)*. In the other Arabic works the reservoir is called *khizāna*, but here the term is *maḥbas al-māʾ* and, instead of *jazʿa*, the outlet is called *majfar*. To graduate the clepsydra it was filled with water while the outlet was blocked, then the outlet was opened and the outflow was allowed to continue for a full day. As the water discharged, marks were made on a vertical line on the inside of the vessel at the passing of each hour. On either side of the vertical line, at right angles to it, horizontal lines were drawn for each month, according to the known length of daylight at the start and finish of each month. There were thus six divisions on each side of the line – the horizontal lines did not coincide as they would have done if the division had been into zodiacal 'houses'.

11. Derek de Solla Price, *Gears from the Greeks* (Science History Publications, New York, 1975), *passim*.

The summer solstice was at the top of these divisions and the winter solstice at the bottom. The months were then subdivided into the correct number of days. The table was engraved on the side of the clepsydra for a length corresponding to a fall in the water-level in a period of fifteen hours, this being the accepted time for the hours of daylight in the latitude of Cordoba at the summer solstice. There was an unmarked section at the bottom of the vessel so that an adequate static head was maintained over the orifice on the shortest day. Every day the clepsydra was filled to the mark corresponding to that day, and in the night to its nadir. (There is a drawing of the divisions on 26 v). On the surface of the water there was a float. This float is also called a *falak*, a designation which might lead to some confusion. It is clearly a float, however, because it was weighted, probably with sand (see al-Jazarī 28)[12], until it was almost submerged in the water. The rope which provided the drive for the mechanisms was tied to a ring on the top of the float.

Clearly, some further refinement was required in order to keep the distance fallen by the float constant throughout the year, while allowing the speed of its descent to vary from day to day. In the clocks of the pseudo-Archimedes, Muḥammad al-Sāᶜātī and al-Jazarī this was achieved by using a float-chamber and flow-regulator, although only al-Jazarī was able to make the system completely accurate. No mention of such a system occurs anywhere in this treatise, but a passage on 15v, although it is somewhat obscure and partly defaced, indicates how the adjustment was made. When the clepsydra had been graduated, the mark for the longest day was taken as the datum, i.e. the start position for the float's travel every day. On the longest day the vessel was filled to this line. The gears and transmission systems were sized so that the time-recording devices tripped twelve times in the period taken by the float to complete its full travel. On other days the water was poured in up to the level corresponding to the day in question on the scale, then raised to the datum line by placing solid heavy objects in the water. There was a set of large objects *(ashkāl)*, and a set of small ones, so that the correct displacement could be obtained for the different periods of daylight by varying the combinations of objects that were placed in the water. In fact, the changes were not made daily, but at intervals of four days, so the number of objects in each set need not have been large. The writer does not specify the exact number, but says 'three, four, or five, or any number we wish'. It would not have been possible to have achieved correct results by reducing the cross-sectional area of the clepsydra over its full working length by different amounts for each four-day period, because such a change in the volume of the

12. Al-Jazarī's monumental clock was accurately reconstructed for the Science Museum, London, for the 1976 World of Islam Festival. It was found necessary to pour sand into the float until its total weight was 9 pounds (about 4 kg.), before the weight of the float was sufficient to drive the mechanisms. The reconstruction of the clock is to be described in a paper for the *Journal of the Antiquarian Horological Society*, by myself and Mr. P. N. Haward, the craftsman who built the clock.

water conflicts with the required change in the hydrostatic formula.[13] Instead the objects must have been below the datum line, a positioning indicated by the writer, who says that they were 'in the water'. As mentioned above, this clock, Model 8, had a set of twelve mirrors, one of which was illuminated every hour — a very similar system to the glass roundels in al-Jazarī's first clock. In the nights the displacing weights were changed, as was the direction of the wheel that uncovered the mirrors, and a lamp was kindled in the box-like structure.

Figure 2 shows Model 10 (17v-18v), whose main release mechanism is of interest since it is a copy of the ball discharging device in the 'Archimedes' clock. A wheel is fixed to a vertical axle and has twelve equally spaced holes, each one fingerbreadth in diameter, bored through its surface near its circumference. Close below this is a static wheel that does not touch the axle; in its surface is a single hole that coincides with the holes in the upper wheel as they pass over it. A channel from this hole leads to the head of a figure. One ball is placed in each of the twelve holes in the upper wheel, so that as this wheel rotates a ball reaches the lower hole every hour and runs through the channel and is ejected from the figure's mouth. Fixed to the same axle, below these two wheels, is a cogwheel 2 spans in diameter; it has a groove on its perimeter to receive the rope which passes over a system of pulleys to the float in the clepsydra. Another cogwheel, also of 2 spans diameter, meshes with the first cogwheel. It operates a mercury balance and a system of pulleys, ropes, and weights for turning the head of the figure of an astrolabist — the precise operation of these mechanisms has not yet been determined. It is apparent however, that these mechanisms were quite sophisticated and contrast markedly with the relative crudeness of the clepsydra itself. It is strange that the engineers whose work Ibn Muᶜādh was describing knew of the ball-release in the 'Archimedes' clock, yet did not adopt the elegant system of outflow control described in the same work.

13. In a vessel of uniform cross-sectional area A, let the datum be height H above the orifice and the end of the scale height H_0. Let the height for a given day on the scale be H_1. If t is the time taken for the water to fall from H to H_0, and t_1 the time from H_1 to H_0 then, assuming the coefficient of discharge and the area of the orifice remain constant:

$$t = k\,A\,(H^{\frac{1}{2}} - H_0^{\frac{1}{2}})$$

and

$$t_1 = k\,A\,(H_1^{\frac{1}{2}} - H_0^{\frac{1}{2}}) \qquad (1)$$

where k is a constant.
If water is poured in up to height H_1 and the cross-sectional area of the vessel is reduced from the bottom to H, to a uniform cross-sectional area of A_1, then

$$A_1 = A \cdot H_1/H. \qquad (2)$$

It is impossible to reconcile (1) with (2), so the displacing weights must have been used to reduce the cross-sectional area of part of the vessel only. It is very probable that the correct times were arrived at by trial and-error, since the formula given in (1) above was almost certainly not known at the time of Ibn Muᶜādh. At least 150 years later, al-Jazarī used empirical methods to obtain the correct rates of discharge for different static heads of water (24-25).

This necessarily brief and sketchy outline of the contents of the treatise naturally omits much of importance, but it is hoped that it fulfils its purpose in bringing this crucial work on machines to the attention of scholars. It seems reasonable to advance the hypothesis that there were two distinct traditions of mechanical technology in medieval times in Europe and Western Asia. The eastern tradition, represented by the Banū Mūsā and al-Jazarī, although it included water-raising machines and large water-clocks, was characterised by the use of delicate mechanisms and subtle hydraulic and pneumatic controls. The western tradition was common to Europe, North Africa and Andalusia — i.e. it was applied both in Christendom and in Islam and placed much more emphasis on large machines with powerful prime movers. Of course, the two traditions were not wholly different, nor was the one unaffected by the other. Nevertheless, the hypothesis of the two traditions may help us to avoid the assumption that the development of a heavy machine technology was mainly a European phenomenon.

Notes for Translation into Arabic

So that there shall be no ambiguities in the translation, the following list of words and phrases, referred to in the pages of the typescript, is provided for guidance.

Page 36 box = *ṣandūq* = صندوق

Page 39 cross = *ṣalīb* = صليب

tube = *qaṣaba* = قصبة

mercury = *zibaq* = زبق أو زئبق

raṭl = رطل

lantern-pinion = *dawlāb dhū ᶜawāriḍ* = دولاب ذو عوارض

tooth = *dandānja* = دندانجة

Page 38 *falak* pl. *aflāk* = فلك ج أفلاك

wheel = ᶜajala = عجله

Sentence about the noria reads in the original (on 5 V) :

فلنعمل فلك شبيه النعورة [وهـو فلك زح] فان كان الماء قليلا كان على شبيـه النـاعورة التي تـدير الأرحى.

wheel = *dawlab* or *dawlāb* = دولب أو دولاب

Page 40 Description of mirrors, of 'fine white glass' is on 15 R. It reads :

[ويكون في كل جزء] مرآة من زجاج أبيض رقيق

concentric siphon = *ka*ᵓ*s al-*ᶜ*adl* = كأس العدل

reservoir = *khizāna* or *maḥbas al-mā*ᵓ = خزانة أو محبس الماء

outlet = *jaz*ᶜ*a* or *majfar* = جزعه أو مجفر

Page 41 objects = *ashkāl* = أشكال

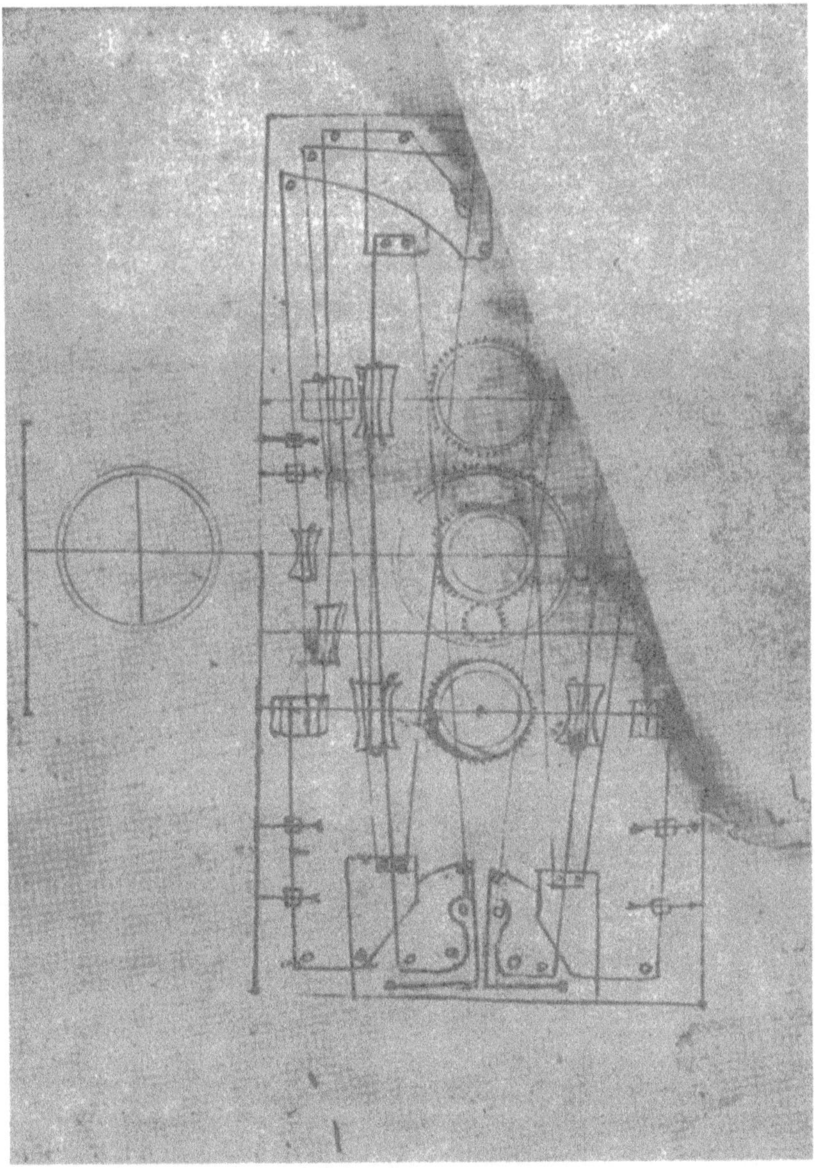

Fig. 1 : *Reproduced by kind permission of the Biblioteca Laurenziana, Florence.*

Fig. 2 : *Reproduced by kind permission of the Biblioteca Laurenziana, Florenca.*

XVIII

Andalusian Technology

There are two distinct types of engineering technology as applied to the Islamic civilisation of Andalusia in the centuries of Muslim occupation. The first is Utilitarian Technology, namely the various types of everyday constructions that helped to maintain and enhance the standards of life. The second, which we may refer to as Fine Technology, was concerned with machines such as water-clocks and automata that were designed as aids for astronomical reckoning or for the amusement and aesthetic pleasure of courtly circles. Of less obvious benefit to society than the first type, the creations of Fine Technology were very important in the history of machine design because of the mechanisms and techniques incorporated in them.

Utilitarian Technology

The most important contribution made by engineers to the society of Andalusia was in the field of hydraulic engineering. The quality of life was sustained by a flourishing agriculture which in turn depended upon irrigation systems, especially because crops introduced from the east—for example, rice, oranges, sugar-cane—required copious amounts of water in the growing season. Irrigation and water supply were achieved by the diversion of rivers into carefully engineered canal systems and by the use of water-raising machines. Irrigation was undoubtedly practised in Roman and Visigothic Spain but the introduction by the Muslims of improvements in dam-construction techniques and of new water-raising machines led to a substantial increase in the irrigated lands, particularly in the province of Valencia and in the valley of the River Guadalquivir.

Cordova was the capital city of Muslim Spain for nearly 300 years and it is here, on the Guadalquivir, that the remains of what is probably the oldest Muslim dam in Spain are to be seen. Standing just below the Roman bridge (the Puente Romano) it follows a zigzag course across the river so that its total length is over 1400 feet though the river is only some 1000 feet wide. It is now very dilapidated and in fact little of the structure remains above water level. It seems to have been built of rubble masonry, but its facing was apparently cut stone, since al-Idrīsī in the twelfth century says that it was made of *qibṭiyya* stone with large marble pillars. Al-Idrīsī adds that below the dam there were three millhouses and in each house were four mills. These millhouses can still be seen. Although the machinery has long since disappeared, the masonry conduits built to conduct the flow of the river in to the millhouses have survived. Also below the dam was a large water-wheel (noria) that was driven by the water impounded behind the dam. The noria lifted water from the Guadalquivir to the level of an aqueduct, which is still in place. There is also a recently-built replica of the noria, enabling visitors to appreciate one of the purposes of the dam. The installation at Cordova is in fact a multi-purpose dam meeting the needs of milling and water supply. Indeed, the Muslims were pioneers in the use of dams for producing hydropower.

The eight dams along a six-mile stretch of the River Turia in Valencia incorporate several points of interest in hydraulic engineering. Each dam was designed to divert part of the river's flow into an irrigation canal. All the dams are similar and the one at Mestella can be regarded as typical. It is 240 feet long and 7 feet high and consists of a core of rubble masonry and mortar with a facing of large masonry blocks. The foundations are very substantial, consisting of masonry supported on wooden piles, the combined depth of masonry and piles being about 25 feet. This seems to be a massive substructure when compared to the superstructure but it must be remembered that the Turia is sometimes subjected to dangerous floods in which the flow is 100 times greater than normal. At such times the dams are submerged to a depth of nearly 20 feet and must resist a battering of water, stones, rocks and trees. Because they are so low and flat and are provided with deep and very firm foundations, the Turia dams have been able to survive these conditions for over a thousand years, and still continue to meet the irrigation needs of Valencia. Another point to be made about these dams is that the canals they serve have between them a total capacity of only slightly less than that of the river. This therefore raises the possibility that the Muslims were able to gauge a river and then design the dams and canals to match.

Another point about the Turia dams and most others elsewhere in Andalusia, is the provision of desilting sluices. One end of each dam is abutted onto a masonry wall which formed one side of the head waters of the irrigation canal. Two sluices were built into this wall, which served two purposes: during normal operations they were used as escapes to allow surplus water to drain back into the river; and occasionally they would both be opened to their full extent in order to desilt the approaches to the canal mouth. Such scouring sluices, closed with planks carried in grooves, are absolutely essential in this type of dam. Without them the canal intakes and the canals themselves would become hopelessly choked and obstructed. They were a Muslim development which later Christian dams were to utilise on a grand scale. In general, Muslim dams indicate that their builders understood well, if empirically, the problems of structural and hydraulic engineering in order to build dams that would stand up to the local conditions.[1] Most lands were irrigated by a network of irrigation canals. The problems to be solved in this ancient technique are basically straightforward: a question of conducting the water from the offtake points on to the fields and arranging for the drainage of the surplus back to the rivers. The *qanāt*, a method for conducting water from an underground aquifer to the place where it is needed, is technically more demanding. It was probably introduced in Assyrian times and was used in the ancient Near East and by the Romans, but its greatly increased spread in North Africa and Spain was due to the Muslims. A *qanāt* is an almost horizontal underground conduit that demands considerable skill in siting and construction. First the landowner, having decided that he wishes water to be supplied at a particular point in his land, engages a skilled surveyor, usually a former *muqanni* (*qanāt* engineer). The surveyor then carefully examines the alluvial fans, looking for traces of seepage and often for a barely noticeable change in vegetation, before deciding where the trial well is to be dug. A winch is set up at the selected point and two skilled *muqanni*s excavate a vertical shaft about three feet in diameter, the spoil being hauled to the surface and deposited around the mouth of the shaft. If the well is successful in striking the aquifer it becomes the "mother well".

The surveyor now determines the route, gradient and precise outlet of the *qanāt*. Having selected the route the surveyor levels from the surface of the water in the mother well to the outlet point, setting the gradient at 1:1000 to 1:1500. The *muqanni*s then begin to excavate the *qanāt* from the outlet point to the mother well. The conduit is about 3 feet wide by 5 feet high and there are vertical shafts at intervals of about 50 yards, both to provide ventilation and to allow for the disposal of the spoil. A characteristic of *qanāt*s, which in modern-day Iran are typically about 10 miles long, is the presence of the spoil heaps at the head of each shaft. *Qanāt*s still supply much of the water of Tehran and other Iranian communities, and are also in use in other parts of the Near East. Two Spanish systems, both built by the Muslims, have been studied. The Madrid system, which still functions, brings water to the city from the springs of the Guadarrama. The system of Crevillente, in the province of Valencia, was some 1500 yards long and had 19 ventilation shafts.[2]

Because of the long distances involved and the almost imperceptible gradients, surveying techniques were at their most demanding in the case of *qanāts*. There were, however, many other demands on the skills of surveyors: in hydraulic engineering, in the setting-out of new building complexes and in the demarcation of boundaries. Several methods of levelling were used, but the most sophisticated was very similar to modern methods, with the obvious omission of optical aids. A sighting tube was mounted on a central pin soldered to a circular metal plate which was suspended vertically. The tube was brought to the horizontal and sighted onto the divisions of a graduated staff held vertically by an assistant. The procedure was then repeated with the staff on another station along the line of the survey; the difference in readings gave the rise or fall of the second station with regard to the first. The instrument was then moved while the position of the staff remained unchanged. A sighting was then made ('back sight'), the staff moved to another station and another reading ('foresight') taken, and so on. The difference in level between two widely separated points could thus be obtained from the algebraic sum of the rises and falls. Large-scale or difficult surveys were undertaken by specialists in land measurement (*muhandis* in Andalusia). Triangulation procedures were carried out with the astrolabe, including the measuring of heights and distances.

In many cases it was necessary to erect water-raising machines to supplement gravity-fed irrigation or to lift supplies of water to communities. Where wells were the only local sources of water, domestic supplies could be provided by using well-head windlasses or even the simple rope-and-bucket method, but these procedures were inadequate for larger-scale applications. The simplest of the traditional machines was the *shādūf* or swape, a pole balanced on a fulcrum with a bucket at the end of the longer arm and a counterweight made of rock or clay at the other end. The bucket was dipped into a well or a watercourse and raised mainly by the action of the counterweight, its contents being discharged into a channel or a head-tank. The *shādūf* was known in ancient times and was in use in the Roman Mediterranean. Even today its use is widespread; it is easy to make and to maintain and is fairly efficient. Its main drawback is that its operation involves back-breaking labour.

Of the other two machines the noria had been in use in Spain in Roman imperial times but its exploitation was greatly increased after the Muslim conquest. The noria is current-driven, and thus is self-acting and requires the presence of neither man nor animal for its operation. It is a large wheel, constructed of timber, mounted on a horizontal axle over a running stream. It is rotated by paddles that dip into the stream. Some types have earthenware pots lashed to the rim while in others the rim itself is divided into compartments. As the wheel rotates, the pots or compartments fill with water at the bottom of the rotation and discharge their contents at the top into a head tank or an aqueduct. The construction of a noria requires substantial capital outlay and considerable skill in manufacture; the names of some of the engineers who constructed norias are known to us. Famous norias in Spain include the one at Cordova, mentioned above, the well-known wheel at La Nora driven by the current of the Aljufia Canal and the large wheel at Toledo, described by al-Idrīsī in the twelfth century. The *sāqiya*, a device still to be seen in many parts of the Iberian Peninsula, was first introduced by the Muslim conquerors in the eighth century. It is a chain-of-pots driven by animal power through a pair of gear wheels meshing at right angles. The machine is constructed from about two hundred separate parts, nearly all of them wooden. It has survived in some areas of the world partly because it is quite an efficient machine and partly because it can be kept in repair by the local carpenter. Importation of fuel is of course unnecessary. The animal, usually a donkey, is harnessed to a shaft which moves a horizontal lantern pinion. This engages teeth set in a vertical wheel which, in turn, raises the water by means of an endless chain of pots fixed to its rim with a double continuous rope. In order to prevent the wheel from going into reverse the machine is provided with a pawl mechanism which acts on the cogs of the potgarland wheel. The pull to which the animal is subjected is exerted by the part of the potgarland wheel carrying the full pots. The pawl is activated in two cases - when the animal is to be unharnessed and in the event of the harness or traces breaking. Without the pawl the machine would turn backwards at great speed and, after one revolution, the drawbar would hit the animal on the head. At the same time many of the pins of the lantern pinion would break and the pots smash.

There is no doubt as to the importance of the *sāqiya* for water supply and irrigation. Apart from their agricultural uses, *sāqiya*s were also associated with royal parks. They were installed, for example, in ʿAbd al-Raḥmān's residence in a park at Madīnat al-Zahrā' outside Cordova, and in Toledo they raised water from the Tagus to supply elegant fountains in al-Maʾmūn's garden. There are references to *sāqiya*s in the works of Arab geographers of the tenth century and there is a full description of the machine in an Arabic encyclopaedia written by Ibn Sīda, who died in 1066. The Andalusian writers suggested practical means by which farmers could increase the efficiency and longevity of the machine. For example, Ibn al-ʿAwwām, writing in the twelfth century, recommended the use of hardwoods for the teeth of the potgarland wheel. He also recommended that the pots be supplied with air vents to prevent breakage as the force of the water pushed each pot against the wall of the well or into the pot behind it.[3] Muslim techniques of hydraulic engineering were transmitted to the Christian part of the Iberian Peninsula, whence they were eventually carried to the New World. Masonry dams were a usual feature of Spanish irrigation systems in Mexico, New Mexico and California. The *sāqiya* was used extensively in Mexico and less widely in South America. The Spaniards also constructed *qanāt*s in the New World. The systems of Tehuacan and Parras in Mexico are almost certainly of Spanish origin. Similar works in Peru and Chile, however, appear to be pre-Columbian.

Further research is required to establish the relationship between imported and indigenous traditions in *qanāt* construction. There is no doubt, however, of the importance, in general, of the introduction of Andalusian hydraulic engineering of Islamic origin in the New World.

It would be impossible, of course, in the space available, to give even summary descriptions of the wonderful water gardens of the Alhambra and the Generalife, both supplied from an aqueduct built by the Nasrids in the thirteenth century. In any case, these gardens have been described many times and are readily available to visitors. The same comments apply to the achievements of Muslim architects such as the Alhambra palace itself and the Great Mosque at Cordova. One foundation, however, is worth mentioning since it is perhaps less well known than the aforementioned and was constructed at the very peak of Muslim power and prosperity in Andalusia. This was Madīnat al-Zahrā', a governmental city of the Umayyad caliphs about 5 km north-west of Cordova. Built by the great ʿAbd al-Raḥmān III, ostensibly at the whim of a favourite slave-girl, it had remained almost unknown until the present century. The city was destroyed at the beginning of the eleventh century, only some 74 years after its foundation, and the ruins were extensively looted to furnish materials for buildings in Cordova. Only in the present century were archaeological explorations begun, the most important excavations and restorations being under the direction of the great master of Hispano-Moorish art Félix Hernández until his death in 1975. Since then the enormous task of restoration has been continued by the architect Rafael Manzano. The Arab chroniclers have given us excellent descriptions of Madīnat al-Zahrā', including the motives for its foundation; the choice of site; the duration of its construction; the number of workmen and pack animals employed; the materials used; the cost of construction; details of the buildings and the uses to which they were put; and finally the decline and destruction of the city. Sufficient has been uncovered up to the present for us to appreciate the high degree of craftsmanship and the exquisite artistic sensibility of all the artisans who worked for the great caliph ʿAbd al-Raḥmān III.[4]

It is known that the water-mill in its several forms had been invented before the start of the Christian era, and that milling installations, some of them on a large scale, had been in widespread use during the later Roman Empire. There are two basic types of water-wheel. The horizontal wheel, the ancestor of the turbine, consisted of a central hub to which either vanes or paddles were attached. The millstones were mounted directly on to the vertical shaft of the wheel - there was no gearing. The vertical mill could be either undershot or overshot. In the former case the motive power was provided by a paddle-wheel which was mounted directly in a running stream and rotated by the velocity of the water. In the overshot wheel the rim was divided into compartments into which the water was directed from above, turning the wheel mainly by its weight.

We are seldom told which type of mill in Andalusia a given report refers to. (The same problem arises elsewhere: nearly 6000 mills are listed in the *Domesday Book*, compiled in England in 1086, without any differentiation between the possible types.) It seems likely that the relatively cheap, low-powered horizontal mills were used by small communities, whereas the vertical types were located on the larger rivers and served cities and major towns. Certainly most of the many references to mills by the Arab geographers were of the vertical variety. Mills mounted on boats or on the piers of bridges could only be powered by vertical wheels. There are reports of large milling installations at Merida, Almería, Jaén and on the banks of the Guadalquivir between Seville and Cordova. In Cordova itself, as we have seen, the mills were powered by the water impounded behind the dam. The ship-mills at Murcia, used for grinding corn, were moved from place to place as required.

The problem of the origins of industrial milling - i.e., mills used for purposes other than the grinding of corn - is not easy to resolve. Industrial mills appeared in Christian Spain, notably in Catalonia, during the twelfth century, the period of expansion of milling technology throughout western Europe. There are frequent citations of fulling mills in Catalonia from 1150 onwards and towards the end of the century water power was applied to the Catalan forges. Paper mills also appear in Catalan documentation during the 1150s. Generally speaking, however, the industrial uses of water power appeared earlier in the Islamic world. The application of water power to paper manufacture began in the Islamic world in the eighth century and there are references to the fulling of cloth in the tenth century and to the crushing of metallic ores in the eleventh. It seems likely that these techniques were introduced into Islamic Spain for the husking of rice and the crushing of sugar-cane when these crops were brought to the peninsula with the advent of Islam.

After its earlier introduction into eastern Islam, paper-making was introduced into Andalusia by the mid-tenth century. The centre of the industry was Játiva, in the province of Valencia. It is assumed that the water-driven mill was used in Islamic Játiva; documentation shows Muslims operating a paper-mill in Játiva. Since the texture and composition of the oldest surviving specimens of medieval Catalan paper are identical to the Andalusian product, the techniques of paper making in both societies can be presumed to have been identical. References to paper mills abound in Catalan documents of the twelfth century and by the thirteenth paper was exported to Sicily from both Barcelona and Valencia. The Valencian exports doubtless originated in Játiva, where the Muslim community received royal support for the continued production of paper, after the Christian reconquest of the region.[5] The balance of probability is that industrial milling was introduced to Christian Spain from al-Andalus, whence it spread to the rest of western Europe. There is at present, however, no conclusive proof for this supposition.

Fine Technology

Our knowledge of water-clocks, automata and similar devices must be derived from literary sources, since the machines themselves were fragile and nearly all of them have long since disappeared. There is, however, ample evidence for their construction. The treatises on machines obviously describe actual constructions and not unrealised concepts, and indeed a number of these devices have been successfully reconstructed by modern craftsmen. Moreover, contemporary reports by geographers and travellers describe water-clocks and other devices that the writers had inspected. Finally, the remains of two monumental water-clocks can still be seen in Fez, Morocco. In a related field, examples of astrolabes and other astronomical instruments have survived, being small, compact and fairly robust. But these belong more properly to the history of astronomy rather than that of technology.[6]

The earliest unequivocal reference to a water-clock in Andalusia, and indeed in the whole world of Islam, occurs in a poem of ʿAbbās ibn Firnās (d. 887), poet and engineer to the Andalusian amīrs ʿAbd al-Raḥmān II and Muḥammad I. No description of the clock is given, but it was almost certainly a water-clock, since it could be used in cloudy weather or at night.

There is a well-known though brief description of two water-clocks built on the banks of the River Tagus about the year 1080, reputedly by the famous astronomer al-Zarqālluh. According to the Spanish Muslim geographer al-Zuhrī (fl. 1137) these clocks each consisted of a basin (probably cylindrical) that was housed in a chamber built into the banks of the river at Toledo. They recorded not only the passage of the hours but also the phases of the moon: for the first 14 days of the month the basins filled at constant speed, then for the second half of the month they emptied at constant speed. According to the description, if water was removed or added - either deliberately or accidentally - the rate of flow was automatically adjusted to bring the water back to the correct level. This is not fanciful; it could have been achieved by using a system invented by Banū Mūsā, three brothers who worked in Baghdad about 850. Both the Toledo clocks were in use until 1133, and one of them after that date. It is possible that the constructor was not al-Zarqālluh, but that the description applies to an earlier machine.

Some of the most important information about Andalusian fine technology, and indeed about Islamic fine technology in general, occurs in a treatise that came to light only in the 1970s. Entitled *The Book of Secrets about the Results of Thoughts*, it was written by a certain Ibn Khalaf al-Murādī, about whom nothing is known except that he worked in Andalusia in the eleventh century. The treatise is part of a unique manuscript in the Biblioteca Medicea Laurenziana in Florence. It was copied in 1266 and is the only known specimen of an Arabic manuscript copied at the Toledan court of Alfonso X of Castile.

Unfortunately, the manuscript incorporates a number of defects that lessen its value as a document for the history of engineering. In the first place, it is in a very poor condition: every page is partly obliterated, either having been cut or affected by dampness. On average about 40 per cent of each page is affected, which means that no page can be read in its entirety. The illustrations are very poor, and their value is further lessened by the fact that all representations of human and animal figures have been omitted from them, doubtless for religious reasons. Furthermore, the descriptions are much too short to do justice to the evident complexity of the machines. Given this combination of defects of the manuscript and the treatise itself, it is impossible to make a detailed study of all the models. Nevertheless, by taking a selection of passages from different models, all describing a similar mechanism or technique, it is sometimes possible to make a plausible reconstruction of what is being described, particularly when the concept is known from the writings of other Muslim engineers.

In his introduction, al-Murādī addresses his kinsmen and friends, telling them that the purpose of his work was to revive a science that was falling into disuse. After this the work is divided into 31 chapters describing various devices. Nos. 1-5 are essentially very large toys, similar to clocks in that automata are caused to move at intervals, but without precise timing. Nos. 6-20 and 27-30 are clocks, all of which record the passage of the temporal hours by the movements of automata. Nos. 21-24 are war-machines in the form of towers that can be raised and lowered by a scissors action like that of 'lazy tongs'. Nos. 25 and 26 are machines for raising water from wells, and no. 31 is a universal sundial.

Although they are in some ways cruder than the machines described by writers in the eastern Islamic world, such as the Banū Mūsā, who have been mentioned above, in other respects they embody important innovations. The machines are larger and more robust than those of their eastern counterparts; we have ropes instead of cords, counterweights of several kilograms instead of a few hundred grams. The most significant features of al-Murādī's machines, however, are the gear-trains used to transmit movements from the water-machinery to the automata. These gears were quite complex and although the state of the text does not permit us to arrive at a detailed understanding of the mechanisms, the gears certainly included segmental wheels, i.e. wheels with teeth on only part of their perimeters in order to permit intermittent operation. It is possible that epicycle gears were also included; they appear to be shown on several of the illustrations, but the manuscript is defective in every place where the gears are described. Complex gear-trains were certainly known at an early time, for example in the Antikythera mechanism, an astronomical calculator dated to the first century B.C. In this and other devices, however, the gears transmitted movements rather than loads, whereas al-Murādī's gear-trains were required to transmit high torques.[7]

The work which gives us the clearest evidence for the existence of a tradition of an Islamic tradition of fine technology in Andalusia is, oddly enough, written not in Arabic but in Castilian. This is the *Libros del Saber de Astronomía*, prepared in 1277 under the direction of Alfonso X of Castile. It is a collection of translations and paraphrases made from Arabic originals, with the declared intention of making Arabic knowledge available to the Christian world. The section that concerns us is at the end of the work and contains descriptions of five timepieces. Three of these have close connections with earlier Arabic constructions. One is a water-clock whose mechanisms and time-recording systems are inferior to those of al-Jazarī, who completed his great book on machines in 1206 in Diyār Bakr. The system for maintaining a constant head is pneumatic, and similar to a method used by the Banū Mūsā. There is a candle-clock, but again this is much inferior to those described by al-Jazarī. The most significant of these clocks, since it incorporated a weight-drive with a means for controlling the speed of its descent, was a clock consisting of a large drum made of walnut or jujube wood tightly assembled and sealed with wax or resin. The interior of the drum was divided into twelve compartments, with small holes between the compartments through which mercury flowed. Enough mercury was enclosed to fill just half the compartments. The drum was mounted on the same axle as a large wheel powered by a weight-drive wound around the axle. Also on the axle was a pinion with six teeth that meshed with 36 oaken teeth on the rim of an astrolabe dial. The mercury drum and pinion made a complete revolution every 4 hours and the astrolabe dial made a complete revolution every 24 hours. Clocks of this type had been known in Islam since the eleventh century - at least 200 years before the first appearance of weight driven clocks in western Europe. In later European versions of this type of clock the mercury was replaced by water and the drums themselves rotated and descended. The falling-drum clepsydra, probably derived ultimately from Islamic prototypes, was extremely popular in

Europe from the sixteenth to the eighteenth century as a cheap and reliable timekeeper, especially in rural areas.[8]

Andalusian fine technology is important in the history of mechanical engineering, particularly for its influence on the development of European horology. The mechanical clock was the most important machine invented in the Middle Ages, for the number of mechanisms that it incorporates and for its successful solution of the problem of controlling the force of gravity. It is certain that the mechanical escapement for controlling the descent of the weight-drive in a clock was invented in western Europe towards the close of the thirteenth century. The evidence points to its inventor having come from the ranks of makers of water-clocks, which was an established craft in Europe by the twelfth century. At this time there was a marked improvement in the techniques of hydraulic timekeeping in western Europe and it is credible to attribute this improvement to Andalusian Islamic influences. All the techniques and mechanisms that were embodied in the mechanical clock, except the mechanical escapement, were already present in Andalusian horology: automata, complex gear-trains including segmental gears, weight-drive and audible signals. This array of ideas became familiar to western European horologists and it required only the addition of the mechanical escapement, admittedly an invention of genius, for the mechanical clock to be born.

It is never easy to trace the transmission of technological ideas from one culture to another, since such transmissions rarely occur by literary means. Nevertheless, a fairly credible conjecture can be made for the passage of Islamic horological ideas into western Europe. First, it has now been established that the first European treatises on the astrolabe were of Arabic inspiration and were produced in the monastery of Ripoll in Catalonia. These early treatises were full of unassimilated, half-understood Arabic words, and it was not until the middle of the twelfth century that properly-organised treatises on the instrument began to appear in Latin. Second, and very significantly, the first known European description of a water-clock was written in the monastery of Ripoll. Knowledge of the astrolabe was probably disseminated into Europe by Gerbert d'Aurillac, later Pope Sylvester II, after he had visited Ripoll about the year 967. Since the church was keenly interested in finding a reliable means for timekeeping, it is likely that information about Arabic water-clocks was also carried to western Europe by a churchman.[9]

It is hoped that the present article has indicated the importance of Andalusia in the history of hydraulic engineering and timekeeping, to name but two fields. It must be remembered, however, that there were several centuries of flourishing Islamic civilisation in the Iberian Peninsula, and this civilisation can be regarded as a splendid flowering of the human spirit in its own right, quite apart from its significance in the history of technology.

Notes

* A Spanish translation, richly illustrated in colour, was published as "Tecnología andalusí", in Julio Samsó *et al.*, eds., *El Legado Científico Andalusí*, Madrid: Ministerio de Cultura, 1992, pp. 157-172 and 298-309 (exhibits). The original English text is here reproduced courtesy of Professor Julio Samsó, University of Barcelona, who kindly provided me with the typescript. A very few minor changes have been included here in this version typed by Wolf-Dieter Wagner, Frankfurt University. (Ed.)

1 Norman Smith, *A History of Dams*, London: Peter Davies, 1971, pp. 75-101.

2 Henri Goblot, *Les Qanats; une technique d'acquisition de l'eau*, Paris: Mouton, 1979, pp. 28-35.

3 Thorkild Schiøler, *Roman and Islamic Water-Lifting Wheels*, Odense: Odense University Press, 1973, *passim*.

4 M. Ocaña Jiménez, article "Madīnat al-Zahrā'", *Encyclopaedia of Islam*, vol. V, pp. 1008-1010.

5 Thomas F. Glick, *Islamic and Christian Spain in the Early Middle Ages*, Princeton N.J.: Princeton University Press, 1979, pp. 241-242.

6 [On this claim see the editorial remarks in the preface. (Ed.)]

7 Donald R. Hill, *Arabic Water-Clocks*, Aleppo: Institute for the History of Arabic Science, 1981, pp. 36-46.

8 Silvio A. Bedini, "The Compartmented Cylindrical Clepsydra", *Technology & Culture*, 3 (1963), pp. 115-141.

9 Donald R. Hill, "Islamic Fine Technology and its Influence on the Development of European Horology", *Al-Abhath* (Beirut), vol. 35, (1987), pp. 8-28. For aspects of mechanics in al-Andalus see also Juan Vernet, *De 'Abd al-Raḥmān I a Isabel II*, Barcelona: University of Barcelona, 1989, pp. 297-351.

XIX

TREBUCHETS

Introduction

This article is concerned with the historical development, construction, performance, and dynamic characteristics of those types of missile-throwing machines that derived their propulsive force from the rotation of beams supported at a fulcrum. The larger part of the original sources used here were Arabic, with some attention to medieval European works, while information on the situation in the Far East was obtained from the works of modern writers. Emphasis is therefore placed upon the Islamic cultural area. This emphasis has its merits, since the Middle East was a focal point for the transmission of inventions in both directions.

The word "trebuchet" has been used for convenience to designate the rotating-beam siege machines, in the full knowledge that other terms were also used in the Middle Ages, and that the question of nomenclature remains unresolved. There were two main types of these machines: one, called hereafter the "traction trebuchet" (fig. 1), was operated by a team of men pulling on ropes attached to the end of the shorter arm of the beam; the other, called hereafter the "counterweight trebuchet" (fig. 2) had motion imparted to the beam by a counterweight fixed to the end of the shorter arm.

The Traction Trebuchet

There are attestations that this machine was known in China in Antiquity and in early medieval times.[1] Needham suggests that its transmission westward may have begun in the seventh century A.D. passing to the Turks of Central Asia and reaching the Western Islamic world by the late seventh century, together with Khurasani or Sogdian artificers.[2] One can test the

[1] S. A. Školjar, "L'artillerie de jet à l'Époque Sung," in *Études Song in memoriam Étienne Balars*, ser. 1, Histoire et institutions 2 (Paris 1971) 123-125.

[2] J. Needham, *Science and Civilization in China*, vol. 5, section 30.2. This section was unpublished at the time of my writing this paper. Dr. Needham kindly allowed me to read the typescript.

probability of this hypothesis by referring to the Arabic historians, although the vocabulary in itself is of no assistance. The expressions *manjanīq* and *ʿarrāda* are used to describe missile-throwing siege machines, and although the *ʿarrāda* may have been the smaller engine, the two expressions often seem to be interchangeable. *Manjanīq* occurs much more frequently than *ʿarrāda*, but their presence at a siege is often confined to a mere mention, without any description of the machines being given. A *manjanīq* is said to have been used when Muḥammad besieged the town of Ṭā'if in A.D. 630. It is very probable that this machine was of the type used by the Greeks and Romans, operated by twisted fibers or by the bending of a large bow.[3] Hence it can be deduced that the expression *manjanīq* could be used for all types of stone-throwing machines, and was not specific to any one type.

But although we cannot derive any information from the terminology by itself, there is nevertheless sufficient information in the Arabic chronicles to give an indication of the type of machines being used. At the siege of Mecca in A.D. 683 there was a *manjanīq* called *Umm Farwa* (Mother of the Hair). This description may well have been derived from the appearance of the ropes hanging down from the end of the beam. A poet added his own description: "swishing its tail like a foaming stallion [camel]."[4] At the siege of Daybul, in Sind, in A.D. 708, the Muslims had a siege machine called *al-ʿArūs* (the Bride). It was operated by 500 men, and was under the control of a skilled operator who took charge of the aiming and shooting.[5] In Khurasan in A.D. 710, there was a machine called *al-Faḥjā'* (the Straddle-legged).[6] Finally, there was a battery of machines at the siege of Baghdad in A.D. 865: men were assigned to every *manjanīq* and *arrāda*, who pulled on ropes to let fly the missiles.[7] The *Mafātīḥ al-ʿUlūm* (Keys of the Sciences), a kind of encyclopedia of technical and scientific terms, was compiled by Abū ʿAbd Allah al-Khuwārizmī towards the end of the tenth century A.D. In one section he describes siege machines—*manjanīq* or *ʿarrāda*. The support structure was the *kursī* (lit. chair): the fulcrum or axle was called the *Khinzīra* which means

[3] Al-Balādhurī, Aḥmad b. Yaḥyā (d. A.D. 892), *Futūḥ al-Buldan* (*Liber Expugnationi Regionum*), ed. M. J. de Goeje (Leyden 1866) 55. For the classical weapons, see E. W. Marsden, *Greek and Roman Artillery* (Oxford 1969).

[4] Al-Ṭabarī, Abū Jaʿfar Muḥammad b. Jarīr, A.D. 839-923, *Annales quos scripsit* etc., ed. M. J. de Goeje (Leyden 1879-1901) ser. 2, 426. The Arabic is *khaṭṭāra mithla al-faniq al-muzbid*. According to E. W. Lane, *An Arabic-English Lexicon* (London 1863-1893) the word *khaṭṭāra* is a synonym for *manjanīq*, "its casting being likened to the action termed *khaṭarān* of the stallion-camel": the stallion camel, when challenging another, raises its tail time after time and strikes its thighs with it. The phrase therefore clearly describes the motion of a traction trebuchet. Huuri (n. 33 below) 142 incorrectly translated this important phrase, thereby missing its significance.

[5] Al-Balādhurī (n. 3 above) 437.

[6] Al-Ṭabarī (n. 4 above) ser. 2, 1230.

[7] Al-Ṭabarī, ser. 3, 1551f.

literally "sow"—the word is also found with the same meaning in the work of Marḍā b. ʿAlī. The word *sahm* is used for the beam, which is described as a long piece of wood, as straight as a palm trunk. The *isṭām* was a piece of iron on the end of the beam to which the missile was attached.[8] Since the earliest date to which we can assign the introduction of the counterweight machine is the end of the twelfth century (see below), the machine described by al-Khuwārizmī must be a traction trebuchet—only the ropes and the sling are missing to complete the description. The other devices identified by al-Khuwārizmī in his book had been in use for many years, and it seems reasonable to suppose that he included only implements and mechanisms whose effectiveness had been established by long use.

The first phase of the Arab conquests began in A.D. 632 and lasted until A.D. 656, when they were brought to a temporary standstill by the dissensions that led up to the death of the caliph ʿUthmān. In this period they conquered Syria, Iraq, Egypt, Mesopotamia, Cirenaica, and most of Iran, and they made their first incursions into Armenia, North Africa, Seistan, and Khurasan. It is quite apparent that, in these early conquests, they had little skill in siege warfare, and that such machines as they possessed were almost completely ineffective. Fortified towns were usually captured by ruses or by treachery, or they surrendered in the aftermath of pitched battles, or because the Arabs were already in possession of the countryside and communications.[9] The first invasion of Khurasan had taken place in A.D. 642, but this was little more than a large-scale raid, and serious attempts to conquer and colonize Khurasan and central Asia did not begin until the Umayyads were firmly established in power. Thereafter the Arabs pushed gradually towards the East, consolidating their hold on Khurasan and establishing themselves in the Oxus valley. The conquests of Qutayba b. Muslim resulted in the bringing of Bukhara under Arab control by A.D. 709 and Sogdiana by A.D. 712. Expeditions were then mounted into the Syr Darya Provinces, bringing the Arabs to the fringes of the Chinese sphere of influence. An Arab trading mission visited China in A.D. 713 or 714.[10]

Considering the evidence as a whole, together with the fact that a regular artillery was first introduced into the Arab armies in Umayyad times,[11] there seems little reason to doubt that the Arabs had adopted the traction trebuchet

[8] Al-Khuwārizmī, Abū ʿAbd Allah Muḥammad b. Aḥmad b. Yūsuf, late tenth century A.D., *Liber Mafātīḥ al-Olūm*, ed., with Latin critical apparatus, G. van Vloten (Leyden 1895) chap. 8, bk. 2. A translation of this chapter is given in: E. Wiedemann, *Aufsätze zur arabischen Wissenschaftsgeschichte* (Hildesheim 1970) 188ff.

[9] D. R. Hill, *The Termination of Hostilities in the Early Arab Conquests, 632-656 A.D.* (London 1971) 133f.

[10] H. A. R. Gibb, *The Arab Conquests in Central Asia* (London 1923) 49-50.

[11] N. Fries, *Der Heerwesen der Araber zur Zeit der Omaijjaden nach Ṭabari* (Tubingen 1921) 45, 55f.

by the end of the seventh century. The references to siege machines, mentioned above, all strongly suggest beam-operated, rather than torsion engines. The actual transmission of the invention must have taken place in central Asia, perhaps soon after the resumption of the Arab offensive in that region in early Umayyad times, that is, after A.D. 660.

We do not have much information about the introduction of this machine into Europe, where it is first mentioned at the beginning of the twelfth century, and in 1147 a battery of these trebuchets were used by northern Crusaders to capture Muslim Lisbon.[12] The counterweight trebuchet came into general use in Europe at the beginning of the thirteenth century, so there is some reason to believe that the traction trebuchet was never such an important weapon in European armies as it was in Islam and China.

Descriptions of the traction trebuchet occur in the works of European, Islamic, and Chinese medieval writers, the last named providing the most detailed reports, from which we learn that the machine consisted essentially of a beam, resting on a fulcrum, which was supported on a timber structure, a tower. This tower was often provided with wheels, to assist in the emplacement and aiming of the weapon. The beam itself might be a single spar, or several such spars bound together to form a composite beam. The trebuchets of the Sung epoch usually had from 1 to 10 spars in the beam, but machines with 13 or even 15 were not unknown. The spars were from 5.60 meters to 8.40 meters long, with diameters at the extremities of 12.5 centimeters and 7 centimeters. At the narrower end was a copper "nest," attached to the spar by iron wire, thus forming a sling. The missile, which could weigh from 2 to 130 pounds, was placed in the copper "nest." At the other end of the beam there was a special attachment to which the ropes were attached, from 40 to 125 in number, between 12.40 meters and 15.50 meters in length, and 16 millimeters in thickness. The team of men who pulled on these ropes to discharge the missile could range in number from 40 to 250 or more. The beam was divided into two arms by the axle—a long arm and a short arm—in the ratio 5:1 or 6:1 for the light machines, 2:1 or 3:1 for the heavy machines.[13] The range of the missiles varied from 85 to 133 yards.[14]

Thus far from Chinese writings. Murḍā b. ʿAlī, in his treatise on weapons, gives us information that does not conflict with the essential features given above. He does not, however, describe beams composed of several spars, but only of single spars, although another Arabic source mentions a machine with 8 spars (*sahm*).[15] He also describes several different models, the Arab, the

[12] Lynn White jr., *Medieval Technology and Social Change* (Oxford 1962) 102-103.

[13] Školjar (n. 1 above) 126.

[14] Needham (n. 2 above).

[15] Al-Bundarī, al-Fatḥ b. Muḥammad, early thirteenth century A.D., *Zubdat al-nuṣrah*, ed. M. T. Houtsma, *Recueil des textes relatif à l'histoire des Seljoucides* 2 (Leyden 1889) 42.

Persian or Turkish, and the Byzantine or Frankish, but these do not differ except in points of detail. He mentions a missile weight of 50 *raṭl*, but since the *raṭl* is a very variable figure, this would be anything from 55 to 350 pounds. The greatest distance attainable, he says, is 60 fathoms (*bāʿ*), the least 40 fathoms: say from 120 yards down to 80 yards. The fulcrum divided the beam in the ratio 6:1. The sling was a cubit (*dhirāʿ*), or a little less than two feet long. The best effects were obtained if the beam was flexible, not rigid, and the best wood was said to be cherry. Murḍā b. ʿAlī assigns an important role to the *rāmī* or "shooter." This man held the pouch containing the missile and pulled it against his chest with all his force. It was important that he held the sling at the correct angle, otherwise there would either be no discharge, or the stone would fly vertically upwards. Presumably he let go of the pouch at the moment before the crew pulled hard on the ropes. The sling release system is shown clearly in figure 3.[16] Of the traction trebuchet, Aegidio Colonna says merely that instead of a counterweight, it has ropes that are pulled by men. It does not throw such large stones as the counterweight machines, but it is quicker to install and can shoot more frequently.[17]

The foregoing descriptions give an adequate idea of the construction and performance of this machine—which is shown diagrammatically in figure 1. It was a light, easily handled weapon, which could discharge stones of up to about 130 pounds in weight for distances somewhat greater than 100 yards. One suspects, however, that the additional resources in men and materials which were required for building and operating machines of high performance, in terms of missile weight and range, were out of all proportion to the increase in military effectiveness, when compared with the lighter machines. Even the smaller traction trebuchets were wasteful in manpower and the pulling team must have been very vulnerable to counterattack from missiles and from sorties.

The dynamic system of the traction trebuchet is discussed at the end of this paper, together with that of the counterweight trebuchet, with which it is compared.

The Counterweight Trebuchet

This machine appears to have been invented somewhere in the Mediterranean area in the late twelfth century, and to have spread outward very rapidly from its point of origin into northern Europe and western Islam. But the ques-

[16] Claude Cahen, "Un traité d'armurie composé pour Saladin," written by Murḍā b. ʿAlī b. Murḍā al-Ṭarsūsī, Arabic text, French translation and notes, in *Bulletin d'études orientales* 12 (1947-1948) 16-18.

[17] Aegidio Colonna (d. 1316), bk. 3, pt 3 of *De regime principium*. Edit. as Aegidio Romani, *De re militari veterum* in *Collectio monumentarum veterum et recentium* 1, by D. S. F. Hahn (Brunswick 1724) 51.

tion of the exact provenance of the invention, whether in Europe or in Islam, is not yet resolved. The earliest unambiguous description of the machine in Europe refers to its use in northern Italy in A.D. 1199.[18] Writing a few years earlier than this, Murḍā b. ʿAlī describes a "Persian" *manjanīq* that is definitely not a traction trebuchet, and yet the description is too confused for us to identify it for certain as a counterweight machine.[19] The first report we have of counterweight trebuchets in use in western Islam refers to the siege of Ḥimṣ in A.D. 1248-1249 where the machine in question is referred to as a *manjanīq maghribī*—a western or a North African trebuchet.[20] Counterweight machines were used in great numbers by the Muslims at the siege of Acre in A.D. 1291.[21] The trade of "trebuchator," however, appears to have been well established in England by the first quarter of the thirteenth century, where we have the names of a number of engineers whose chief responsibility was the construction of trebuchets.[22] Although we cannot be quite certain that these were all counterweight machines, this seems probable. A certain Gerard was employed on fortification works in Northumberland in A.D. 1244 and stones for the engines he made were cut from a form and a mold supplied by him.[23] Such close standardization of ammunition implies a standardization of weapons, and this would never have been possible with traction trebuchets, where the strength of the men and the resilience of the wood are such uncertain factors. Even so, the evidence is far from conclusive and on present data we can only locate the origin of the counterweight trebuchet somewhere in Mediterranean Christendom or western Islam, towards the end of the twelfth century. Almost certainly, however, it passed into the Far East from Islam. Two Muslim engineers, ʿAlā' al-Dīn and Ismāʿīl, are honored by a biography in the official history of the Yuan dynasty. They constructed the machines for Kubilai for the siege of Fan-chhêng towards the end of A.D. 1272. The counterweight trebuchet acquired the name of "the Muslim *phao*."[24]

References to the design and performance of this machine are not infrequent, and many illustrations of it occur in medieval and later works.[25] It seems to

18 White (n. 12 above) 102-103.

19 Cahen (n. 16 above) fig. 14.

20 Abū-l-Fidā', Ismāʿīl b. ʿAlī (A.D. 1273-1331), *Mukhtasar ta'rikh al-bashar*, Recueil des historiens des croisades, *Historiens orientaux* 1 (1872) 125.

21 Al-Maqrīzī, Aḥmad b. ʿAlī (A.D. 1364-1442), *al-sulūk li maʿrifat duwal al-Mulūk*, ed. Quatremère, *Histoire des sultans Mamlouks* (Paris 1837-1842) 2.125.

22 J. Harvey, *English Mediaeval Architects* (London 1954) 22f., 91, 150f.

23 *Ibid.* 111f.

24 Needham (n. 2 above). It should be observed, however, that, contrary to Chinese records, Marco Polo claimed that the counterweight trebuchet was introduced into China by a German artisan and a Nestorian in A.D. 1276. See White (n. 12 above) 103 n. 1.

25 Apart from works cited elsewhere in this paper, one can find scattered references and drawings in Leonardo da Vinci's notebooks. Professor Ladislao Reti of Milan was kind enough to send me some photographs of manuscript pages, including one from the Cod.

have been found particularly fascinating by the European mind. (The standard machine is shown in fig. 2). According to Aegidio Colonna[26] the *trabuchium* had a beam (*virga*) to which a sling (*funda*) was attached. It had a fixed counterweight in the form of a box filled with stones and sand, or with lead, or with other heavy bodies. Colonna also describes another machine, which he calls a *biffa*. This had a moving counterweight that rotated around the beam. He says that the *biffa* had a greater range than the *trabuchium* but was less accurate. No drawing of this machine is shown here, because the details of its construction are unknown. When the beam was pulled down the closed counterweight box had to be rotated so that it was above the end of the beam, and held there until the moment of discharge. This could have been done by a winch mounted on a superstructure above the main towers. It is inadmissible to suppose, as some modern writers have done, that the "fixed" counterweight meant a load rigidly attached to the beam, while a "moving" weight simply meant that it was suspended from the beam as in figure 2. Between two such machines there would be no difference either in accuracy or in range. This point is returned to in the section dealing with dynamics.

Some of the drawings of counterweight trebuchets in medieval and renaissance works are highly fanciful and the machines as depicted would have been useless.[27] Perhaps the best drawing is in *Bellifortis* by Conrad Kyeser.[28] This shows a trebuchet of workmanlike appearance, constructed of timber. The beam is supported on an axle that rests between triangular supports. The drawing is dimensioned, the total length of the beam being 54 feet, with a short arm of 8 feet and a long arm of 46 feet, giving a ratio of 5.75:1. The sling, carrying a round stone, rests in a wooden channel in the base, the channel being horizontal and in the same vertical plane as the beam; the sling is as long as the complete beam.

The missiles could be very heavy. During the fourteenth-century sieges of Tlemcen the trebuchets were capable of bombarding the town with balls

---

Atl., fol. 57 va, showing a machine in which the counterweight operates a large wheel set on the fulcrum. But, as Professor Reti says, the trebuchet was in decline by this time, and Leonardo was more interested in the mechanics of the machine than in its potential as a weapon.

Many of the medieval works that deal with weaponry are either unedited, or are difficult of access, but copies of original drawings will be found in Payne-Gallwey (n. 31 below), Huuri (n. 33 below), Cahen (n. 16 above) etc. Also of interest is Francis Grose, *Military Antiquities* 2 (London 1788) 285-302. There are several illustrations between pages 302 and 303. One shows a counterweight trebuchet, said to be from a carving in ivory dating from the reign of Edward II.

[26] Colonna (n. 17 above) 50f.

[27] For instance: Agostino Ramelli, *Le diverse et artificiose machine* (Paris 1588). The trebuchet shown on 322, which has no sling, would hardly have got the missile into the air.

[28] Conrad Kyeser von Eichstätt, A.D. 1405, *Bellifortis*, ed. Götz Quarg (Düsseldorf 1967) vol. 1, pl. 30a.

made of marble, some of which have been found there, the largest with a circumference of two meters and weighing 230 kilograms.[29] At the siege of Ḥimṣ in A.D. 1248, a *manjanīq* was erected which would throw a stone weighing 140 Syrian *raṭls*. A Syrian *raṭl* was about 4 pounds in weight,[30] hence the missile weighed 560 pounds. In general, however, missiles were somewhat smaller than this and may have ranged from 100 to 200 pounds. We do not have any really firm information about ranges attained by the counterweight machine, although the figure of 300 yards is probably of the right order.[31] Nor do we have any reliable information about the masses of the counterweights: calculations based upon the approximate dimensions of the counterweight boxes shown in illustrations indicate loads of 10,000 to 30,000 pounds, depending upon the material used for ballast.

The Dynamics of Trebuchets

A knowledge of the basic operating principles of trebuchets is necessary to a full understanding of their historical development. A noticeable example of a faulty deduction due to ignorance of mechanics is Payne-Gallwey's refusal to believe that the traction trebuchet ever existed, because "however many men pulled at the arm of a trebuchet they would not apply nearly the force that would be conveyed by the terrestrial gravitation of a heavy weight."[32] The comparison is quite irrelevant, because one must consider the whole system of forces and resistances.

A body set upon an axle will remain at rest if the masses on either side of the axle are in balance. If it is held by an external restraint when the masses are out of balance, and the restraint is then removed, it will accelerate to a certain maximum velocity, decelerate to a momentary standstill, and then continue to oscillate in this way until brought to rest by friction or by an external restraint. The maximum velocity that can be attained is determined by two factors (neglecting friction and air resistance), one accelerating factor and one retarding factor. The acceleration is due to the fall of the out-of-balance masses under gravitational pull; retardation is due to the Inertia of the whole body. The Moment of Inertia of a body is a measure of its resistance to change of angular velocity. A particle of mass m at a distance r from the fulcrum has a Moment of Inertia of mr^2, and the total Moment of Inertia for a body is the mathematical sum of mr^2 for all the particles in the body.

[29] G. S. Colin in "Ḥiṣār," *Encyclopaedia of Islam*, ed. 2, 470.

[30] E. Wiedemann and F. Hauser, *Über die Uhren in Bereich der Islamischen Kultur*, Nova acta, Abh. der Kaiserl. Leop. Carol Deutschen Akademie der Naturforscher 100.5 (Halle 1915) 47.

[31] Sir Ralph Payne-Gallwey, *The Crossbow, with a treatise on the Balista and Catapult of the Ancients* (London 1903) 309. See also the section on Dynamics in this paper.

[32] Payne-Gallwey 309 n. 1.

In a counterweight trebuchet, for example, doubling the mass of the counterweight doubles its Moment of Inertia, while doubling its distance from the fulcrum quadruples its Moment of Inertia. Exactly the same calculations can be made for the missile if it is resting on the end of the beam, but the sling makes an enormous difference, as we shall see later. Calculations for the beam are somewhat more complex. In general, however, the same considerations apply: increase in its mass increases its Moment of Inertia in direct proportion; increases in the length bring the square relationship into play. An understanding of the basic concept of Inertia and of the interdependence of all parts of a rotating body is essential to an understanding of the way trebuchets worked. The three main components—counterweight, beam, and missile—are all parts of the same body. They combine to provide the out-of-balance mass that is acted upon by gravity to produce acceleration, *and they also combine to produce the total Inertia that resists acceleration.* Both the traction trebuchet and the counterweight trebuchet were compound pendulums but they operated under quite different dynamic conditions. The first required greater skill in handling, the second in design.

Although it is easy to describe the dynamic system of the traction machine, it is impossible to derive a clear idea about its design and performance from theoretical considerations because there are too many unknown factors, notably the resilience of the beam, the tractive force exerted, and the average angle at which it was applied. The acceleration was provided mainly by an external force—the pull on the ropes. The gravitational pull on the out-of-balance masses must have had a negative, decelerating effect, since the Inertia of the missile and the long arm of the beam were greater than that of the short arm. Both beam and missile had therefore to be fairly light, otherwise much of the energy of the pull would have been counteracted by the Inertia of the machine, and the velocity of discharge would have been too low. There was, of course, no Inertia due to a counterweight and the stresses due to the jerk on the ropes were much less than those induced by a counterweight. This is largely because the beam was in motion as soon as the jerk was applied and bending stresses were therefore small. Even if the jerk produced an equal force to that of the dead load of a counterweight, the stress would still be much less, and hence the beam could be much lighter. Another important point is that men can only apply a rapid, powerful, vertical pull for a short distance, perhaps about three feet. Huuri attempted to compare the power of the two machines by imagining the pulling crew loaded into the counterweight box, but rejected this line of reasoning because he felt that the weight of a man does not necessarily equal the pull that he can exert.[33] But this is

[33] K. Huuri, *Zur Geschichte des Mittelalterlichen Geschützwesens aus orientalischen Quellen* (Helsinki 1941) 15f. Despite his faulty mechanics, Huuri's work remains indispensable for a study of siege weapons in the Islamic world, particularly for his information on source material.

the least of the reasons that make this argument absurd: the lightness of the beam due to the absence of a counterweight, the absence of Inertia due to a counterweight, and the shortness of the pull are the essential factors that distinguish the traction trebuchet from the counterweight trebuchet.

It must have taken many years of trial and error to arrive at the correct dimensions for a given performance, that is, the length and cross section of the beam, the ratio between its two arms, and the number of men in the pulling crew. One had to ensure that the "whip" in the beam was fully effective and that discharge occurred at the correct angle, and at the greatest possible velocity. Given these considerations one can see that the dimensions cited by Školjar make sense. A small machine with a beam 5.6 meters long, divided by the fulcrum in the ratio 5:1 would have a short arm of .93 meters. If the pull was taken with the short arm somewhat above the horizontal and discharge occured when the beam was at 45 degrees, then the pull would be applied throughout the rotation, and the Inertia of the beam and missile would not decelerate the machine. With the large machines the ratio of long arm to short arm was reduced to 3:1 or 2:1. Again, this appears to be dynamically right for a heavier beam, since the impulsive torque of the pull would be increased, and the Inertia of beam and missile reduced. The sources give figures of 250,[34] 500,[35] and even 1,200[36] men for the size of the pulling crew. With such large numbers, there must have been a high degree of training and discipline.

At first sight, it may seem surprising that the counterweight trebuchet was not developed, as an effective engine of war, before the end of the twelfth century. After all, not only did military engineers have the example of the traction trebuchet as a spur to their inventiveness, but they could also have been inspired by watching the action of the *s̲h̲ādūf*—the counterweighted irrigation device. The traction trebuchet, however, is a simpler machine, while the *s̲h̲ādūf* does not operate at high velocity. What is in fact surprising, when one comes to consider the dynamics of the counterweight trebuchet, is that it ever became a useful engine of war at all. The engineer, Muslim or Christian, who constructed the first effective machine of this type deserves our unqualified respect.

It will be simplest to consider first the machine without the sling, as if the missile were resting in a cup at the end of the beam, and then take the sling into account later. We now have an ordinary compound pendulum, which is to be released from an out-of-balance condition and allowed to swing —apart from friction, which we ignore for the moment, there are no external

[34] Školjar (n. 1 above) 127.
[35] Al-Balād̲h̲urī (n. 3 above) 437.
[36] Al-Bundarī (n. 15 above) 42.

forces. The gravity acting on the out-of-balance masses produces the acceleration, and this is resisted by the Inertia of missile, beam, and counterweight.

The beam of the counterweight trebuchet, unlike that of the traction machine, has to carry the full load of the counterweight while it is raised and then held in position. The beam must therefore be strong enough to resist the bending stresses produced by the counterweight. For example, if we consider the machine depicted by Kyeser, with a counterweight arm 8 feet long, and assume a counterweight of 10,000 pounds, we can calculate that the cross-section of the beam at the fulcrum must be 9 inches wide by 15 inches deep.[37] Admittedly, the long arm of the beam can be tapered from fulcrum to tip, but not drastically so, because this arm has to be strong enough to take the pull of the ropes from the spanning winch. The Moment of Inertia of such a beam is high, considerably greater than that of a missile weighing 200 pounds at a radius of 46 feet. The range of the machine, without a sling, would therefore have been ridiculously low—well under 100 yards, if allowance is made for frictional resistance.

It is quite incorrect to assume that the range can be increased in direct proportion by some straightforward means.[38] If, for instance, the counterweight is made heavier, this will increase the gravitational pull, but the beam must be made stronger and heavier, and the Inertia of beam and counterweight will be increased. If the counterweight is moved nearer the fulcrum, the beam can be lighter; Inertia of beam and counterweight decreases, but gravitational pull lessens. Using a lighter missile will not have much effect, because the Inertia of beam and counterweight will be unchanged. And so on—all the factors are interrelated and when any change is made, a gain in one direction is offset by a loss in another. It is, in fact, quite impossible to design a counterweight trebuchet, having no sling or only a short one, that would have a militarily effective range, however many permutations of design parameters one tries. This is probably the reason why it took so long to develop an efficient counterweight trebuchet, because this had to wait until someone hit upon the idea of using a very long sling. The sling greatly increased the range of the machine in two main ways. The mass of the missile did not have a direct retarding effect on them achine from the moment the beam was released, since beam and counterweight swung through a large arc before the missile was lifted off the ground. Admittedly, there was some frictional resistance as the missile was dragged along its channel, but this was relatively slight. The beam and counterweight had therefore accelerated to a high velocity

[37] See Appendix.

[38] Thus Payne-Gallwey (n. 31 above) 314 appears to assume that the power of the machine could be indefinitely increased simply by increasing the counterweight. Huuri (n. 33 above) 13 nn. 1, 2 makes a similar error by assuming a direct proportion between counterweight and performance.

before the missile load was exerted. The sling was then rotated upwards. Since it provided a virtually weightless extension to the beam, the retardation effect of sling and missile was quite small. A high linear velocity of discharge was therefore achieved.

Furthermore, the sling may have been accelerating, relative to the beam, at the moment of discharge, and this would have further increased the range. If the artificer could also have contrived that the missile was launched at the correct angle for maximum range—about 45 degrees—then the machine would have been at its peak of performance.

The trebuchet with a rotating counterweight was probably a late development. It had the advantage of increasing the gravitational pull without a corresponding increase in Inertia. It cannot have been easy, however, to have devised a suitable means for raising the counterweight, or to have built a supporting structure that was rigid enough to withstand the sudden severe rotational stresses. Colonna is undoubtedly correct in stating that this machine was more powerful but less accurate than the simple counterweight trebuchet.[39] Accuracy would have been impaired, because it must have been almost impossible to ensure that the release of counterweight and beam always occurred at the same instant.

Returning to the standard machine discussed above, having no sling, it is apparent that the design limitations still apply when the sling is taken into account. All the dimensions and proportions of the trebuchet's construction were interdependent and critical. The sling makes one significant difference, however: without it, no combination of design factors gives a range that is anything but derisory; with it, there appears to be an optimum design that gives maximum efficiency for a given missile/counterweight ratio. The writer has made a series of calculations that seem to support this view, and these calculations are felt to be admissible as evidence, because any inaccuracies in the assumptions upon which they are based will affect all the results in the same direction. Before considering the results of these calculations, we should examine the experimental data, such as it is.

There have been occasional attempts to assess the performance of counterweight trebuchets by the use of replicas, of which the best known is the experiment associated with Napoleon III and reported by I. Favé.[40] A full size trebuchet was constructed, having a beam 10.30 meters long, and a counterweight of 4,500 kilograms. Unfortunately the experiment was not well controlled and the report omits several vital details, such as the weight of the beam, the exact weights of the missiles, and the approximate angles of discharge.

[39] Colonna (n. 17 above) 50f.

[40] Louis Napoléon Bonaparte and I. Favé, *Études sur le passé et l'avenir de l'artillerie* (Paris 1846-1871) 2.26ff.

After some initial difficulties with the sling-release mechanism the following results were obtained:

Missile	Range
boulet de 24	175 meters
bombe de 22 centimetres remplie de terre	145 meters
bombes de 27 et 32 centimetres remplies de terre	120 meters

The machine was not robust enough to allow the counterweight to be increased to 8,000 kilograms, as Napoleon had required.

Although the experiment was carried out and reported in a manner that was scientifically slipshod, it at least demonstrated that respectable ranges could be achieved with a comparatively light machine, without its operators having the refined skills of the medieval engineers.

A series of experiments was carried out by B. Rathgen[41] using a small model with a beam 1 meter long, divided in the ratio 1:5. From the results he obtained Rathgen tried to predict the performance of a full-size trebuchet by direct arithmetical proportion. Experiments of this kind with small-scale models are not only valueless, they are misleading. It is impossible to ensure that the range obtainable by the model is exactly, or even approximately, proportional to the range of the full-size machine. There are too many imponderables. To take an important example: the behavior of a sling on the end of a rotating arm has not been studied, but it is certain that its rotational acceleration depends upon the actual rotational velocity of the beam at each instant. The angle turned through by the sling before discharge is an important determinant of the range. But the actual terminal velocity of the model's beam will be much less than that of the real machine, and hence the model's sling will behave quite differently from that of the full-size machine.

We are therefore left with the evidence of the sources together with anything that can be deduced from dynamic calculations based upon reasonable working assumptions. (The assumptions made in these calculations, and the design specifications, are summarized in the appendix.) From the calculations it would appear that the correct proportions for the two arms of the beam are 1:5 or 1:6. We can then postulate the design and performance of two machines, one light, one heavy, as follows:

1 Missile weight 100 pounds. Counterweight 10,000 pounds.
Total length of beam 36 feet, throwing arm 30 feet.
Sling length 15 feet. Likely range: 245 yards.

2 Missile weight 500 pounds. Counterweight 20,000 pounds.
Total length of beam 57.6 feet. Throwing arm 48 feet.
Sling length 30 feet. Likely range: 285 yards.

[41] B. Rathgen, *Das Geschütz in Mittelalter* (Berlin 1927) 631f.

Conclusion

At the siege of Acre in A.D. 1291 there were ninety-two siege machines.[42] Probably these included light machines for raining missiles into the city, and heavy machines for battering the walls. It must have been extremely unpleasant to endure bombardment from such a battery, with stones and burning missiles falling continuously into the area behind the walls while the heavier stones crashed against the walls themselves.

The trebuchet was succeeded by the cannon, which had the advantages of more rapid rate of fire, ease of transport, higher missile velocity, greater range and accuracy. Against this must be set the fact that the trebuchet could often be fabricated from materials available near the beleaguered strongpoint, whereas the cannon foundry might be hundreds of miles away. Moreover, the higher velocity of the cannonball was partly offset by the greater weight of the trebuchet missile. For a long time after the introduction of cannon the trebuchet must have been the superior weapon in some types of siege warfare. Convenience, and perhaps questions of prestige, rather than military effectiveness, gave victory to the cannon before its due time.

APPENDIX

Method of Calculating Ranges for Counterweight Trebuchets

Assumptions and Design Data

1. The missile is laid with its sling in a channel between the supporting towers.
2. The short arm is at an angle of 45 degrees above the horizontal before release, and 45 degrees below the horizontal at the instant when the missile is discharged. At this instant the sling is in line with the beam and traveling at the same velocity, without tangential acceleration.
3. The point of impact is in the same horizontal plane as the point of discharge.
4. The sling is weightless and unextensible.
5. On the advice of Mr. M. A. Sawyer of the Timber Research Association it has been assumed that the timber used for the beam was a 50-grade timber with a moisture content of 18 percent. Its weight is assumed as 35 pounds per square foot, and its extreme working fiber stress at 3,000 pounds per square inch.

[42] Al-Maqrīzī (n. 21 above) 125.

Calculations

1. For a given design of machine the required cross-section of the beam at the fulcrum was calculated by working out the bending moment caused by the counterweight, using an extreme fiber stress of 3,000 pounds per square inch. A reasonable taper was given to the long arm of the beam.
2. An average radius for the missile, between lift-off and discharge, was calculated; also the mean position of the center of gravity of the whole system during this period.
3. Moments of Inertia of beam, counterweight, and missile were calculated.
4. The machine was treated as a compound pendulum using the energy equation:

$$1/2\ \mathrm{Mk}^2 \left(\frac{d\,\Theta}{d\,t}\right)^2 + \mathrm{Mgh}\,(\mathrm{I} - \cos.\ \Theta) = \text{constant}$$

   This was used twice: first for beam and counterweight alone, to give the angular velocity at the instant the missile leaves the ground; second, for the whole system to give the angular velocity at discharge, and hence the linear velocity.
5. Since the angle of discharge was assumed to be 45 degrees, and since air resistance was neglected, the range was a maximum and is given by:

$$\text{Range} = \frac{V^2}{g} \quad \text{where V is the linear velocity at discharge.}$$

6. The theoretical range of the Napoleon and Favé machine was also calculated by the above method and compared with the range actually achieved. This gave an efficiency of about 0.56. Loss of efficiency was doubtless due mainly to friction in the machine. Efficiency could also be adversely affected by air resistance and incorrect angle of discharge. The performance would be improved if the point of discharge were significantly higher than the target, or if the sling were accelerating at the moment of discharge.

   In view of the fact that the medieval engineers were much more experienced than Favé and his team it was decided to work with an efficiency of 0.70.

Results

Two types of machine were investigated, a light and a heavy, as follows:

Light Machine

Missile weight 100 pounds; counterweight 10,000 pounds; throwing arm 30 feet; sling length 15 feet.

Design no.	Ratio of arms	Length of short arm (Feet)	Cross-section of beam at axle (inches)*	Range (yards)	Range × 0.70 (yards)
1	1:3	10.0	16 × 9	243	170
2	1:4	7.5	15 × 8	295	207
3	1:5	6.0	16 × 6	350	245
4	1:6	5.0	14 × 6	350	245
5	1:7	4.28	13 × 6	347	243
6	1:8	3.75	12 × 6	347	243

*Depth given first.

Heavy Machine

Missile weight 500 pounds; counterweight 20,000 pounds; throwing arm 48 feet; sling length 30 feet.

Design no.	Ratio of arms	Length of short arm (feet)	Cross-section of beam at axle (inches)*	Range (yards)	Range x 0.70 (yards)
7	1:3	16.0	22 × 12	307	215
8	1:4	12.0	25 × 9	420	294
9	1:5	9.6	23 × 9	407	285
10	1:6	8.0	22 × 8	410	287
11	1:7	6.86	20 × 8	396	277
12	1:8	6.0	19 × 8	370	259

*Depth given first.

Fig. 1. Traction trebuchet. This is taken from Školjar, fig. 5, page 130. It is one of a number of different designs shown in the Chinese sources. The various types differ in the construction of the supporting towers, and in the number of spars to a beam. The basic principle, however, is the same in all the patterns.

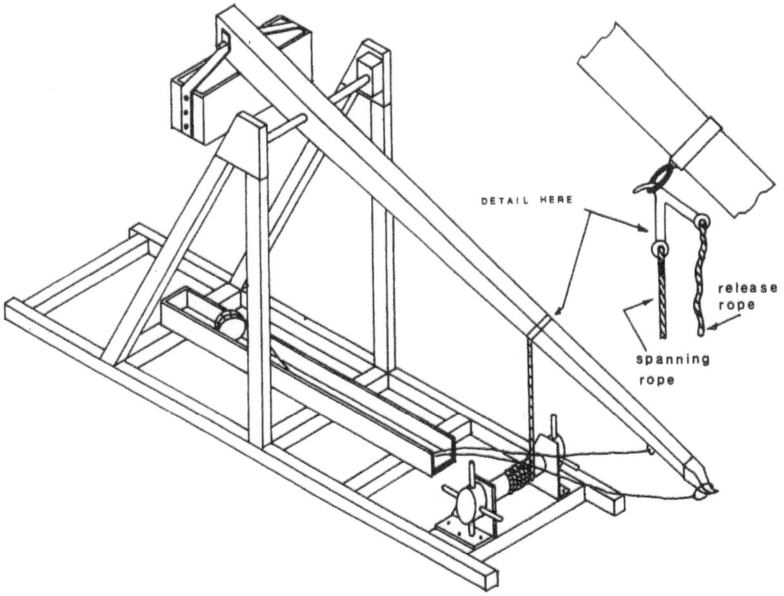

Fig. 2. Counterweight trebuchet. This follows Kyeser's Drawing (Plate 30a) fairly closely. The double hook, for spanning and release, is taken from Payne-Gallwey; although such a device is not shown clearly on any of the old drawings, it is a practical design for the dual purpose. Fittings, fixings and anchorages are not shown in detail, lateral and longitudinal bracing is omitted for the sake of clarity. Kyeser does not show bracing either, although this would have been essential for the stability of the machine.

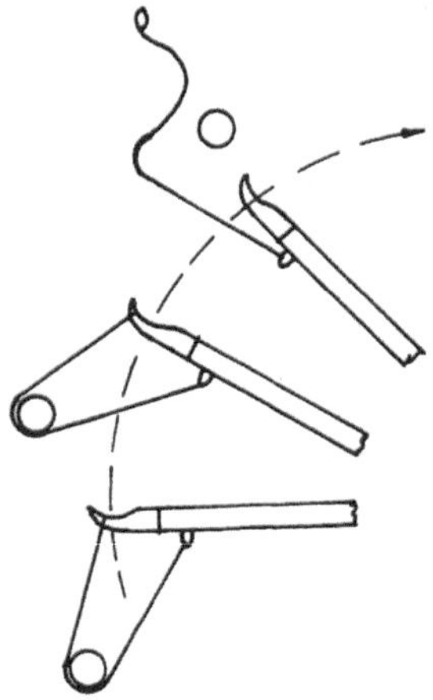

FIG. 3. Sling-release device. The sling-release worked in an exactly similar way on both the traction trebuchet and the counterweight trebuchet.

XX

The Role of the Camel and the Horse in the Early Arab Conquests[1]

I. THE CAMEL

THE most frequently used name for the camel in the Arabic sources is *ba'īr*. The noun is of feminine gender, but in the twentieth century the Badw use this term for 'camel' regardless of sex. Other names which occur commonly are: *nāqa*, an adult female; *ibl*, a collective noun; *jamāl*, a male camel; *rāḥila*, a female riding camel. The term *rikāb* usually means a riding camel and in the Qur'ān (LIX, 6) it is used for camels as distinct from horses. Occasionally other terms are used, including *nāḍiḥ*, for a camel used to draw water, but also sometimes ridden, and *najība*, a thoroughbred riding camel, while both *dhalūl* and *matīy* are other names for riding camels. A *bakr* was a young animal. Sometimes the term *dāba* is used, but this word can mean any riding animal or pack animal. Women were carried in camel-litters, a kind of cage made of wooden poles. This was known as the *ẓa'īna* and the same term was used, by extension, for the women who carried the rein.

The poverty of resources in early Islam is indicated by the scarcity of camels on the march to Badr, where there were only seventy camels to three hundred men. On this expedition, and on others, men took turns in riding the camels. Even the Meccans at Badr and at the Battle of the Ditch had fewer mounts than men. More rarely, on operations outside Arabia, there were occasions when the number of camels was insufficient.

Nevertheless, the situation must have improved gradually, since camels were often taken as booty, and this, together with the adherence to Islam of camel-riding nomads, tended to increase the proportion of camels to warriors. Long-distance travel in the desert could hardly be contemplated without the camel—even the short pursuit to Ḥamrā al-Asad after Uḥud was considered a hard journey on foot. It seems likely that the great majority of Muslims had their own camels by the time the foreign conquests were in progress.

The sources rarely state specifically that journeys were undertaken on camel-back; they merely say 'they went to such a place', even for journeys of several hundred miles. The many incidental references to camels make it certain, however, that the Arab forces in Arabia, Syria, Iraq, and Mesopotamia

[1] *General.* The standard Arabic sources for the early Islamic period have been used, notably Ṭabarī and Balādhurī (Futūḥ al-Buldān), both in the de Goeje editions, together with Ibn 'Abd al-Ḥakam (ed. Torrey), al-Ya'qūbī (ed. Houtsma), Ibn Isḥāq (ed. Wüstenfeld) and al-Wāqidī (ed. Jones).

used the camel as their means of transport. The army of ʿAmr b. al-ʿĀṣ for the invasion of Egypt is called a cavalry force in one passage in Ṭabarī,[1] and John of Nikiou always refers to the Muslims as mounted on horses, never on camels.[2] Camels are mentioned elsewhere, however, both in Egypt and in North Africa. When they entered Egypt, the Muslims were expecting to take camels as booty and, since the dromedary had long been domesticated in Egypt and North Africa, no doubt they were able to meet their needs.[3] ʿAmr's force may have included a higher proportion of cavalry than the Muslim armies in other war theatres, but it is likely that the invasion forces and the expeditions to the Maghrib relied mainly on the camel for transport.

In all the desert and steppe regions which have unbroken land communications with Arabia, with no sea barriers or large mountain ranges intervening, the dromedary had been domesticated for centuries before the Arab invasions. This continuity of terrain and of camel-culture, together with the evidence of the sources, makes it virtually certain that the methods of warfare and transport in these areas remained very similar to those used in Arabia, with the dromedary as the indispensable means of long-distance transport.

The accounts of the warfare in the highlands of Iran and Asia Minor, however, contain only a few references to camels. It is probable that these were the Bactrian variety—one source says so[4]—and the two-humped type is a pack-animal, which can be ridden at a walk, not at speed. Couriers bound for Medina used dromedaries, but they obtained their mounts in Kūfa or Basra. In the highland regions the use of mules, pack-horses, and Bactrian camels must have altered the traditional pattern of Badw warfare and movement. As opposed to the situation in the desert lands, most soldiers would no longer have had mounts and there must have been a transformation into the Persian style of warfare, with squadrons of cavalry forming an elite, while as a general rule the mass of the fighting men were on foot.

There is no evidence to indicate the method of transport in the arid lands of the Great Desert and Khurāsān. The first raids into Khurāsān took the desert route, where water is scarce and conditions are not suitable for horses and mules. The use of the dromedary had been known in these areas for centuries before the Muslim invasion. It was used, for instance, by Alexander the Great in his march from Persepolis to Khurāsān.[5] It is highly likely, therefore, that dromedaries were available to the Arabs for transport and riding in the deserts and steppes to the east and north-east of the Iranian plateau.

[1] Cf. Ṭabarī, I. 2592.

[2] Cf. Jean Évêque de Nikiou, *Chronique*, ed. and trans. M. H. Zotenberg (Paris, 1883).

[3] For the historical development of camel-culture and camel-nomadism cf. *EI*², s.v. *Badw*. Also of value are *L'antica società beduina*, ed. F. Gabrieli (Rome, 1959) (a collection of essays by several authors), and V. Monteil, *Essai sur le chameau* (Ifan (Mauretania), 1952).

[4] Cf. Ṭabarī, I. 2704.

[5] Cf. C. Ritter, *Vergleichende Erdkunde von Arabien* (Berlin, 1847), ii. 640.

The Camel and the Horse 34

The dromedary is used by the Arabs as a riding animal, as a beast of burden, and for agricultural work. It is not easy to fight from the back of a camel and its use as a cavalry mount in the conquests was very infrequent. The standard-bearer of the Hawāzin at Ḥunayn fought with the lance from camel-back and there is mention of a party of infantrymen on camel-back at Qādisiyya.[1] Such reports are rare, however, whereas attestations for dismounting are much more numerous. The camels were left in the camp and guarded by slaves—at the Yarmūk they were hobbled and stationed around the encampment. The warriors who had camels but not horses must therefore be considered as mounted infantry.

March speeds varied considerably, depending upon the nature of the journey and the composition of the force. The fastest travellers were, of course, couriers who were unencumbered by baggage or by women and children. The journey from Mecca to Medina, a distance of over 250 miles, took three or four days, while al-Muthannā b. Ḥāritha is said to have ridden from Medina to Ḥīra in ten days, a distance of about 670 miles. March times given by modern writers agree reasonably well with these figures. Thesiger covered 450 miles in nine days,[2] while Musil quoted a journey by two couriers who rode 94 miles in twenty-four hours,[3] and another occasion when a man rode 160 miles from sunrise to sunset, slept the night, and performed the return journey on the following day, also from sunrise to sunset.[4] Doughty says that a riding camel (*dhalūl*) in good condition may do 70 miles a day for short journeys, 60 to 65 miles a day for a week, and 50 miles a day for a fortnight.[5] It can be assumed, therefore, that the speed of couriers during the conquests was similar to these estimates: a rider could cover about 100 miles in one day, 400 miles in a week, and 700 miles in a fortnight.

March times for raiding parties, although slower than those of couriers, were still impressive. It is calculated that the speed of march to and from Tabūk, and of Usāma's force to and from the Balqā, was in both cases about 40 miles a day. The waterless stage of Khālid b. al-Walīd's journey from Iraq, between Qurāqir and Suwā, was accomplished at an average speed of 35 miles a day.

A march such as that of Sa'd b. Abī Waqqās and his troops to Qādisiyya was affected by considerations other than the endurance of men and mounts. Many of the recruits were from the Yemen and came with their women, children, and movable possessions. The advance was made in stages, both to ensure that water and pasture were available in sufficiency and to recruit additional warriors from the local tribes. The speed of march cannot have been more than 20 miles a day.

[1] Cf. Ṭabarī, I. 1660, 2309.
[2] Cf. W. Thesiger, *Arabian Sands* (London, 1959), 17.
[3] Cf. A. Musil, *Arabia Deserta* (New York, 1927), 120.
[4] Cf. A. Musil, *Northern Negd* (New York, 1928), 145.
[5] Cf. C. M. Doughty, *Travels in Arabia Deserta* (London, 1936), ii. 553.

II. THE HORSE

References to horses and cavalry abound in the sources; and although the camels, particularly in the early days, must have greatly outnumbered the horses, the references to horses are much more numerous. The horse is nearly always called *faras* and the rider is called *fāris*. Muḥammad called female horses *faras* and the word is of female gender, but among the modern Badw it has come to mean any horse, regardless of sex, because with a few exceptions only mares are bred.

Individual animals are often described in detail: by colour, markings, temperament, accoutrements, and by name. The general name for cavalry is *khayl* and is widely used for describing horse formations in various circumstances. The word *kurāʿ* is used as a collective term meaning mounts, never as the designation of a battle formation. Other names were used for cavalry units according to the function they performed:

1. The *katība*

   This was the equivalent of a cavalry squadron.

2. The *ṭalīʿa*

   This was usually a small spy troop or reconnaissance patrol sent out to obtain information and its size was between one and ten men, but the expression is also used for a raiding party, and in later years came to be used for a unit of the *taʿbiya* formation.

3. The *sarīya*

   The term can have a similar connotation to the *ṭalīʿa*, a patrol sent out for information or for forage. In the *Maghāzi* of Wāqidī, however, *sarīya* is used to distinguish an expedition in which Muḥammad did not take part from an expedition led by Muḥammad.

4. The *jarīda*

   Usually this term was used in the phrase *jarīda khayl* and described a horse troop which was acting independently, e.g. on long-distance raids into enemy territory.

5. The *mujarrada*

   This term was used of the cavalry arm of an army in battle formation.

6. The *rābiṭa*

   When a town or a region had been conquered, it was often necessary to station a mounted garrison in the territory as a mobile guard force to put down revolts and discourage insurrection: this was the *rābiṭa*. The term *murābiṭa* has the same meaning.

7. The *kurdūs*

   The account, by Sayf b. ʿUmar, of the battle of the Yarmūk is the only place where this term has been found. He divides the cavalry into many *karādīs* and gives the name of the leader of each one.[1]

The number of horses available to the Muslims in the early days was very small, but must have been augmented by the accession of Mecca, since the richer Meccans could always put an effective cavalry force into the field. At Badr and at the expulsion of the Banū Qurayẓa there were horses in the booty, but the numbers were small; no horses were mentioned in the booty taken at Ḥunayn, although 6,000 camels were captured. Horses were bought from the spoils of the Banū Qurayẓa and the Banū Naḍīr; and in the peace treaty with Najrān it was stipulated that the Muslims be provided with 30 horses in case of war. The figure of 10,000 horses at Tabūk is certainly a gross exaggeration, especially as the journey was undertaken at a time of great heat and many thousands of camels would have been needed merely to keep the horses alive. It is certain that the number of horsemen with the Arab forces remained small until the victories of the Yarmūk and Qādisiyya opened up greater opportunities for taking horses as booty. We are told, for example, that there were horses in the spoils of Qādisiyya and Jalūlā'. There seems also to have been a more systematic attempt to provide horses for the Muslim armies. 'Umar I had horses collected from all areas for use in military operations. There were 4,000 of them available in Kūfa[2] and there was a horse pasture for Muslim mounts in northern Syria.[3] A certain Nāfi' b. al-Ḥārith was the first to wean a colt in Basra and he was granted a fief in Basra by 'Umar[4] for the purpose of rearing horses. Although we are told that all the Muslims were mounted after Qādisiyya, this is extremely unlikely and is not borne out by the statement that the cavalry which forded the river at al-Madā'in numbered 600; this was probably the total number of horsemen with the Arabs at that time. Both in north Syria and in Iran the cavalry is mentioned as distinct from the main body of the army under the commander-in-chief. Thus the cavalry of Ḥabīb b. Maslama in Armenia roamed about, subduing villages. Cavalry was sent against a town and the main body of the army marched, until they rejoined the cavalry before the town. In mountain regions, when the army was united, mules were ridden and the horses were led at the sides. This is, of course, an extension of the Badw technique, when on a raid, of leading the horses by the side of the camels, and then changing to horseback immediately before launching the attack. It is a very old practice and was employed by the Nabataean allies of the Romans in their attack on Jerusalem in A.D. 67.[5]

In the early days the battles were fought on the Muslim side almost entirely by infantry, but the Meccans possessed sufficient horses for a cavalry force. At Badr there is little mention of cavalry, although the Meccans had

[1] Cf. Ṭabarī, I. 2093 ff.
[2] Cf. Ṭabarī, I. 2499.
[3] Cf. Balādhurī, 191.
[4] Cf. Balādhurī, 350 ff.
[5] Cf. *EI*², s.v. *Badw*.

100 horses with them, and it is significant that no attempt was made to assault the Muslim ranks with a cavalry charge. Nor was the cavalry effective at the Battle of the Ditch, being unable to gain a foothold on the Muslim side of the trench. At Uḥud, however, the horsemen under Khālid b. al-Walīd probably played a decisive role in the Muslim defeat, but again it is interesting to note that no frontal attack was made by the cavalry on the Muslim defensive position along the slopes of Mount Uḥud. Instead Khālid adopted the sound tactics of attempting to turn the Muslim flank by going round Mount Uḥud and advancing westward along the Wādi Qanat, until he was halted by a company of archers stationed, to the south of the wādi, on Mount Rumat. He was unable to make any headway against them, as long as they maintained their position; here, as in later medieval battles, archers were a serious threat to mounted troops. It was only when most of the bowmen left their posts that the cavalry broke through and fell upon the Muslims, whose ranks were disorganized in the quest for plunder. The horsemen were then operating in conditions favourable to cavalry, with room to manœuvre and to wield the lance.[1]

By the beginning of the seventh century the Byzantine armies had already been reorganized to deal with mobile enemies such as the Berbers and the cavalry had become the best-armed and most numerous of the Byzantine troop formations. The Persians also, with their horse-mounted knighthood, the *Asāwira*, could presumably put much larger cavalry forces into the field than the Arabs. The type of open, level country best suited to cavalry is well described by the author of a *strategikon*.[2]

The mobility of the Arabs, so superior to that of their enemies for movement over long distances in desert or steppe, became inferior when they were faced by their opponents on the battlefield. The evidence suggests strongly that the Arabs, for the pitched battles, chose positions which would favour infantry, supported by archers, and afford little scope for the effective use of heavy cavalry. Such was the position at the Yarmūk, with the deep ravine of the river in front of the Muslims and with an easy line of retreat to Arabia in case of defeat.[3] Although the battle of Qādisiyya was fought on a plain, the Muslim position was behind a channel called ʿAtīq, which the Persians had to fill with soil, reeds, and pack-saddles in order to make the crossing. It was the Arab intention to retreat to the desert, if they were defeated, and to await a more favourable opportunity for renewing the assault. At Nihāwand,

[1] Cf. M. Hamidullah, 'Les champs de bataille au temps du prophète', in *Revue des Études Islamiques*, xiii (1939), 5–8.

[2] Cf. *L'Armée byzantine à la fin du VIe siècle d'après le Strategikon de Maurice*, trans. F. Aussaresses (Paris, 1909), 9.

[3] Equally, the Byzantines may have regarded the Yarmūk line as a defensive barrier against the renewal of the Arab invasion. This does not alter the fact that the battle was eventually fought and won by the Arabs on terrain which was much more suitable for infantry than for cavalry. See J. B. Glubb, *The Great Arab Conquests* (London, 1963), 173 ff.

although we have little information about the topography of the battlefield, the greater part of the Arab army seems to have been composed of infantry (they were instructed to tighten their sandal-straps before the battle was joined).

There seems to be little doubt that infantry played an important part in these victories, but it is difficult to ascertain from the sources the role played by the archers. There are many references to archers in the accounts of Muḥammad's campaigns, so it seems likely that this branch of warfare was already well understood by the Arabs. A Persian who fought at Qādisiyya is reported as saying that the Arab archery was more effective than the Persian and that the Muslim arrows pierced armour and cuirasses.

In addition to considerations of tactics and topography, the Arab conduct of battles may have been influenced by another factor. Their armies were usually composed both of nomadic tribesmen and of townsmen, although the relative numbers of these two groups varied between one army and another. Now, it is evident that Badw methods of fighting are markedly different from those of settled communities, so much so, indeed, that serious contests between Badw and settlers, decided solely by force of arms, must have been of rare occurrence. The Badw method is the *ghazw*, which is dependent upon mobility, surprise, and the swift evasion of pursuit. The operations of al-Muthannā b. Ḥāritha along the Euphrates, after the departure of Khālid b. al-Walīd, were of this nature:[1] attacks on settlements were made from the desert at dawn after a night march and the raiders escaped back to the desert, as soon as they had obtained their spoils. Only gold, silver, and other valuables were taken, so that the party was not slowed down by carrying weighty articles of booty.

If the raiding party was sufficiently large compared with the potential opposition, the Arabs used another method of attack to attain their objectives. A horse-guard would be stationed at the gates of the town, to prevent sorties, and the raiders then menaced the livelihood of the inhabitants by carrying off livestock, by pasturing their mounts in the standing corn, and by damaging or threatening to damage crops and orchards. The townspeople usually capitulated and complied with the demands of the raiders, which were not, in general, exorbitant. This method was used against Ḥīra and other towns along the Euphrates and it may also have been the means by which the permanent conquest of Mesopotamia was achieved. Here, of course, the Arab forces were more than mere Badw raiders; and there was no regular army to put an end to their depredations. Inside Arabia, however, and on the margins of the desert, before the Byzantine and the Persian resistance had been broken, the balance of power was more equal. Although the townspeople did not come out to challenge the nomads in the open desert, neither were the nomads able to break into the settlements and take them by assault. The same

[1] Cf. Ṭabarī, I. 2202 ff.

situation obtained until quite recently. Doughty wrote: 'In Taymà mud walls were sufficient protection—no-one had the ingenuity to use powder, or a palm trunk as a battering ram.'[1]

There were thus two distinct elements in the Arab armies, each with its own method of waging war. The nomads were mobile, skilful at surprise attacks and evasion of pursuit, effective as light cavalry when horses were available, but unreliable as infantry in a stubbornly contested battle. The men from the towns and oases were unskilled at *ghazw* warfare, but steadier and more dependable when fighting on foot. No doubt the contrast was blurred, as was the distinction between the two groups, and probably neither had much skill in siege warfare, while both were probably able to adapt themselves to act as cavalry when horses were in greater supply. It is considered, however, that the distinction was sufficiently real, in the early years of the conquests, for it to have had an effect upon the conduct of the warfare.

At the battle of Yamāma the brunt of the fighting was said to have been borne by the Companions of the Prophet, the tribesmen being largely ineffectual.[2] At 'Ayn Shams, 'Amr b. al-'Āṣ complained that the 'men of Yemen' were not playing their part and we are told that he had to rely upon the Companions for victory.[3] These data support the hypothesis that the fighting core of the armies was formed from the sedentary population of the towns and oases and that it was they, as infantry, who broke the power of the Byzantines and the Persians in the pitched battles.

When cavalry was used in combat, its position in the order of battle varied: sometimes it was on the wings, sometimes in front of the main body of troops. A discharge of missiles usually preceded the cavalry attack, but at Qādisiyya the cavalry charge (*muṭārada*) was said to have been first and to have been repeated thirty times. It is hard to believe that the cavalry at Qādisiyya was very effective on the battlefield, particularly as their numbers were limited.

Some details are given of fighting from horseback. There is frequent mention of foot-soldiers being attacked by cavalry. At Uḥud, Khālid b. al-Walīd is described as riding among the Muslims, when their ranks were disorganized, killing them with his lance. Duels were also fought between horsemen in a style similar to that of the joust familiar in later medieval times.

III. THE USES OF MOBILITY

The conduct of the Arabs in pitched battles must not be confused with the tactics used on raiding expeditions. Such raids were frequently mounted both before and after Nihāwand—as, for example, the raids during the

[1] Cf. Doughty, *Arabia Deserta*, ii. 329. [2] Cf. Ṭabarī, I. 1945 ff.
[3] Cf. Ṭabarī, I. 2592.

conquest of Mesopotamia and the first incursions into Adharbayjān, Ṭabaristān, and Khurāsān. There are several indications that these expeditions were of a *ghazw* nature; and, with the exception of Mesopotamia, no permanent conquests were made at this time. Forts were left alone, if they resisted. Towns were said to have rebelled, after having been conquered in this period, and to have been subdued again later; but it is probable that these early raiders had merely exacted a single payment of tribute, before continuing their march. Later, during the time of 'Uthmān, troops from Kūfa were stationed in the frontier regions, the *thughur*, and there were 10,000 men in these garrisons. The number of fighting men in Kūfa was 40,000, so that 'each man got a chance for booty once every four years'.[1] In the same period, when al-Walīd raided Adharbayjān, Jīlān, and Mūqān, he returned to Kūfa after his expedition. There is a marked contrast, therefore, between such expeditions, where a maximum use was made of mobility for long-range transport, and the pitched battles of Badr, Uḥud, the Yarmūk, and Qādisiyya.

It must be emphasized that the mobility of the Arabs was most effective when they were in their natural environment—the open desert. Even in Arabia there were natural obstacles that riders found difficulty in negotiating. One of these was lava (*ḥarra*), which was a barrier to riders. Medina, at the Battle of the Ditch, needed no additional protection on the sides guarded by lava, as this terrain was held to be impassable. The ditch itself was a simple enough device, yet it completely nonplussed the Meccan cavalry, used as they were to the open desert. *En route* to Ḥudaybīya the Muslim army had difficulty on the march because of the broken nature of the country, intersected by many ravines (*shi'b*). Wādis could be obstacles to movement, when the beds had become soft after rain; and they became impassable, when they were converted into torrents after downpours. The desert itself was feared, when it was unknown and waterless: a poet describes the difficulties of travel in the northern deserts, where strong camels become feeble and where lie the skeletons of fallen animals; a force mounted on camels suffered greatly from thirst in crossing the narrow Dahnā' desert.

But these were difficulties that the Arabs were accustomed to encounter and surmount. Once out of the familiar surroundings of Arabia, however, each new obstacle presented itself as a formidable barrier, the overcoming of which demanded a mental effort at least as great as the physical one. This apprehension is reflected in the attention which is paid to such episodes in the sources. Several pages are devoted by Ṭabarī to the crossing of the Tigris at al-Madā'in and the whole enterprise is described as fraught with difficulty and danger. On another occasion the Arabs were unable to follow the Persians who were in boats, because of the intervening water. The early Muslim invaders had no skill in bridge building and had to enlist the services of local

[1] Cf. Ṭabarī, I. 2805.

inhabitants to do this work for them.[1] In North Africa and in the Syrian and 'Irāqī deserts the mobility of camel-mounted troops could be used to the full. Conditions in the Syrian desert, in particular, where there are fewer obstacles than in many parts of Arabia, are especially favourable for the movement of large forces on camel-back. In parts of Egypt and Iraq, in Persia, and in Asia Minor movement is hampered by swampy conditions and rivers in the riparian regions and by mountains and rivers in the highlands. In these areas the Arabs had to adapt themselves to a new form of warfare, without the ease of movement which they had enjoyed in the desert. Their mobility in such conditions cannot have been greater than that of their enemies.

In the early years of warfare in these regions, moreover, the Arabs were almost certainly inferior to their enemies in the use of heavy cavalry, both technically and numerically. Their body armour and horse accoutrements, especially stirrups, were probably of inferior design, so that the Arab cavalrymen were unable to meet their opponents on equal terms. Nor were they able to use the bow effectively from horseback. On one occasion at Isfahān a troop of Arab horse was unable to close with a Persian one, because the Persians held them off by threatening to use their bows.[2] More than once in Persia, the Arabs only avoided defeat by taking up strong defensive positions, presumably in order to reduce the effectiveness of the enemy cavalry. Al-Aḥnaf b. Qays, for example, when hard-pressed in Lower Tukharistān, made his stand in a narrow pass, with the river Murghāb on his right flank and a mountain on his left.[3]

Not only were the Arabs inferior in heavy cavalry, but they also had little skill in siege warfare. It is beyond the scope of this paper to discuss the way in which the Arabs adapted their methods of warfare to meet the new conditions, an adaptation which included the use of non-Arab cavalry forces. This development, which may have begun as early as the first campaign in Khuzistān (A.H. 17), eventually produced Muslim armies that were markedly different in composition and technique from the armies which had triumphed in Syria, Iraq, and Egypt.

If, therefore, one is to examine the manner in which the Arabs gained an advantage over their enemies by the use of superior mobility, conferred on them by the dromedary, nothing of value can be obtained by considering the campaigns in Persia, Armenia, and Asia Minor. Nor can any advantage to the Arabs from superior mobility be discovered by a study of the decisive pitched battles. Where superior mobility played a weighty role was in the strategical conduct of the campaigns in Syria, Iraq, and Egypt.

This superiority made itself felt in several ways, but all were linked to the central fact that the Arabs were able to cover long distances rapidly and

[1] Cf. Balādhurī, 251 ff.; also Ibn 'Abd al-Ḥakam, 73; and Jean de Nikiou, 559.
[2] Cf. Balādhurī, 312 ff.
[3] Cf. Ṭabarī, I. 2901; also Balādhurī, 407.

to use the desert for passage, for a raiding base, and as a refuge. Thus al-Muthannā b. Ḥāritha was able to raid the Euphrates towns with impunity and to evade pursuit by retreating to the desert. Even after the Muslim defeat at the Battle of the Bridge, the Persians made no attempt to consolidate their success by pursuing the Arabs into the desert. Before Qādisiyya, it was the Arabs' declared strategy to fight the Persians on the border of the desert and the town, so that in the event of defeat they could retreat into the desert without fear of pursuit and await a more favourable opportunity for attack. Their rear, with its lines of communications, was thus safe from enemy interference, the sending of reinforcements was without hazard, and the families were left in safety at desert watering-places. The Arab victory, when it came, meant the end of the Sasanid Empire, whereas a Persian victory would have been a hollow one.

Even more important was the ability of the Arabs to concentrate their forces swiftly at the point of greatest danger. Movements like the transfer of Khālid b. al-Walīd and his troops from Iraq to Syria, the swift movement of the scattered Arab forces to meet the Byzantine threat at Ajnādayn, and the dispatch of al-Zubayr with 12,000 men to reinforce 'Amr b. al-'Āṣ in Egypt, all had a major influence upon the outcome of the campaigns. These dramatic events are reasonably well documented, but there were numerous minor incidents which had their effect on the final result and about which we know little. One has in mind the disruption of enemy communications, the isolation of fortified cities, the surprise attacks upon the encampments of Christian Arabs—all creating a situation which must have been as damaging to enemy morale as it was destructive of his war potential.

It is difficult to assess the effect of good communications upon the course of these conquests and upon their immediate aftermath. No doubt part of the ability to concentrate forces effectively was dependent upon the easy passage of information between commanders, who might be separated by many miles. At a different level, the exchange of dispatches between field commanders and the caliph is frequently mentioned in the sources. The role of 'Umar I in directing operations is frequently exaggerated, in that he is often said to have given instructions about tactical matters and local objectives, which would only have been decided in the theatre of war by the army commanders. 'Umar I, however, was certainly kept informed of all major events; and there can be little doubt that his orders were accepted on important strategical matters, on the appointment of governors, and on the administrative arrangements for the conquered territories.

The superior mobility of the Arabs was, therefore, one of the most important factors in ensuring their success in the early conquests. In a wider context it can be stated with confidence that, without the domestication of the dromedary, the conquests could never have been undertaken at all. Because the way of life in Arabia was nomadic and based on camel transport, the Arabs were able to mobilize sufficient forces to invade the lands of the Byzantine and Persian Empires with some hope of success. Had the prevailing culture been limited to oasis agriculture in widely scattered, mutually isolated areas, no combination of forces and no mass movement would have been possible.

INDEXES

Note: In the ordering of entries the Arabic definite article "al-" has been overlooked

Names of Persons

Modern Scholars

Titles of Books

Topics

Technical Terms